ADVANCES IN CELL AGING AND GERONTOLOGY

VOLUME 12

Membrane Lipid Signaling in Aging and Age-Related Disease

ADVANCES IN CELL AGING AND GERONTOLOGY

VOLUME 12

Membrane Lipid Signaling in Aging and Age-Related Disease

Volume Editor:

Mark P. Mattson,
PhD, National Institute on Aging,
NIH
Baltimore, MD
USA

2003

ELSEVIER

Amsterdam – Boston – London – New York – Oxford – Paris
San Diego – San Francisco – Singapore – Sydney – Tokyo

ELSEVIER SCIENCE B.V.
Sara Burgerhartstraat 25
P.O. Box 211, 1000 AE Amsterdam, The Netherlands

First edition 2003

British Library Cataloguing in Publication Data

Membrane lipid signaling in aging and age-related disease.
  (Advances in cell aging and gerontology; v. 12)
  1. Aging 2. Membrane lipids 3. Cellular signal transduction
  I. Mattson, Mark Paul
  612.6'7

ISBN 0444512977
Library of Congress Cataloging in Publication Data
A catalog record from the Library of Congress has been applied for.

ISBN: 0-444-51297-7
ISSN: 1566-3124 (Series)

Transferred to digital print 2007
Printed and bound by CPI Antony Rowe, Eastbourne

TABLE OF CONTENTS

Chapter 10
Phospholipase A$_2$ in the Pathogenesis of Cardiovascular Disease 177

Chapter 11
Retinal Docosahexaenoic Acid, Age-Related Diseases, and Glaucoma 205

PREFACE

The lipids that comprise membranes of cells, be they phospholipids, sphingolipids, glycolipids or even cholesterol, serve many functions beyond their role in the structural organization of the membranes. In response to stimulation by various agonists, and environmental insults such as oxidative stress, several different membrane lipids are cleaved either enzymatically or non-enzymatically. Products are thereby released that act as important signals mediating physiological responses to the stimuli. Among the most intensively studied signaling pathways is the inositol phospholipid signaling, which leads to the production of IP3 and diacylglycerol with consequent calcium release and protein kinase activation. Another example of a prominent signaling pathway involves sphingomyelin hydrolysis resulting in the production of ceramide, which acts as an important signal in both physiological and pathophysiological settings. Recently there have been a flurry of studies characterizing so called lipid rafts, which are complex microdomains in membranes in which sphingolipids and cholesterol are concentrated, together with various receptors and transducing proteins for an array of growth factor, cytokine and neurotransmitter signals. In this issue of *Advances in Cell Aging and Gerontology* entitled **Membrane Lipid Signaling in Aging and Age-Related Disease** experts in the fields of lipid signaling and aging provide timely reviews of specific aspects of membrane lipid signaling from the perspectives of aging and diseases of aging including cardiovascular disease, cancers and neurodegenerative disorders. It is becoming quite clear that alterations in sphingolipid, inositol phospholipid and cholesterol metabolism occur in a variety of tissues throughout the body during aging, and that specific abnormalities in these signaling pathways play roles in many different age-related diseases.

The book begins with a chapter by Chris Fielding which provides an overview of signaling in the microdomains called lipid rafts and caveolae. The localization of these membrane microdomains in strategic positions within cells provides spatial control over signaling. This spatial control is particularly critical in structurally complex cells such as neurons. Kathleen Montine and colleagues then provide a review of how oxidative stress effects membranes and how the generation of lipid peroxidation products plays a role in aging and diseases of aging. Oxidative stress is considered an important factor in aging and membrane lipid peroxidation has been implicated in a variety of disorders including atherosclerosis, cancers and Alzheimer's disease. Studies of invertebrate systems where genetic manipulations can be readily performed to identify genes of interest have made major contributions to our understanding of the genes that control lifespan. Cathy Wolkow reviews studies that have identified lipid related signaling pathways that appear to play a major role in controlling lifespan. In particular, inositol phosphate signaling pathways activated by insulin and insulin-related signals appears to play an important role in coupling environmental stress signals to energy metabolism and thereby may regulate cellular aging rate.

Cancers continue to be a major killer in the United States and other industrialized countries. Ruvolo et al. review the evidence that altered sphingolipid metabolism and ceramide production play a role in the pathogenesis of cancers by altering cell cycle and cell death pathways. Altered sphingolipid metabolism also plays a role in the pathogenesis of cardiovascular disease. Atherosclerosis is a complex process involving damage to vascular endothelial cells and inflammatory processes. Sphingolipid signaling appears to play important roles in several different steps in the process of atherosclerosis, and studies of sphingolipid signaling in atherosclerosis are revealing novel targets for therapeutic intervention. In the nervous system sphingomyelin signaling plays important roles in regulating development and plasticity of neuronal circuits. Mattson and Cutler describe the roles of sphingomyelin and ceramide signaling in brain aging and the pathogenesis of neurodegenerative disorders such as Alzheimer's disease and amyotrophic lateral sclerosis. It appears that increased oxidative stress in these disorders leads to excess production of ceramides and cholesterol esters which can trigger the degeneration and death of neurons.

Arachidonic acid is a product of phospholipase A_2 activation that plays important roles in inflammatory responses and also has a broad array of functions in cells in many different organ systems including the nervous system. Hari Manev and Tolga Uz describe eicosanoid pathway changes during aging of the nervous system with a focus on the regulation of cyclooxygenases and lipoxygenases. It appears that inflammation-like processes occur during brain aging and in neurodegenerative disorders, and a better understanding of arachidonic acid metabolism is therefore likely to lead to new approaches for preventing and treating neurodegenerative disorders as well as other diseases. Alterations in cholesterol metabolism are well known to play important roles in the pathogenesis of cardiovascular disease. However, recent findings have demonstrated important roles of cholesterol metabolism in modulating various signal transduction pathways and suggest that perturbed cholesterol metabolism plays roles in many different age-related diseases. John Incardona describes how studies of inherited disorders of cholesterol and sphingolipid metabolism have provided insight into the functions of lipid signaling in human disease. He describes Niemann–Pick C disease and Smith–Lemli–Opitz syndrome as two examples that have increased our understanding of the various roles of membrane cholesterol and lipid rafts in cellular biology and disease. Suzana Petanceska et al. review the emerging evidence that alterations in cholesterol metabolism play a role in the pathogenesis of Alzheimer's disease. Interesting relationships between cholesterol metabolism and the production of amyloid beta peptide, the main component of plaques in the brains of Alzheimer's disease patients, have emerged from epidemiological and experimental studies in animal models of Alzheimer's disease.

Eva Hurt-Camejo and colleagues describe reactions of phosholipases A_2 in cellular and metabolism with a focus on a role of these enzymes in the pathogenesis of cardiovascular disease. These enzymes exist both within cells and outside cells and have actions in both locations that may contribute to alterations in lipoprotein and lipid metabolism and the activation of inflammatory pathways in the process of atherosclerosis. Finally, Nicolas Bazan provides a concise review of the role of docosahexaenoic acid in aging and age-related disease with a specific focus on glaucoma, a major cause of blindness in elderly individuals. The production and metabolism of docosahexaenoic acid display unique features. Of particular interest is the enrichment of docosahexaenoic acid in excitable membranes of the retina and brain. Several messengers are derived from docosahexaenoic acid including neuroprostanes and hepoxillins; these lipid mediators may have important roles in aging and age-related disease.

Collectively the information contained in the chapters of this volume of *Advances in Cell Aging and Gerontology* provide the reader with a clear picture of the very important roles of membrane lipid signaling in the regulation of various physiological processes throughout the body. It is quite clear that alterations in several lipid-signaling pathways occur during aging and in age-related diseases. At least in some cases these alterations may be early and pivotal events in disease processes as suggested by recent studies documenting beneficial effects of interventions that target lipid signaling. We expect that this book will be a valuable resource for investigators in fields of aging and age-related disease and that the information contained within these pages will foster new experiments aimed at unraveling the roles of lipid signaling in cellular physiology and disease.

MARK P. MATTSON, PhD

**Advances in
Cell Aging and
Gerontology**

Overview: Spatial control of signal transduction by caveolae and lipid rafts

Christopher J. Fielding

*Cardiovascular Research Institute and Department of Physiology,
University of California, San Francisco, CA 94143, USA.
E-mail address: cfield@itsa.ucsf.edu*

Contents

1. Introduction

The cell surface is heterogenous, consisting of microdomains with varying lipid and protein composition. Well-recognized microdomains include coated pits (sites of the internalization of many macromolecules) and the microvilli that identify the absorptive surface of many epithelial cells. An additional class (DRMs: detergent-resistant microdomains) has been distinguished by its insolubility in neutral detergents. This fraction is rich in free cholesterol (FC) and sphingolipids, signaling proteins, GPI-anchored proteins and FC-binding caveolin proteins. More recent research has shown that this fraction represents a mixture of two distinct species of microdomains, lipid rafts and caveolae (Fielding and Fielding, 2002).

Lipid rafts are planar microdomains, typically 50–350 nM in diameter, rich in GPI-anchored proteins but deficient in caveolins, the structural proteins of caveolae. Caveolae are invaginated domains, whose openings at the cell surface are typically 60–120 nm in diameter. These microdomains, rich in caveolin, are deficient in GPI-anchored proteins (Iwabuchi et al., 1998; Abrami et al., 2001; Sowa et al., 2001).

Caveolae and lipid rafts probably coexist in most cells, though the relative number of each varies widely in different cells. Caveolae are enriched in terminally differentiated cells such as adipocytes, vascular smooth muscle and endothelial cells. Their numbers are very low in blood leucocytes, which are rich in lipid rafts. As a result, leucocytes have been used for much of the published characterization of lipid rafts (Matko et al., 2002). Under basal conditions some cancer and transformed cell lines lack caveolae and caveolin entirely (Koleske et al., 1995; Lee et al., 1998). However, even these cells may express large numbers of caveolae under some experimental conditions. For example, human MCF-7 cancer cells express caveolin and caveolae concomitant with the upregulation of the p-glyco-protein transporter following incubation with adriamycin (Lavie et al., 1998).

The expression of caveolin is necessary but not sufficient for the development of morphological caveolae. In addition to caveolae, caveolin is also present in several intracellular pools, including the *trans*-Golgi network, a chaperone complex carrying newly synthesized FC, and recycling endosomes (Fielding and Fielding, 1996; Uittenbogaard et al., 1998; Gagescu et al., 2000). While the presence of caveolin in cell and cell membrane fractions does not necessarily imply the presence of cell surface caveolae, in most primary cells the plasma membrane contains the bulk of caveolin. In transformed cells, what caveolin is present may be largely in the form of intracellular vesicles (Sowa et al., 2001).

Caveolin binds FC via a central domain which also interacts with a "scaffold" sequence present on many transmembrane signal kinases, as well as a number of signal intermediate proteins, such as *ras*, and protein kinases A and C (Smart et al., 1999). Protein–protein association is in many cases dependent on or facilitated by post-translational lipid modifications (*N*-palmitoylation, *N*-myristoyl-ation) (Sowa et al., 1999). As discussed more fully below, transport of caveolin to the cell surface and caveolar assembly may depend on these modifications.

Signaling proteins and GPI-anchored proteins bound to lipid rafts usually lack a recognizable scaffold-binding site. Their association with rafts depends both on *N*-acylation and on the presence of amino acid sequences recognize lipids (Fielding and Fielding, 2002). FC levels in lipid rafts are reported to be lower than those in caveolae (Iwabuchi et al., 1998). Ganglioside GM1, once considered a marker for caveolae, is present in both caveolae and lipid rafts. The same distribution may hold for other lipids involved in signal transduction, such as ceramide, phosphatidic acid and diglyceride. Enzymes which generate these lipids, including sphingomyelinase and phospholipase, are present in caveolae and lipid rafts. This suggests that lipid signals may be generated in caveolae. Lipid signals may in some cases extend the effect of protein-mediated signal transduction (Igarashi et al., 1999).

There is considerable difference in the lifetime of lipid rafts and caveolae at the cell surface. Lipid rafts are labile, with a measured half-life of only a few minutes, similar to the time required for signal transduction (Sheets et al., 1997; Pralle et al., 2000). They may be stabilized by the presence of FC, GPI-anchored and other proteins. In contrast, caveolae (with the possible exception of those in endothelial cells) remain at the cell surface over periods of many hours (Thomsen et al., 2002). However, while caveolae are stable, their association with signaling molecules is dynamic (Liu et al., 2000; Fielding et al., 2002).

Although caveolae and lipid rafts have a similar lipid composition, there is little to suggest that they are interconvertible. They appear to play different roles in the cell. However, crosstalk between caveolae and lipid rafts has been identified (Abrami et al., 2001). Overexpression of caveolin and caveolae was associated with a decrease in GPI-anchored proteins in rafts. Loss of GPI-anchored proteins was coupled with an upregulation of caveolin and caveolae. While under basal conditions individual signaling proteins may be associated mainly with lipid rafts, or mainly with caveolae, a change in the proportions of rafts and caveolae can lead to a redistribution of these proteins, though modification of their kinetic properties (Sowa et al., 2001; Vainio et al., 2002).

2. Functions of rafts and caveolae – signaling and FC homeostasis

These roles have been studied in detail for both classes of microdomains. Signaling via both lipid rafts and caveolae is FC-dependent. Unlike the situation with caveolae, rafts may play only a minor or negligible role in whole cell FC homeostasis. Additional roles for caveolae have been postulated in the uptake of folate via its receptor (FR), a GPI-anchored protein; in the uptake and release of certain bacteria and viruses; and (in endothelial cells) in the transcytosis of albumin. The first two of these functions now seem more likely to depend on lipid rafts, not caveolae. Caveola-mediated transcytosis has not been described except in endothelial cells. The present overview focuses on the signaling and FC-dependent roles of caveolae and lipid rafts, both because these appear to be the best established, and because recent evidence indicates them to be linked.

Of signaling pathways involving caveolae, that mediated by the platelet-derived growth factor receptor (PDGF-R) is among the best characterized. In primary cells (vascular smooth muscle cells, fibroblasts) PDGF-R is present mainly in caveolae (Liu et al., 2000; Fielding et al., 2002). Following PDGF binding, and dimerization and autophosphorylation of PDGF-R in a complex that includes the exchange factor *sos* and the small GTPase *h-ras*, independent phosphorylation cascades are mediated via the p38, PI3 kinase, ERK1/2 and other pathways. The primary sequences of PDGF-R and *ras* and caveolin itself include caveolin-binding "scaffold" sequences (Smart et al., 1999). Caveolin is also irreversibly *N*-palmitoylated (Parat and Fox, 2001). *H-ras* is both myristoylated and palmitoylated and these modifications play a key role in its association with caveolae (Prior et al., 2001). In continuous cell lines, where the expression of caveolae is often much reduced, an association of PDGF-R with lipid rafts instead of caveolae has been reported (Matveev and Smart, 2002). While initial studies suggested that caveola-associated signaling proteins represented an inactive reservoir (Smart et al., 1999), subsequent findings were made of active signaling from caveolae (Gustavsson et al., 1999; Zhu et al., 2000). However, caveolar FC content was a major regulatory factor in all cases.

Caveolae are the terminus within the plasma membrane for newly synthesized FC as well as recycling lipoprotein-derived FC. In fibroblasts and vascular smooth muscle cells, FC loading was associated with both induction of caveolin synthesis as well as increased expression of caveolae and caveolar FC at the cell surface (Fielding et al., 1997; Thyberg et al., 1998). In aortic endothelial cells, caveolin levels, already high, were not further increased by FC loading, but increased amounts of FC became caveola-associated (Zhu et al., 2000). A decrease in cellular FC was associated with a decrease in cell surface caveolae, and down-regulation of caveolin synthesis (Hailstones et al., 1998). Caveolae were identified in primary fibroblasts and vascular smooth muscle cells as sites from which FC is preferentially transferred out of the cell (Fielding and Fielding, 1995; Fielding et al., 2002). It is not yet clear if this results directly from a destabilization of caveolar FC, for example following signal transduction, or if it involves the activity of caveolar ancillary proteins such as p-glycoprotein or SR-BI, to facilitate FC exchange between biological membranes (Liscovitch and Lavie, 2000; Liu et al., 2002). Together these data suggest a model for FC homeostasis in peripheral cells, in which a rise in cellular FC leads to an increase in both caveolin synthesis and the expression of cell surface caveolae. This in turn leads to an increase in FC efflux, and thus a decrease in cell FC, followed by downregulation of caveolin expression (Fielding et al., 1997).

Initiation of signal transduction from PDGF-R by PDGF is associated with a major, rapid decrease in the level of FC associated with caveolae. As much as 80% of initial FC content was lost within 5 min, at least in part via a transient 4-fold stimulation of FC efflux, when the physiological acceptor apolipoproptein A-1 (apo A-1) was present in the medium. Apo A-1 or lipid-poor prebeta-migrating HDL are the major acceptors of FC derived from caveolae. That this FC was directly derived from caveolae was shown by balance studies. The presence of apo A-1 (and stimulation of the loss of FC from caveolae) was associated with a 2–4 fold stimulation of protein kinase activity (Fielding et al., 2002). These data illustrate the dynamic relationship that exists between signal transduction and FC homeostasis in caveolae (Fig. 1). Whether other signals originating in caveolae are stimulated by FC efflux in a similar way remains to be determined.

The structure and regulation of raft-associated signal complexes have been determined mainly in blood leucocytes, where these make up the major proportion of FC-rich cell surface domains. A well-studied example is the T-cell immuno-globulin receptor. This signaling complex includes the proteins *lyn* and *lck*. Both are acylated, and the nonacylated proteins are inactive in signal transduction (Kovarova et al., 2001; Hawash et al., 2002; Lang et al., 2002). Raft integrity, and specifically the presence of FC, is also required. Unlike the case with caveolae, there is no evidence at present that the FC content of rafts is dynamically regulated. Their short lifetime probably limits such a role (Fig. 1).

To summarize, the structure, properties and functions of lipid rafts and caveolae, while significantly different, show some common features. The three-dimensional structure of caveolae permits the possibility of increased specificity in the efflux of FC, and amplification and dynamic control of signal transduction.

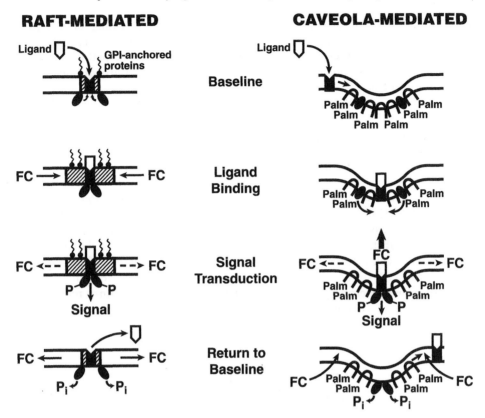

Fig. 1. Models for raft-mediated (left) and caveolae-mediated (right) signal transduction. The transient nature of raft complexes is compared with the semi-permanent existence of caveolae. The figure also shows the ability of caveolae to rapidly transfer FC to the extracellular space, in contrast to the lateral transfer in the plane of the membrane suggested for lipid rafts. The horseshoe shaped symbols represent caveolin; palm, palmitoylated caveolin.

3. Caveolae in cell division and aging

Cell division requires a doubling of cell-associated FC within a period of a few hours. In subconfluent primary skin fibroblasts, where the doubling time is 24–28 h, such an increase was complete within 8 h following S-phase, and preceded mitosis. This was achieved without any significant increase in cholesterogenesis above the minimal rates characteristic of quiescent cells (Fielding et al., 1999). The increase in cellular FC mass is the result both of an increase of the uptake of preformed lipoprotein FC, and a decrease in the rate of FC efflux. At the same time the expression of caveolae is reduced, and caveolin synthesis downregulated at the level of transcription. These effects are dependent on transcription factors E2F and p53. p53 acts as a tumor suppressor in many cell lines. Entry into E2F-dependent cell cycling depends on p53 downregulation, while p53 expression is associated with cell arrest. Chronic p53 overexpression leads to apoptosis.

The finding that the caveolin gene was p53-dependent (Fielding et al., 2000; Galbiati et al., 2001) may indicate that the downregulation of caveolae and caveolin expression represents a significant cell cycle regulatory pathway effective at the S/G2 interface, potentially involving both resistance to growth factors and retention of cellular FC.

In the aging cell, these relationships appear to be greatly modified. While the expression of p53 and caveolin are both increased, these changes were not associated with the expected increase in cell surface caveolae. High levels of total caveolin, but an absence of cell-surface caveolae, were reported in senescent human fibroblasts (Wheaton et al., 2001). These cells also lacked caveola-associated tyrosine kinases including the receptor proteins for several growth factors. In a second study, an increase in caveolin protein was found in aging cells, together with an increase in cytoplasmic vesicles that though identified as caveolae, obviously differed in structure and location from normal caveolae (Park et al., 2000). Instead, most caveolin in aging cells appeared to be present in intracellular vesicles. Here it could play no role in the stimulation of FC efflux or signaling via cell surface kinases. The aging cell is generally refractory to the effects of growth factors, probably at least in part associated with a reduction in the expression of caveolae. Together these data suggest that the induction of

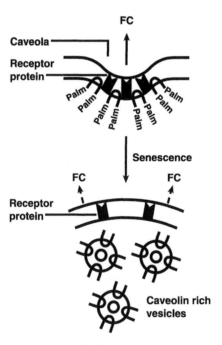

Fig. 2. Effects of aging/senescence on the distribution of caveolae and caveolin. The model shown suggests that the inability of senescent cells to palmitoylate caveolin may be central to the inability of caveolin to form cell surface caveolae, and the disruption of normal cell-surface tyrosine kinase-based signaling complexes.

caveolin expression is normal in aging cells, but that the conversion of this caveolin to caveolae functional at the cell surface is not.

Caveolin at the cell surface is palmitoylated at residues 133, 143 and 156. Early experiments did not distinguish between vesicles containing caveolin and cell surface caveolae, and reported that mutant, palmitoylation-defective caveolin was recovered, like wild-type caveolin, in the detergent-resistant membrane fraction (Dietzen et al., 1995) while observing that caveolin polymerization was stabilized by acylation (Monier et al., 1996). Functional studies of caveola-dependent signaling now show that in the absence of palmitoylation, normal caveolae are not formed (Uittenbogaard and Smart, 2000).

Oxidative stress is a significant factor in cellular aging, and stress induced in endothelial cells was associated with an inhibition of caveolin palmitoylation (Parat et al., 2002) and trafficking (Kang et al., 2000; Parat et al., 2002). This suggests that in aging cells, a decrease in cell surface caveolae, and the accumulation of intracellular caveolin vesicles mediated by oxidative stress, could be responsible for abnormalities in both signaling by receptor kinases, and potentially, in FC homeostasis (Fig. 2).

4. Summary

Several lines of investigation connect FC homeostasis and signaling via caveolae and lipid rafts. Caveolae appear to represent a stabilized, three-dimensional evolution of rafts, that can control their FC content dynamically, both during signal transduction and in response to lipid loading. In the normal dividing cell, this system limits signaling from the cell surface except to initiate the cell cycle, and retains FC as required for cell membrane synthesis, despite the presence of extracellular lipoprotein FC acceptors. In the aging cell, these mechanisms are deranged and while caveolin accumulates in intracellular vesicles, rather than in cell surface caveolae, and caveolin palmitoylation is inhibited. A detailed study of the relationship between palmitoylation status and caveolar function would be of great interest. Studies of the molecular basis of these effects should also offer many opportunities to understand the role of the plasma membrane in aging.

Acknowledgments

Research by the author by supported through the National Institutes of Health via Grants in Aid HL 57976 and HL67294.

References

Abrami, L., Fivaz, M., Kobayashi, T., Kinoshita, T., Parton, R.G., van der Groot, F.G., 2001. Cross-talk between caveolae and glycosylphosphatidylinositol-rich domains. J. Biol. Chem. 276, 30729–30736.

Bist, A., Fielding, C.J., Fielding, P.E., 2000. p53 regulates caveolin gene transcription, cell cholesterol and growth by a novel mechanism. Biochemistry 39, 1966–1972.

Dietzen, D.J., Hastings, W.R., Lublin, D.M., 1995. Caveolin is palmitoylated on multiple cysteine residues. Palmitoylation is not necessary for localization of caveolin to caveolae. J. Biol. Chem. 270, 6838–6842.

Fielding, C.J., Bist, A., Fielding, P.E., 1997. Caveolin mRNA levels are upregulated by free cholesterol and downregulated by oxysterols in fibroblast monolayers. Proc. Natl. Acad. Sci. USA 94, 3753–3758.

Fielding, C.J., Bist, A., Fielding, P.E., 1999. Intracellular cholesterol transport in synchronized human skin fibroblasts. Biochemistry 38, 2506–2513.

Fielding, P.E., Fielding, C.J., 1995. Plasma membrane caveolae mediate the efflux of cellular free cholesterol. Biochemistry 34, 14288–14292.

Fielding, P.E., Fielding, C.J., 1996. Intracellular transport of low-density lipoprotein-derived free cholesterol begins at clathrin-coated pits and terminates at cell surface caveolae. Biochemistry 35, 14932–14938.

Fielding, C.J., Fielding, P.E., 2002. Relationship between cholesterol trafficking and signaling in rafts and caveolae. Biochim. Biophys. Acta (in press).

Fielding, P.E., Russel, J.S., Spencer, T.A., Hakamata, H., Nagao, K., Fielding, C.J., 2002. Sterol efflux to apolipoprotein A-1 originates from caveolin-rich microdomains and potentiates PDGF-dependent protein kinase activity. Biochemistry 41, 4929–4937.

Gagescu, R., Demaurex, N., Parton, R.G., Hunziker, W., Huber, L.A., Gruenberg, J., 2000. The recycling endosome of Madin-Darby canine kidney cells is a mildly acidic compartment rich in raft components. Mol. Biol. Cell 11, 2775–2791.

Galbiati, F., Volonte, D., Liu, J., Capozza, F., Frank, P.G., Zhu, L., Pestell, R.G., Lisanti, M.P., 2001. Caveolin-1 expression negatively regulates cell cycle progression by inducing G(0)/G(1) arrest via a p53/p21(WAF-1/Cip1)-dependent mechanism. Mol. Biol. Cell 12, 2229–2244.

Gustavsson, J., Parpal, S., Karlsson, M., Ramsing, C., Thorn, H., Borg, M., Lindroth, M., Peterson, K.H., Magnusson, K.E., Stralfors, P., 1999. Localization of the insulin receptor in caveolae of adipocyte plasma membrane. FASEB J. 13, 1961–1971.

Hailstones, D., Sleer, L.S., Parton, R.G., Stanley, K.K., 1998. Regulation of caveolin and caveolae by cholesterol in MDCK cells. J. Lipid Res. 39, 369–379.

Hawash, I.Y., Hu, X.E., Adal, A., Cassady, J.M., Geahlen, R.L., Harrison, M.L., 2002. The oxygen substituted palmitic acid analogue, 13-oxypalmitic acid, inhibits Lck localization to lipid rafts and T cell signaling. Biochim. Biophys. Acta 1589, 140–150.

Igarashi, J., Thatte, H.S., Prabhakar, P., Golan, D.E., Michel, T., 1999. Calcium-independent activation of endothelial nitric oxide synthase by ceramide. Proc. Natl. Acad. Sci. USA 96, 12583–12588.

Iwabuchi, K., Handa, K., Hakamori, S., 1998. Separation of glycosphingolipid signaling domain from caveolin-containing membrane fraction in mouse melanoma B16 cells and its role in cell adhesion coupled with signaling. J. Biol. Chem. 273, 33766–33773.

Kang, Y.S., Ko, Y.G., Seo, J.S., 2000. Caveolin internalization by heat shock or hyperosmotic shock. Exp. Cell Res. 255, 221–228.

Koleske, A.J., Baltimore, D., Lisanti, M.P., 1995. Reduction of caveolin and caveolae in oncogenically transformed cells. Proc. Natl. Acad. Sci. USA 92, 1381–1385.

Kovarova, M., Tolar, P., Arunchandran, R., Draberova, L., Rivera, J., Draber, P., 2001. Structure-function analysis of Lyn kinase association with lipid rafts and initiation of early signaling events after Fcepsilon receptor I aggregation. Mol. Cell Biol. 21, 8318–8328.

Lang, M.L., Chen, Y.W., Shen, L., Gao, H., Lang, G.A., Wade, T.K., Wade W.F., 2002. IgA Fc receptor (FcalphaR) cross-linking recruits tyrosine kinases, phosphoinoside kinases and serine/threonine kinases to glycolipid rafts. Biochem. J. 364, 517–525.

Lavie, Y., Fiucci, G., Liscovitch, M., 1998. Up-regulation of caveolae and cavoelar constituents in multidrug-resistant cancer cells. J. Biol. Chem. 273, 32380–32383.

Lee, S.W., Reimer, C.L., Oh, P., Campbell, D.B., Schnitzer, J.E., 1998. Tumor cell growth inhibition by caveolin re-expression in human breast cancer cells. Oncogene 16, 1391–1397.

Liscovitch, M., Lavie, Y., 2000. Multidrug resistance: a role for cholesterol efflux pathways? Trends Biochem. Sci. 25, 530–534.

Liu, P., Wang, P., Michaely, P., Zhu, M., Anderson, R.G.W., 2000. Presence of oxidized cholesterol in caveolae uncouples active platelet-derived growth factor receptors from tyrosine kinase substrates. J. Biol. Chem. 275, 31648–31654.

Liu, T., Krieger, M., Kan, H.Y., Zannis, V.I., 2002. The effects of mutations in helices 4 and 6 of apo A-1 on scavenger receptor class B type I (SR-BI)-mediated cholesterol efflux suggest that formation of a productive complex between reconstituted high density lipoprotein and SR-BI is required for efficient lipid transport. J. Biol. Chem. 277, 21576–21584.

Matko, J., Bodnar, A., Vereb, G., Bene, L., Vamosi, G., Szentesi, G., Szollosi, J., Gaspar, R., Horejsi, V., Waldmann, T.A., Damjanovich, S., 2002. GPI-microdomains (lipid rafts) and signaling of the multi-chain interleukin-2 receptor in human lymphoma/leukermia T cell lines. Eur. J. Biochem. 269, 1199–1208.

Matveev, S.V., Smart, E.J., 2002. Heterologous desensitization of EGF receptors and PDGF receptors by sequestration in caveolae. Am. J. Physiol. 282, C935–C946.

Monier, S., Dietzen, D.J., Hastings, W.R., Lublin, D.M., Kurzchalia, T.V., 1996. Oligomerization of VIP21-caveolin *in vitro* is stabilized by long-chain fatty acylation or cholesterol. FEBS Lett. 388, 143–149.

Parat, M.O., Fox, P.L., 2001. Palmitoylation of caveolin-1 in endothelial cells is posttranslational but irreversible. J. Biol. Chem. 276, 15776–15782.

Parat, M.O., Stachowicz, R.Z., Fox, P.L., 2002. Oxidative stress inhibits caveolin-1 palmitoylation and trafficking in endothelial cells. Biochem. J. 361, 681–688.

Park, W.Y., Park, J.S., Cho, K.A., Kim, D.I., Ko, Y.G., Seo, J.S., Park, S.C., 2000. Up-regulation of caveolin attenuates epidermal growth factor signaling in senescent cells. J. Biol. Chem. 275, 20847–20852.

Pralle, A., Keller, P., Florin, E.L., Simons, K., Horber, J.K., 2000. Sphingolipid-cholesterol rafts diffuse as small entities in the plasma membrane of mammalian cells. J. Cell Biol. 148, 997–1008.

Prior, I.A., Harding, A., Yan, J., Sluimer, J., Parton, R.G., Hancock, J.F., 2001. GTP-dependent segregation of H-ras from lipid rafts is required for biological activity. Nature Cell Biol. 3, 368–375.

Sheets, E.D., Lee, G.M., Simson, R., Jacobson, K., 1997. Transient confinement of a glycosylphosphatidylinositol-anchored probe in the plasma membrane. Biochemistry 36, 12449–12458.

Smart, E.J., Graf, G.A., McNiven, M.A., Sessa, W.C., Engelman, J.A., Scherer, P.E., Okamoto, T., Lisanti, M.P., 1999. Caveolins, liquid-ordered domains, and signal transduction. Mol. Cell Biol. 19, 7289–7304.

Sowa, G., Liu, J., Papapetropoulos, A., Rex-Haffner, M., Hughes, T.E., Sessa, W.C., 1999. Trafficking of endothelial nitric oxide synthase in living cells. Quantitative evidence supporting the role of palmitoylation as a kinetic trapping mechanism limiting membrane diffusion. J. Biol. Chem. 274, 22524–22531.

Sowa, G., Pypaert, M., Sessa, W.C., 2001. Distinction between signaling mechanisms in lipid rafts vs. caveolae. Proc. Natl. Acad. Sci. USA 98, 14072–14077.

Thomsen, P., Roepstorff, K., Stahlhut, M., van Deurs, B., 2002. Caveolae are highly immobile plasma membrane microdomains, which are not involved in constitutive endocytic trafficking. Mol. Biol. Cell 13, 238–250.

Thyberg, J., Calara, F., Dimayuga, P., Nilsson, J., Regnstrom, J., 1998. Role of caveolae in cholesterol transport in arterial smooth muscle cells exposed to lipoproteins *in vitro* and *in vivo*. Lab. Invest. 78, 825–837.

Uittenbogaard, A., Ying, Y., Smart, E.J., 1998. Characterization of a cytosolic heat-shock protein-caveolin chaperone complex. Involvement in cholesterol trafficking. J. Biol. Chem. 273, 6525–6532.

Uittenbogaard, A., Smart, E.J., 2000. Palmitoylation of caveolin-1 is required for cholesterol binding, chaperone complex formation, and rapid transport of cholesterol to caveolae. J. Biol. Chem. 275, 25595–25599.

Vainio, S., Heino, S., Mansson, J.E., Fredman, P., Kuismanen, E., Vaarala, O., Ikonen, E., 2002. Dynamic association of human insulin receptor with lipid rafts in cells lacking caveolae. EMBO Rep. 3, 95–100.

Wheaton, K., Sampsel, K., Boisvert, F.M., Davy, A., Robbins, S., Riabowol, K., 2001. Loss of functional caveolae during senescence of human fibroblasts. J. Cell Physiol. 187, 226–235.

Zhu, Y., Liao, H.L., Wang, N., Yuan, Y., Ma, K.S., Verna, L., Stemerman, M.B., 2000. Lipoprotein promotes caveolin-1 and Ras translocation to caveolae: role of cholesterol in endothelial signaling. Arterio. Thromb. Vasc. Biol. 20, 2465–2470.

Advances in
Cell Aging and
Gerontology

Overview: Membrane lipid peroxidation

Kathleen S. Montine*, Joseph F. Quinn, and Thomas J. Montine

University of Washington, Department of Pathology, Harborview Medical Center, Box 359791, Seattle, WA 98104, USA.
Correspondence address: Tel: + 1-206-341-5244; E-mail address: kmontine@u.washington.edu

Contents

Abbreviations

AD: Alzheimer's disease; AGE: Advanced glycosylation endproduct; ALS: Amyotrophic lateral sclerosis; Apo: apopoliprotein; Aβ: Amyloid beta; CSF: Cerebrospinal fluid; EAAT2: Excitatory amino acid transporter; Gpx1: Glutathione peroxidase gene; HHE: 4-hydroxy-2-hexenal; HNE: 4-hydroxy-2-nonenal; L$^\bullet$: Carbon-based lipid radical; LH: Fatty acyl chain; LDL: Low density lipoprotein; LO$^\bullet$: Lipid alkoxyl radical; LOO$^\bullet$: Lipid peroxyl radical; LOOH: Lipid hydroperoxide; MDA: Malondialdehyde; MPTP: 1-methyl-4-phenyltetrahydropyridine; NFT: Neurofibrillary tangles; PAF: Platelet activating factor; PD: Parkinson's disease; RAGE: Receptor for AGE's; ROS: Reactive oxygen species; SOD1: Cu/Zn-superoxide dismutase; TBARS: Thiobarbiturate reactive substances.

Advances in Cell Aging and Gerontology, vol. 12, 11–26

1. Introduction

Free radical-mediated injury, termed oxidative damage, is thought to be a central contributor to processes of aging as well as the pathogenesis of several age-related diseases. The biochemistry of free radical-mediated injury is complex and its sequellae and the response to such injury are protean. In this chapter we will review the biochemical mechanisms of lipid peroxidation, one of the major outcomes of free radical-mediated injury to tissue, and the possible contributions of lipid peroxidation to common age-related diseases.

2. Biochemistry of lipid peroxidation

2.1. Peroxidation of fatty acyl groups

Peroxidation of fatty acyl groups, mostly in membrane phospholipids, has three phases: initiation, propagation, and termination (Fig. 1). Initiation occurs when a hydrogen atom is abstracted from a fatty acyl chain (LH), leaving behind a carbon-based lipid radical (L$^{\bullet}$). Hydrogen atoms can be abstracted by carbon-, nitrogen-, oxygen-, or sulfur-based radicals. Among the oxygen-based radicals, $^{\bullet}$OH is the most effective at hydrogen atom abstraction. Allylic hydrogens are most labile to

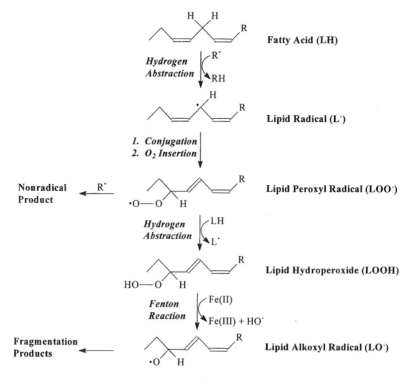

Fig. 1. Chemical diagram of the steps in lipid peroxidation.

abstraction because their carbon–hydrogen bond is made more acidic by the adjacent carbon–carbon double bond. Therefore, polyunsaturated fatty acids are the most vulnerable to lipid peroxidation. Following rearrangement to form a conjugated diene, propagation of lipid peroxidation begins with the reaction of L• with molecular oxygen to form a lipid peroxyl radical (LOO•). This extremely reactive species can abstract a second hydrogen atom from a nearby fatty acyl chain (L'H) to generate a lipid hydroperoxide (LOOH), thus forming a new carbon-based lipid radical (L'•) that can initiate another round of hydrogen atom abstraction. LOOH's may undergo Fenton-type reaction to generate lipid alkoxyl radicals (LO•) that then fragment to form a variety of hydrocarbons and reactive aldehydes. Finally, termination of lipid peroxidation occurs when two radical species react with each other to form a nonradical product. Overall, lipid peroxidation is a self-propagating process that will proceed until substrate is consumed or termination occurs. Cellular antioxidant systems may intercede at either the first or second phase. Superoxide dismutases, catalases or paramagnetic ion chelators prevent initiation of lipid peroxidation by eliminating radicals or sources of radicals before they react with lipid. In contrast, ascorbate, α-tocopherol, and reduced glutathione act by suppressing propagation of lipid peroxidation.

There are two broad outcomes to lipid peroxidation, *viz.*, structural damage to membranes and generation of bioactive secondary products. Structural membrane damage derives from the generation of fragmented fatty acyl chains, lipid–lipid crosslinks, and lipid-protein crosslinks (Farber, 1995). In addition, lipid hydroperoxyl radicals can undergo endocyclization to produce novel fatty acid esters that may disrupt membranes. Two classes of cyclized fatty acids are the isoprostanes and neuroprostanes, derived *in situ* from free radical-mediated peroxidation of arachidonyl or docosahexadonyl esters, respectively (Morrow and Roberts, 1997; Roberts et al., 1998). In total, these processes combine to produce changes in the biophysical properties of membranes that can have profound effects on the activity of membrane-bound proteins.

2.2. *Lipid peroxidation products*

Bioactive secondary products of lipid peroxidation can be generated by a number of mechanisms. Fragmentation of LOOH's, in addition to producing abnormal fatty acid esters, also liberates a number of diffusable products, some of which are potent electrophiles (Esterbauer et al., 1991; Porter et al., 1995). The most abundant diffusable products of lipid peroxidation are chemically reactive aldehydes such as malondialdehyde (MDA), acrolein, 4-hydroxy-2-nonenal (HNE) from ω-6 fatty acyl groups, and 4-hydroxy-2-hexenal (HHE) from ω-3 fatty acyl groups (Esterbauer et al., 1991) (Fig. 2). Other highly chemically reactive products are also formed but these presumably remain esterified to lipid; examples of these are the isoketals and neuroketals (Roberts et al., 1999; Bernoud-Hubac et al., 2001). Alternatively, hydrolysis of oxygenated fatty acyl groups generated during lipid peroxidation can liberate novel products from damaged lipid. For example, free isoprostanes are easily detectable in

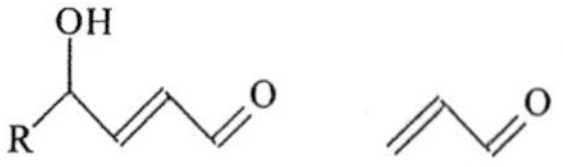

HNE: R = C$_5$H$_{11}$ Acrolein
HHE: R = C$_2$H$_5$

Fig. 2. Chemical diagram of some of the major reactive products of lipid peroxidation.

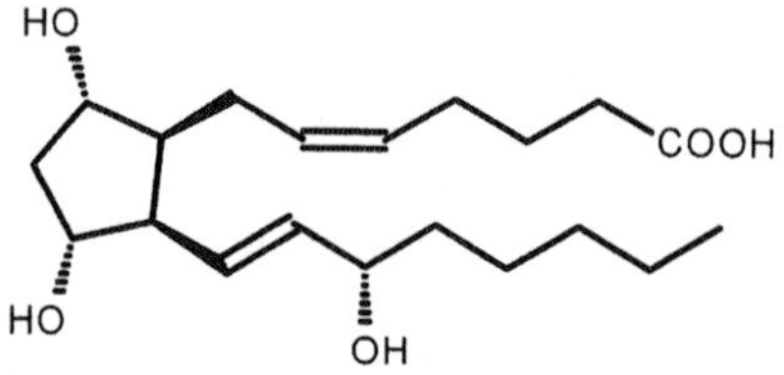

Fig. 3. Chemical diagram of 15-F$_{2t}$-isoprostane.

plasma, urine, and cerebrospinal fluid (CSF) (Morrow et al., 1997; Montine T.J. et al., 1998, 1999a,b; Roberts et al., 1998).

Some of these secondary products are thought to contribute significantly to the deleterious effects of lipid peroxidation in tissue. Reactive aldehydes from lipid peroxidation adduct a number of cellular nucleophiles, including proteins, nucleic acids, and some lipids (Esterbauer et al., 1991). Indeed, many of the cytotoxic effects of lipid peroxidation can be reproduced experimentally using electrophilic lipid peroxidation products such as HNE (Farber, 1995). These include depletion of glutathione, dysfunction of structural proteins, reduction in enzyme activities, and induction of cell death. Chemically stable products of lipid peroxidation may also contribute to the pathogenesis of lipid peroxidation through receptor-mediated signaling. For example, peroxidation and fragmentation of polyunsaturated fatty acyl groups in phosphatidylcholines can generate platelet activating factor (PAF) analogues that stimulate the PAF receptor (McIntyre et al., 1999). Also, at least one isomer of the F-ring isoprostanes, 15-F$_{2t}$-IsoP (Fig. 3), has been shown to be a potent vasoconstrictor and platelet activator, likely through receptor-mediated mechanisms (Audoly et al., 2000).

Advanced glycosylation endproducts (AGE's) are complex post-translational modifications of protein derived from reducing sugars and related carbohydrates and have been studied extensively as contributors to most, if not all, sequellae of adult-onset diabetes mellitus (Booth et al., 1997). AGE formation follows protracted elevation of glucose concentrations, a condition classified as substrate stress. Importantly, recent data demonstrate that AGE's may also form from lipid peroxidation in the absence of substrate stress (Fu et al., 1996). Therefore, both substrate stress and lipid peroxidation may converge in a common pathway of AGE formation. Like other products of lipid peroxidation, AGE's contribute to tissue injury by inactivation of critical proteins; however, AGEs also lead to reactive oxygen species (ROS) production either through direct chemical means or by stimulation of the receptor for AGE's (RAGE) that leads to free radical production via cell signaling pathways (Smith M.A. et al., 1995).

2.3. Quantification of lipid peroxidation

Quantification of lipid peroxidation has been pursued by measuring several of the secondary products mentioned above. A very important distinction to consider is whether the assay is measuring lipid peroxidation occurring *in vitro* or *in vivo*. Assays such as those for thiobarbiturate reactive substances (TBARS) or chromatography for specific secondary products accurately measure lipid peroxidation in a controlled *in vitro* system where further metabolism of the lipid peroxidation products does not occur. However, in more complicated model systems such as cell culture or and *in vivo*, extensive metabolism of electrophilic lipid peroxidation products compromises the accuracy of these assays (Gutteridge and Halliwell, 1990; Moore and Roberts, 1998). One solution to the problem of accurately quantifying lipid peroxidation *in vivo* has been the measurement of a specific class of isoprostanes, the F_2-isoprostanes, which are chemically stable products of free radical-mediated damage to arachidonyl esters that are not extensively metabolized *in situ* (Morrow et al., 1997). Indeed, tissue, plasma, urine, and CSF F_2-isoprostanes are now widely used as highly accurate *in vivo* biomarkers of lipid peroxidation in experimental models and patients.

3. Lipid peroxidation in age-related diseases

Increased lipid peroxidation has been correlated to varying extents with common age-related diseases: adult onset diabetes mellitus, atherosclerosis, some types of cancer, and some forms of neurodegeneration. Two major questions remain in all of these diseases: does lipid peroxidation contribute to disease initiation or progression as opposed to being a consequence of disease, and if so, by what mechanisms? There is a very large and rapidly expanding literature on lipid peroxidation in age-related diseases. We will briefly review the findings on the role of lipid peroxidation in atherosclerosis and cancer, and then focus on the contributions of lipid peroxidation to neurodegeneration.

3.1. Atherosclerosis

Atherosclerosis is a complex process that involves a pathologic interaction among several components including blood, endothelial cells, macrophages and vascular smooth muscle cells. The role of lipid peroxidation in atherosclerosis has focused on oxidative damage to low density lipoprotein (LDL) by radicals thought to be produced from macrophages or endothelial cells. Oxidized LDL has not only modified lipid from the structural alterations caused by lipid peroxidation, but has also modified protein, apolipoprotein (apo) B, which is adducted with secondary products of lipid peroxidation like HNE. Indeed, the modifications of apoB by lipid peroxidation have been studied in great detail. The pathologic consequences of exposing vascular cells to oxidatively modified LDL are many and include (i) avid incorporation by macrophage scavenger receptors, in contrast to the apoB/E receptor for unmodified LDL, and the promotion of foam cell formation, (ii) chemotaxis, (iii) stimulation of cytokine and growth factor release, (iv) cytotoxicity to endothelial cells and vascular smooth muscle cells, and (v) humoral immune response.

It has been clearly shown that LDL is oxidized *in vivo* and that lipid peroxidation products and modified LDL do accumulate in atherosclerotic lesions (Palinski et al., 1989). For example, HNE-apoB adducts and F-ring isoprostanes have been immunolocalized to atherosclerotic plaques (Jurgens et al., 1993; Pratico et al., 1998b). Moreover, the results of experimental studies outlined above indicate that the oxidized LDL has many opportunities to promote the progression of atherosclerosis. However, large randomized clinical trials with α-tocopherol at doses that presumably suppress lipid peroxidation had mixed results (Stephens et al., 1996; Yusuf et al., 2000). Major questions that remain to be resolved are what is the source(s) of free radicals responsible for LDL oxidation *in vivo* and which of the many pathologic effects of oxidized LDL demonstrated in experimental models is actually contributing to atherogenesis *in vivo* (Chisolm and Steinberg, 2000).

3.2. Carcinogenesis

ROS-mediated damage to DNA can be direct, resulting in single- and double-strand breaks, abasic sites, and a number of chemically modified nucleotides including thymine and cytosine glycols, 5-hydroxycytidine, 8-hydroxyguanine, and 8-hydroxyadenine (Preston and Hoffman, 2001). In addition to direct attack by ROS, reactive products of lipid peroxidation also modify DNA chemically to form a variety of species including cyclic propano adducts (Burcham, 1998). Indeed, recent advances in analytical methods have shown that such modifications account for a significant subset of endogenous DNA damage. This may be due in part to the self-progating nature of lipid peroxidation compared to a one-time attack of ROS on DNA. The magnitude of oxidative damage to DNA in humans has been estimated in one study that quantified urinary concentrations of 8-hydroxyguanine and thymine glycol; these investigators estimated that approximately 20,000 bases in DNA per cell per day are damaged by ROS.

Base excision is thought to be especially important in repair of oxidative damage to DNA. Mutations can occur with misrepair or nonrepair, the common ones being C → T substitutions, G → C transversions, and G → T transversions. However, it must be stressed that the common oxidative damage-induced mutations are not specific because they too are produced by errors in DNA polymerase copying of undamaged DNA (Jackson and Loeb, 2001). Thus, while there is substantial experimental evidence that oxidative damage-induced mutations can contribute to initiation or progression of cancer, this lack of specificity confounds definite demonstration of oxidative damage as cause of malignancy.

3.3. Neurodegeneration

3.3.1. Alzheimer's disease

There is compelling evidence that the magnitude of lipid peroxidation in the brains of Alzheimer's disease (AD) patients examined *post mortem* exceeds that in age-match control individuals. Seminal experiments demonstrated significantly increased TBARS in diseased regions of AD brain compared to age-matched control individuals (Lovell et al., 1995). Others have measured free HNE and acrolein in AD brain tissue and have shown that both are elevated in diseased regions of AD brain compared to controls (Audoly et al., 2000).

F_2-isoprostane levels are elevated in frontal lobe and hippocampus of AD patients compared to controls with short *post mortem* intervals (Pratico et al., 1998a; Montine T.J. et al., 1999c). In addition, F_2-isoprostanes are elevated in the cerebral cortex of aged homozygous apoE gene-deficient mice (Montine T.J. et al., 1999d; Pratico et al., 1999). A class of free radical-generated products analogous to the F_2-isoprostanes, but generated from docosahexaenoic rather than arachidonic acid, have been described and termed F_4-neuroprostanes (Roberts et al., 1998). Because docosahexaenoic acid is more labile to peroxidation than arachidonic acid and because docosahexaenoic acid is highly enriched in brain, it was proposed that F_4-neuroprostanes may be more sensitive markers of brain oxidative damage than F_2-isoprostanes. Indeed, F_4-neuroprostanes are significantly more abundant than F_2-isoprostanes in the cerebral cortex of aged homozygous *apoE* deficient mice (Montine T.J. et al., 1999d). One group has reported that F_4-neuroprostanes (called F_4-isoprostanes in their publication) are elevated in temporal and occipital, but not parietal lobes of AD patients compared to controls, and that levels of F_4-neuroprostanes are greater than those of F_2-isoprostanes in these regions (Nourooz-Zadeh et al., 1999); however, interpretation of data from this study is limited by long *post mortem* intervals (Nourooz-Zadeh et al., 1999). Finally, others have reported a significant increase in cerebral F_2-isoprostane levels in transgenic mice that express a mutant form of the human amyloid precursor protein gene (Tg2567), and a correlation between F_2-isoprostane and $A\beta$ peptides levels up 18 months of age (Pratico et al., 2001); however, we have been unable to reproduce this finding in the same mice aged up to 20 months (Fig. 4). It is worth considering that Tg2567 mice, although an excellent model of cerebral $A\beta$ amyloidogenesis, are a limited model of AD since they exhibit minimal neuronal loss as well as lack of other features of AD. Whether or not there is increased lipid peroxidation in the cerebrums of these mice may be a critical issue in determining the potential contribution of lipid peroxidation to neurodegeneration. If these mice do experience increased cerebral lipid peroxidation in the absence of significant neuronal loss, then one is left with the conclusion that increased lipid peroxidation, even when combined with cerebral $A\beta$ peptide deposition, is not sufficient to cause significant neuronal loss. Alternatively, if these mice do not have significantly increased lipid peroxidation, then a role for increased lipid peroxidation in neurodegeneration remains a possibility.

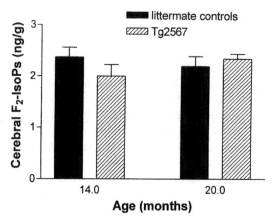

Fig. 4. Cerebral F_2-isoprostane levels in Tg2567 mice that express a mutant form of human amyloid precursor protein gene and that accumulate Aβ peptides in brain compared to their littermate wild type controls.

In contrast to quantification, several groups have also studied the localization of lipid peroxidation products in AD brain. These studies have used immuno-chemical detection of protein covalently modified by lipid peroxidation products or displaying protein carbonyls. There has been broad agreement among these studies. Consistent with the quantitative studies described above, hippocampus and cerebral cortex from AD patients display protein modifications that are undetect-able or barely detectable in the corresponding brain regions from age-matched control individuals (Montine K.S. et al., 1997a,b, 1998; Sayre et al., 1997; Smith, M.A. et al., 1998b; Calingasan et al., 1999). Another study found proteins modified by lipid peroxidation products in diseased regions of AD brain that are not present in uninvolved brain regions. Within diseased regions of AD brain, neuronal cytoplasm and neurofibrillary tangles (NFT's) are the major focus of protein modification; however, none of these studies observed modified proteins in or adjacent to neuritic plaques. This stands in sharp contrast to what is seen in genetically modified mice expressing mutant human amyloid precursor protein, where increased HNE-protein adduct immunoreactivity and AGE immuno-reactivity are localized adjacent to or within amyloid deposits (Smith et al., 1998a). One group observed that the tissue distribution of HNE-protein adducts in *post mortem* human brain varies with *APOE* genotype. In these studies, one of the chemical forms of HNE-protein adducts, the 2-pentylpyrrole adduct, co-localized with NFT's and was significantly associated with homozygosity for the $\varepsilon 4$ allele of *APOE* (Montine, K.S. et al., 1997a,b). The most abundant chemical form of HNE-protein adducts, the Michael adduct, was observed in both pyramidal neuron and astrocyte cytoplasm of AD patients with an $\varepsilon 3$ allele of *APOE*, but only in pyramidal neuron cytoplasm of AD patients homozygous for $\varepsilon 4$ (Montine et al., 1998). These authors have suggested that the tissue distribution and biochemical structure, not the apparent quantity, of HNE-protein adducts in

AD is influenced by *APOE* genotype, perhaps related to the essential role of apoE in central nervous system lipid trafficking.

Studies of cultured hippocampal and cortical neurons, of cortical synaptosome preparations, and of rodent models relevant to the pathogenesis of AD have provided considerable role support for lipid peroxidation and HNE in the synaptic dysfunction and neuronal degeneration that occurs in AD. Exposure of neurons or synaptosomes to amyloid beta-peptide results in lipid peroxidation and HNE production (Mark et al., 1997a; Keller et al., 1997). HNE, in turn, impairs the function of ion-motive ATPases, glucose and glutamate transporters, and GTP-binding proteins resulting in dysregulation of cellular ion homeostasis and increased vulnerability of neurons to excitotoxicity and apoptosis (Mark et al., 1997a,b; Keller et al., 1997; Kruman et al., 1997; Blanc et al., 1997, 1998). Moreover, HNE can induce changes in the microtubule-associated protein tau similar to those present in the neurofibrillary tangles of AD patients (Mattson et al., 1997). Finally, infusion of iron or HNE into the basal forebrain of rats causes depletion of acetylcholine and impairment of learning and memory (Bruce-Keller et al., 1998). Interestingly, apolipoproteins E2 and E3 bind more HNE than does apolipoprotein E4, and this may account for the superior abilities of E2 and E3 to protect neurons against oxidative stress-induced death (Pedersen et al., 2000). Collectively, these findings suggest a major role for lipid peroxidation and HNE in the pathogenesis of neuronal dysfunction and degeneration in AD.

The above studies of brain tissue all used material collected *post mortem*. AD patients undergoing *post mortem* examination typically have advanced disease and an average duration of dementia of 8–12 years. Therefore, a serious limitation to analysis of tissue obtained *post mortem* is that the increased brain lipid per-oxidation in AD patients might be a late stage consequence of disease. Obviously, a late stage consequence of AD would be a less attractive therapeutic target than a process contributing to disease progression at an earlier stage. More recently, CSF has been investigated as an accessible source of central nervous system tissue for the assessment of brain lipid peroxidation in earlier stages of AD. Initial studies measured CSF obtained from the lateral ventricles *post mortem* to determine the feasibility of this approach. One study measured free HNE in CSF obtained from the lateral ventricles *post mortem* and showed that its concentration is significantly elevated in AD patients compared to age-matched controls (Lovell et al., 1997). A few studies have determined the concentration of F_2-isoprostanes, and in one case F_4-neuroprostanes, in CSF obtained from the lateral ventricles *post mortem* and have shown significant elevations in AD patients compared to age-matched controls (Montine, T.J. et al., 1998, 1999c). Importantly, CSF F_2-isoprostane concentrations in AD patients are significantly correlated with decreasing brain weight, degree of cerebral cortical atrophy, and increasing Braak stage, but not with *APOE* genotype or the tissue density of neuritic plaques or NFT's (Montine, T.J. et al., 1999c).

More recent studies have focused on probable AD patients early in the course of dementia. F_2-isoprostane levels are reproducibly increased in CSF obtained *intra vitam* from the lumbar cistern of AD patients early in the course of their dementia

(Montine, T.J. et al., 1999a; Pratico, D. et al., 2000) compared to age-matched controls. In contrast, the literature on peripheral F_2-isoprostanes in AD has been conflicting. Four small studies have reported plasma or urine F_2-IsoP levels in patients with mild to moderate dementia and in age-matched controls. Two concluded that neither plasma nor urine F_2-IsoPs are elevated in AD patients compared to controls (Feillet-Coudray et al., 1999; Montine, T.J. et al., 2000), one concluded that plasma but not urine F_2-IsoPs are elevated in AD patients versus controls (Waddington et al., 1999), and the last concluded that both plasma and urine F_2-IsoPs are elevated in AD patients compared to controls (Pratico et al., 2000); however, this study was unique among the four in that it did not exclude individuals who smoke, a behavior known to elevate plasma and urine F_2-isoprostanes (Morrow et al., 1995). We recently completed the largest study of 56 AD patients and 34 controls (smokers excluded) and observed no difference in urine levels of F_2-isoprostanes or their major metabolite. In addition, urine and CSF F_2-isoprostanes levels in 32 AD patients did not correlate. Supporting these conclusions, elevated rat cerebral F_2-isoprostanes and F_4-neuroprostanes following systemic exposure to kainic acid were not associated with a significant change in plasma or urine levels from the same animals. These results show that plasma and urine F_2-IsoPs and F_4-NeuroPs do not accurately reflect CNS levels of these biomarkers and are not reproducibly elevated in body fluids outside of CNS in AD patients.

In combination, these data indicate that regionally increased lipid peroxidation in brain is a feature of AD that correlates with disease severity at end stages and that is increased early in the course of this disease. Lack of a reproducible increase in lipid peroxidation markers in peripheral body fluids suggests that oxidative damage to AD is not systemic but rather focused in the central nervous system. Finally, it remains unclear what role lipid peroxidation may be playing in mouse models of $A\beta$ cerebral amyloidogenesis. As with atherosclerosis, possible sources and mechanisms for excess free radical generation in diseased regions of AD abound in model systems, but none has yet been proven as a dominant pathway *in vivo*. In addition, what role lipid peroxidation and its products play in the initiation of progression of AD is not clear.

3.3.2. Parkinson's disease

Idiopathic Parkinson's Disease (PD) is the second most common age-related neurodegenerative disease. Similar to AD, there is compelling evidence from human *post mortem* tissue that increased oxidative stress occurs in the midbrain of patients with PD compared with age-matched controls (Coyle and Puttfarcke, 1993; Cohen and Werner, 1994; Jenner and Olanow, 1998). However, the mechanisms of oxidative stress may differ somewhat from AD. For example, depletion of nigral-reduced glutathione is proposed to be a relatively early and specific event in PD (Jenner, 1994). Lipid peroxidation also appears to be a component of nigral degeneration in PD when examined *post mortem*. Compared with controls, the midbrains of PD patients have elevated TBARS and LOOH's as well as immunochemically detectable HNE-protein adducts; free HNE is elevated in CSF of patients with PD compared to controls (Dexter et al., 1989, 1994; Yoritaka et al., 1995; Shelley, 1998). Interestingly, HNE-protein adducts were present in several midbrain nuclei in PD patients, not just those in the *substantia nigra* (Yoritaka et al., 1995), a pattern similar to 8-hydroxyguanosine immunoreactivity in midbrain from PD patients (Zhang et al., 1999). Despite the *post mortem* evidence that associates PD with increased midbrain lipid peroxidation, negative results with high dose α-tocopherol supplementation in the large clinical trials have questioned the significance of lipid peroxidation earlier in the course of PD (DATATOP, 1989).

In humans and some mammals including mice, exposure to 1-methyl-4-phenyltetrahydropyridine (MPTP) produces selective degeneration of dopaminergic neurons in the central nervous system. MPTP-induced nigral dopaminergic degeneration in animals has been utilized widely as a model of PD (Langston and 1994) and is thought to involve mitochondrial dysfunction and oxidative damage. Indeed, mice lacking both alleles of the Cu/Zn-superoxide dismutase gene (*sod1*) or the glutathione peroxidase gene (*gpx1*) are significantly more vulnerable to MPTP-induced dopaminergic neurodegeneration than littermate controls (Zhang et al., 2000). Moreover, brainstems of mice systemically exposed to MPTP show an acute 8-fold increase in HNE concentrations, a 50% reduction in reduced glutathione levels, and a 6-fold increase in the concentration of HNE-glutathione adducts within 24 hours of exposure (Shelley, 1998).

While these data suggest that some lipid peroxidation products may participate in the pathogenesis of MPTP-induced dopaminergic neurodegeneration and demonstrate increased lipid peroxidation in late stage PD, the role of nigral lipid peroxidation in earlier stages of PD is not clear.

3.3.3. Amyotrophic lateral sclerosis

Similar to AD and PD, amyotrophic lateral sclerosis (ALS) has both familial and sporadic forms. Research into the role of oxidative damage in ALS has been fueled by the discovery that mutations in the gene for Cu/Zn-SOD (*SOD1*) are the cause of a subset of familial ALS (Rosen et al., 1993). Indeed, some lines of mice expressing ALS-linked *SOD1* mutations ALS develop a disease phenotype that closely mimics familial ALS and that is thought to derive in part from increased free radical-mediated damage (DalCanto and Gurney, 1994; Gurney et al., 1994; Tu et al., 1997). *Post mortem* examination of patients with sporadic ALS has shown increased protein carbonyl formation and other signatures of lipid peroxidation in motor cortex, although TBARS were not elevated, and immunoreactivity for MDA-protein adducts in spinal cord, which was absent in controls (Ferrante et al., 1997). Others have shown that lumbar spinal cord obtained *post mortem* from sporadic ALS patients has immunohistochemically detectable HNE-protein adducts in the anterior horn (Pederson et al., 1998). These investigators also showed by immunoprecipitation that one of the modified proteins is an excitatory amino acid transporter (EAAT2). Interestingly, familial ALS patients, either with or without mutations in *SOD1*, do not have these motor cortex changes seen in sporadic ALS, but do have similar spinal cord changes (Ferrante et al., 1997).

Two studies have examined CSF from living ALS patients for evidence of increased lipid peroxidation earlier in the disease. One study measured free HNE in CSF obtained from the lumbar cistern of sporadic ALS patients at initial diagnosis and prior to therapy (Smith et al., 1998). HNE levels were measured using high performance liquid chromatography with fluorescence detection following derivatization of CSF. This study demonstrated a significant elevation in HNE levels in CSF from sporadic ALS patients compared to patients with several other neurodegenerative diseases, but not when compared to patients with Guillain-Barre syndrome or chronic inflammatory demyelinating polyneuropathy (Smith et al., 1998). In a separate, smaller study, F_2-isoprostanes in CSF from sporadic ALS patients and age-matched controls were measured using chromatography followed by mass spectrometry; there was no significant difference between the two groups (Montine et al., 1999a). The average disease duration in this group of ALS patients was approximately 2 years and many of the patients had already initiated therapy.

In summary, data from *post mortem* studies consistently associate ALS with increased lipid peroxidation in spinal cord and point to the modification of a specific protein by lipid peroxidation products as a mechanism whereby lipid peroxidation products could contribute to disease progression. Moreover, a closely related animal model of familial ALS indicates that free radical-mediated damage may contribute to disease progression. Finally, some but not all data from CSF support a role for increased lipid peroxidation early in the course of sporadic ALS.

4. Summary

Lipid peroxidation has been associated with many pathological conditions, especially age-related diseases. While this association is becoming tighter and the body of evidence strongly supporting a role for lipid peroxidation in the pathogenesis of atherosclerosis, cancer, and neurodegeneration, significant questions remain. What are the major sources of free radicals in each of these diseases? Answers to this question will have important therapeutic implications. What are the mechanisms by which lipid peroxidation contributes to the initiation or progression of disease? Answers to this question will deepen our understanding of the pathogenesis of age-related diseases.

References

Audoly, L.P., Rocca, B., Fabre, J.E., Koller, B.H., Thomas, D., Loeb, A.L., Coffman, T.M., FitzGerald, G.A., 2000. Cardiovascular responses to the isoprostanes $iPF_{2\alpha}$-III and iPE_2-III are mediated via the thromboxane A_2 receptor *in vivo*. Circulation 101, 2833–2840.

Bernoud-Hubac, N., Davies, S.S., Boutaud, O., Montine, T.J., Roberts II, L.J., Morrow, J.D., 2001. Formation of highly reactive gamma-ketoaldehydes (neuroketals) as products of the neuroprostane pathway. J. Biol. Chem. 276, 30964–30970.

Blanc, E.M., Kelly, J.F., Mark, R.J., Mattson, M.P., 1997. 4-hydroxynonenal, an aldehydic product of lipid peroxidation, impairs signal transduction associated with muscarinic acetylcholine and metabotropic glutamate receptors: possible action on Gaq/11. J. Neurochem. 69, 570–580.

Blanc, E.M., Keller, J.N., Fernandez, S., Mattson, M.P., 1998. 4-hydroxynonenal, a lipid peroxidation product, inhibits glutamate transport in astrocytes. Glia 22, 149–160.

Booth, A.A., Khalifah, R.G., Todd, P., Hudson, B.G., 1997. *In vitro* kinetic studies of formation of antigenic advanced glycation end products (AGEs). J. Biol. Chem. 272, 5430–5437.

Bruce-Keller, A.J., Li, Y.-J., Lovell, M.A., Kraemer, P.J., Gary, D.S., Brown, R.R., Markesbery, W.R., Mattson, M.P., 1998. 4-hydroxynonenal, a product of lipid peroxidation, damages cholinergic neurons and impairs visuospatial memory in rats. J. Neuropathol. Exp. Neurol. 57, 257–267.

Burcham, P.C., 1998. Genotoxic lipid peroxidation products: their DNA damaging properties and role in formation of endogenous DNA adducts. Mutagenesis 13, 287–305.

Calingasan, N.Y., Uchida, K., Gibson, G.E., 1999. Protein-bound acrolein: a novel marker of oxidative stress in Alzheimer's disease. J. Neurochem. 72, 751–756.

Chisolm, G.M., Steinberg, D., 2000. The oxidative modification hypothesis of atherogenesis: an overview. Free Rad. Biol. Med. 28, 1815–1826.

Cohen, G., Werner, P., 1994. Free Radicals, Oxidative Stress, and Neurodegeneration, In: Calne, D.B. (Ed.). Neurodegenerative diseases. WB Saunders, Philadelphia, PA, pp. 139–162.

Coyle, J.T., Puttfarcken, P., 1993. Oxidative stress, glutamate, and neurodegenerative disorders. Science 262, 689–695.

Dal Canto, M.C., Gurney, M.E., 1994. Development of central nervous system pathology in a murine transgenic model of human amyotrophic lateral sclerosis. Am. J. Pathol. 145, 1271–1280.

DATATOP Study Group, 1989. Effect of deprenyl on the progression of disability in early Parkinson's disease. N. Engl. J. Med. 321, 1364–1371.

Dexter, D.T., Carter, C.J., Wells, F.R., Javoy-Agid, F., Agid, Y., Lees, A., Jenner, P., Mardsen, C.D., 1989. Basal lipid peroxidation in substantia nigra is increased in Parkinson's disease. J. Neurochem. 52, 381–389.

Dexter, D.T., Cholley, A.E., Flitter, W.D., Slater, T.F., Wells, F.R., Daniel, S.E., Lees, A.J., Jenner, P., Marsden, C.D., 1994. Increased levels of lipid hydroperoxides in the Parkinsonian substantia nigra: an HPLC and ESR study. Movement Dis. 9, 92–97.

Esterbauer, H., Schaur, R.J., Zollner, H., 1991. Chemistry and biochemistry of 4-hydroxynonenal, malonaldehyde and related aldehydes. Free Radic. Biol. Med. 11, 81–128.

Farber, J.L., 1995. Mechanisms of cell injury. In: Craighead, J.E. (Eds.), Pathology of Environmental and Occupational Disease. Mosby-Year Book Inc., St Louis, pp. 287–302.

Feillet-Coudray, C., Tourtauchaux, R., Niculescu, M., Rock, E., Tauveron, I., Alexandre-Gouabau, M.C., Rayssiguier, Y., Jalenques, I., Mazur, A., 1999. Plasma levels of 8-epiPG(F_2alpha), an *in vivo* marker of oxidative stress, are not affected by aging or Alzheimer's disease. Free Rad. Biol. Med. 27, 463–469.

Ferrante, R.J., Browne, S.E., Shinobu, L.A., Bowling, A.C., Baik, M.J., MacGarvey, U., Kowall, N.W., Brown, R.H., Jr., Beal, M.F., 1997. Evidence of increased oxidative damage in both sporadic and familial amyotrophic lateral sclerosis. J. Neuorchem. 69, 2064–74.

Fu, M.X., Requena, J.R., Jenkins, A.J., Lyons, T.J., Baynes, J.W., Thorpe, S.R., 1996. The advanced glycation end product, N epsilon-(carboxymethyl)lysine, is a product of both lipid peroxidation and glycoxidation reactions. J. Biol. Chem. 271, 9982–9986.

Gurney, M.E., Pu, H., Chiu, A.Y., Dal Canto, M.C., Polchow, C.Y., Alexander, D.D., Caliendo, J., Hentati, A., Kwon, Y.W., Deng, H.W. 1994. Motor neuron degeneration in mice that express a human Cu,Zn superoxide dismutase mutation. Science 264, 1772–1775.

Gutteridge, J.M., Halliwell, B., 1990. The measurement and mechanism of lipid peroxidation in biological systems. Trends Biochem. Sci. 15, 129–35.

Jackson, A.L., Loeb, L.A., 2001. The contribututution of endogenous sources of DNA damage to the multiple mutations in cancer. Mutation Res. 477, 7–21.

Jenner, P., 1994. Oxidative damage in neurodegenerative disease. Lancet 344, 796–798.

Jenner, P., Olanow, C.W., 1998. Understanding cell death in Parkinson's disease. Ann. Neurol. 44, S72–S84.

Jurgens, G., Chen, Q., Esterbauer, H., Mair, S., Ledinski, G., Dinges, H.P., 1993. Immunostaining of human autopsy aortas with antibodies to modified apolipoprotein B and apoprotein(a). Arteriosclerosis and Thrombosis 13, 1689–1699.

Keller, J.N., Pang, Z., Geddes, J.W., Begley, J.G., Germeyer, A., Waeg, G., Mattson, M.P., 1997. Impairment of glucose and glutamate transport and induction of mitochondrial oxidative stress and dysfunction in synaptosomes by amyloid β-peptide: role of the lipid peroxidation product 4-hydroxynonenal. J. Neurochem. 69, 273–284.

Kruman, I., Bruce-Keller, A.J., Bredesen, D.E., Waeg, G., Mattson, M.P., 1997. Evidence that 4-hydroxynonenal mediates oxidative stress-induced neuronal apoptosis. J. Neurosci. 17, 5089–5100.

Langston, J.W., Irwin, I., 1994. Organic neurotoxicants, In: D.B. Calne (Eds.), Neurodegenerative Diseases, W B Saunders, Philadelphia, PA, pp. 225–240.

Lovell, M.A., Ehmann, W.D., Butler, S.M., Markesbery, W.R., 1995. Elevated thiobarbituric acid-reactive substances and antioxidant enzyme activity in the brain in Alzheimer's disease. Neurology 45, 1594–1601.

Lovell, M.A., Ehmann, W.D., Mattson, M.P., Markesbery, W.R., 1997. Neurobiol Aging 18, 457–461.

Mark, R.J., Lovell, M.A., Markesbery, W.R., Uchida, K., Mattson, M.P., 1997a. A role for 4-hydroxynonenal in disruption of ion homeostasis and neuronal death induced by amyloid β-peptide. J. Neurochem. 68, 255–264.

Mark, R.J., Pang, Z., Geddes, J.W., Uchida, K., Mattson, M.P., 1997b. Amyloid β-peptide impairs glucose uptake in hippocampal and cortical neurons: involvement of membrane lipid peroxidation. J. Neurosci. 17, 1046–1054.

Mattson, M.P., Fu, W., Waeg, G., Uchida, K., 1997. 4-hydroxynonenal, a product of lipid peroxidation, inhibits dephosphorylation of the microtubule-associated protein tau. NeuroReport 8, 2275–2281.

McIntyre, T.M., Zimmerman, G.A., Prescott, S.M., 1999. Biologically active oxidized phospholipids. J. Biol. Chem. 274, 25189–25192.

Montine, K.S., Kim, P.J., Olson, S.J., Markesbery, W.R., Montine, T.J., 1997a. 4-hydroxy-2-nonenal pyrrole adducts in human neurodegenerative disease. J. Neuropathol. Exp. Neurol. 57, 415–425.

Montine, K.S., Olson, S.J., Amarnath, V., Whetsell, W.O., Jr., Graham, D.G., Montine, T.J., 1997b. Immunohistochemical detection of 4-hydroxy-2-nonenal adducts in Alzheimer's disease is associated with inheritance of APOE4. Am. J. Pathol. 150, 437–443.

Montine, K.S., Reich, E., Neely, M.D., Sidell, K.R., Olson, S.J., Markesbery, W.R., Montine, T.J., 1998. Distribution of reducible 4-hydroxynonenal adduct immunoreactivity in Alzheimer disease is associated with APOE genotype. J. Neuropathol. Exp. Neurol. 56, 866–871.

Montine, T.J., Markesbery, W.R., Morrow, J.D., Roberts II, L.J., 1998. Cerebrospinal fluid F2-isoprostane levels are increased in Alzheimer's disease. Ann. Neurol. 44, 410–413.

Montine, T.J., Beal, M.F., Cudkowicz, M.E., O'Donnell, H., Margolin, R.A., McFarland, L., Bachrach, A.F., Zackert, W.E., Roberts II, L.J., Morrow, J.D., 1999a. Increased CSF F2-isoprostane concentration in probable AD. Neurology 52, 562–565.

Montine, T.J., Beal, M.F., Robertson, D., Cudkowicz, M.E., Biaggioni, I., O'Donnell, H., Zackert, W.E., Roberts II, L.J., 1999b. Cerebrospinal fluid F2-isoprostanes are elevated in Huntington's disease. Neurology 52, 1104–1105.

Montine, T.J., Markesbery, W.R., Zackert, W., Sanchez, S.C., Roberts II, L.J., Morrow, J.D., 1999c. The magnitude of brain lipid peroxidation correlates with the extent of degeneration but not with density of neuritic plaques or neurofibrillary tangles or with APOE genotype in Alzheimer's disease patients. Am. J. Pathol. 155, 863–868.

Montine, T.J., Montine, K.S., Olson, S.J., Graham, D.G., Roberts II, L.J., Morrow, J.D., Linton, M.F., Fazio, S., Swift, L.L., 1999d. Increased cerebral cortical lipid peroxidation and abnormal phospholipids in aged homozygous apoE-deficient C56Bl/6J mice. Exp. Neurol. 158, 234–241.

Montine, T.J., Shinobu, L., Montine, K.S., Roberts II, L.J, Kowall, N.W., Beal, M.F., Morrow, J.D., 2000. No difference in plasma or urinary F2-isoprostanes among patients with Huntington's disease or Alzheimer's disease and controls. Ann. Neurol. 48, 950.

Moore, K., Roberts II, L.J., 1998. Measurement of lipid peroxidation. Free Radical Res. 28, 659–671.

Morrow, J.D., Frei, B., Longmire, A.W., Gaziano, J.M., Lynch, S.M., Shyr Y., Strauss, W.E., Oates, J.A., Roberts II, L.J., 1995. Increase in circulating products of lipid peroxidation (F_2-isoprostanes) in smokers. Smoking as a cause of oxidative damage. New Engl. J. Med. 332, 1198–1203.

Morrow, J.D., Roberts II, L.J., 1997. The isoprostanes: unique bioactive products of lipid peroxidation. Prog. Lipid Res. 36, 1–21.

Nourooz-Zadeh, J., Lie, E.H., Yhlen, B., Anggard, E.E., Halliwell, B., 1999. F_4-isoprostanes as specific marker of docosahexaenoic acid peroxidation in Alzheimer's disease. J. Neurochem. 72, 734–740.

Palinski, W., Rosenfeld, M.E., Yla-Herttuala, S., Gurtner, G.C., Socher, S.S., Butler, S.W., Parthasarathy, S., Carew, T.E., Steinberg, D., Wtiztum, J.L., 1989. Low density lipoprotein undergoes oxidative modification *in vivo*. Proc. Natl. Acad. Sci. USA 86, 1372–1376.

Pedersen, W.A., Fu, W., Keller, J.N., Markesbery, W.R., Appel, S., Smith, R.G., Kasarskis, E., Mattson, M.P., 1998. Protein modification by the lipid peroxidation product 4-hydroxynonenal in the spinal cords of amyotrophic lateral sclerosis patients. Ann. Neurol. 44, 819–824.

Pedersen W.A., Chan, S.L., Mattson, M.P., 2000. A mechanism for the neuroprotective effect of apolipoprotein E: isoform-specific modification by the lipid peroxidation product 4-hydroxynonenal. J. Neurochem. 74, 1426–1433.

Porter, N.A., Caldwell, S.E., Mills, K.A., 1995. Mechanisms of free radical oxidation of unsaturated lipids. Lipids 30, 277–290.

Pratico, D., Lee, V.M., Trojanowski, J.Q., Rokach, J., Fitzgerald, G.A., 1998a. Increased F_2-isoprsotanes in Alzheimer's disease: evidence for enhanced lipid peroxidation *in vivo*. FASEB J. 12, 1777–1784.

Pratico, D., Tangirala, R.K., Rader, D.J., Rokach, J., FitzGerald, G.A., 1998b. Vitamin E suppresses isoprostane generation *in vivo* and reduces atherosclerosis in ApoE-deficient mice. Nature Med. 4, 1189–1192.

Pratico, D., Rokach, J., Tangirala, R.K., 1999. Brains of aged apolipoprotein E-deficient mice have increased levels of F_2-isoprostanes, *in vivo* markers of lipid peroxidation. J. Neurochem. 73, 736–741.

Pratico, D., Clark, C.M., Lee, V.M., Trojanowski, J.Q., Rokach, J., FitzGerald, G.A., 2000. Increased 8,12-iso-iPF$_{2\alpha}$-VI in Alzheimer's disease: correlation of a noninvasive index of lipid peroxidation with disease severity. Ann. Neurol. 48, 809–812.

Pratico, D., Uryu, K., Leight, S., Trojanowski, J.Q., Lee, V.M., 2001. Increased lipid peroxidation precedes amyloid plaque formation in an animal model of Alzheimer amyloidosis. J. Neurosci. 21, 4183–4187.

Preston, R.J., Hoffmann, G.R., 2001. Genetic toxicology In: Klaassen, C.D. (Ed.), Casarett and Doull's Toxicology. McGraw-Hill, New York, pp. 321–350.

Roberts II, L.J., Montine, T.J., Markesbery, W.R., Tapper, A.R., Hardy, P., Chemtob, S., Dettbarn, W.D., Morrow, J.D., 1998. Formation of isoprostane-like compounds (neuroprostanes) *in vivo* from docosahexaenoic acid. J. Biol. Chem. 273, 13605–13612.

Roberts II, L.J., Salomon, R.G., Morrow, J.D., Brame, C.J., 1999. New developments in the isoprostane pathway: identification of novel highly reactive gamma-ketoaldehydes (isolevuglandins) and characterization of their protein adducts. FASEB J. 13, 1157–1568.

Rosen, D.R., Siddique, T., Patterson, D., Figlewicz, D.A., Sapp, P., Hentati, A., Donaldson, D., Goto, J., O'Regan, J.P., Deng, H.X., 1993. Mutations in Cu/Zn superoxide dismutase gene are associated with familial amyotrophic lateral sclerosis. Nature 362, 59–61.

Sayre, L.M., Zelasko, D.A., Harris, P.L., Perry, G., Salomon, R.G., Smith, M.A., 1997. 4-Hydroxynonenal-derived advanced lipid peroxidation end products are increased in Alzheimer's disease. J. Neurochem. 68, 2092–2097.

Shelley, M.L. (E)-4-hydroxy-2-nonenal may be involved in the pathogenesis of Parkinson's disease. 1998. Free Rad. Biol. Med. 25, 169–174.

Smith, M.A., Sayre, L.M., Monnier, V.M., Perry, G., 1995. Radical AGEing in Alzheimer's disease. Trends Neurosci. 18, 172–176.

Smith, M.A., Mirai, K., Hsiao, K., Pappolla, M.A., Harris, P.L., Siedlak, S.L., Tabaton, M., Perry, G., 1998a. Amyloid-beta deposition in Alzheimer transgenic mice is associated with oxidative stress. J. Neurochem. 70, 2212–2215.

Smith, M.A., Sayre, L.M., Anderson, V.E., Harris, P.L., Beal, M.F., Kowall, N., Perry, G., 1998b. Cytochemical demonstration of oxidative damage in Alzheimer disease by immunochemical enhancement of the carobnyl reaction with 2,4-dinitrophenylhydrazine. J. Histochem Cytochem. 46, 731–735.

Smith, R.G., Henry, Y.K., Mattson, M.P., Appel, S.H., 1998. Presence of 4-hydroxynonenal in cerebrospinal fluid of patients with sporadic amyotrophic lateral sclersosis. Ann. Neurol. 44, 696–699.

Stephens, N.G., Parsons, A., Schofield, P.M., Kelly, F., Cheesman, K., Mitchinson, M.J., 1996. Randomised controlled trial of vitamin E in patients with coronary disease: Cambridge Heart Antioxidant Study (CHAOS). Lancet 347, 781–786.

Tu, P.H., Gurney, M.E., Julien, J.P., Lee, V.M., Trojanowski, J.Q., 1997. Oxidative stress, mutant SOD1, and neurofilament pathology in transgenic mouse models of human motor neuron disease. Lab. Invest. 76, 441–56.

Waddington, E., not in medline 1999. Alzheimer's Reports 2, 277–282.

Yoritaka, A., Hattori, N., Uchida, K., Tanaka, M. Stadtman, E.R., Mizuno, Y., 1995. Immunohisto-chemical detection of 4-hydroxynonenal protein adducts in Parkinson disease. Proc. Natl. Acad. Sci. USA 93, 2696–2701.

Yusuf, S., Dagenais, G., Pague, J., Bosch, J., Sleight, P., 2000. Vitamin E supplementation and cardiovascular events in high-risk patients. The Heart Outcomes Prevention Evaluation Study Investigators. New Engl. J. Med. 342, 154–160.

Zhang, J., Perry, G., Smith, M.A., Robertson, D., Olson, S.J., Graham, D.G., Montine, T.J., 1999. Parkinson's disease is associated with oxidative damage to cytoplasmic DNA and RNA in substantia nigra neurons. Am. J. Pathol. 154, 1423–1429.

Zhang, J., Graham, D.G., Montine, T.J., Ho, Y.S., 2000. Enhanced N-methyl-4-phenyl-1,2,3,6-tetra-hydropyridine toxicity in mice deficient in CuZn-superoxide dismutase or glutathione peroxidase. J. Neuropathol. Exp. 59, 53–61.

Advances in
Cell Aging and
Gerontology

Regulation of invertebrate longevity by inositol phosphate signaling

Catherine A. Wolkow

National Institute on Aging, Gerontology Research Center, Laboratory of Neurosciences,
5600 Nathan Shock Drive, Baltimore, MD 21224
E-mail address: wolkowca@grc.nia.nih.gov

Contents

1. Introduction

Phosphoinositide lipid signaling pathways function as effectors from a variety of growth factor receptors, most notably for the purposes of this review, the insulin and IGF-I receptors. Phosphoinositide lipid signaling pathways have been conserved through metazoan evolution, and have important roles in the growth and development of diverse organisms. Thus, the basic components of this signaling pathway are evolutionarily ancient. However, studies of this signaling pathway in a variety of organisms suggest that the outputs of this pathway can be different among species. Such plasticity may target the signaling pathway to accommodate unique needs of each organism.

Advances in Cell Aging and Gerontology, vol. 12, 27–46

One basic function of phosphoinositide lipid signaling that appears to be conserved in evolutionarily diverse organisms is control of cell growth and metabolism. In flies, phosphoinositide lipid signaling directly controls cell growth. The same pathway in nematodes signals that environmental conditions are adequate for growth and reproduction. Phosphoinositide lipid signaling in mammals is a major effector for signaling downstream of the insulin- and IGF-I receptors that control cellular metabolism and growth.

In addition to metabolic outputs, insulin-like signaling controls longevity in fruitflies and nematodes. In nematodes, phosphoinositide lipid signaling is linked to increased cellular stress resistance. The extent to which mammalian longevity is controlled by phosphoinositide lipid signaling is also currently under active investigation. The purpose of this article is to compare and contrast the components and functions of phosphoinositide lipid signaling in nematodes, fruitflies and mammals. The links between phospholipid signaling and longevity in invertebrates will be examined and the possibility that such pathways also control mammalian longevity will be discussed. Deciphering the pathways by which phospholipid signaling impinges on invertebrate longevity is important for understanding the strategies that nature has adopted for lifespan control.

2. Activation of PI3K by insulin/IGF-I receptor signaling

The membrane lipid, phosphatidylinositol (PtdIns), can be phosphorylated on the D-3, D-4 and/or D-5 positions of the inositol ring. In response to growth factor signaling, levels of PtdIns-3,4,5-P_3 rise dramatically as a consequence of class I_A PI3K activation. Class I_A PI3K are heterodimeric enzymes consisting of a 110 kDa catalytic subunit and an SH2-domain containing adaptor, or regulatory, subunit (for an in-depth review, see Fruman et al., 1998).

Three p110 catalytic subunits have been identified in mammalian genomes, p110 α, β, and δ (Domin and Waterfield, 1997). The p110 amino terminus contains domains that mediate binding to the adaptor subunit and *ras* (Fig. 1A). The catalytic domain is located in the carboxyl terminus. Catalytic activation requires that the p110 catalytic subunit bind to an adaptor subunit, also referred to as a regulatory subunit. The mammalian PI(3)K adaptor subunits are p85α, p55α, p50α p85β and p55γ (Fig. 1B). The p85 isoforms differ from the p55/p50 proteins by the presence of an amino terminal extension containing SH3 and BH (breakpoint cluster region homology) domains. The p85/p55/p50α proteins are all transcribed as splice variants from the same gene (Fruman et al., 1996). All class I_A PI3K adaptors contain a pair of SH2 domains that target the protein to phosphotyrosines on activated growth factor receptors or linker proteins, including IRS. The adaptor subunit SH2 domains bind preferentially to phosphotyrosines within the YXXM motif. YXXM sites are found on the vertebrate insulin and IGF-I receptors, the insulin receptor-like proteins of *Drosophila* and *C. elegans*, and within the IRS linker proteins of both vertebrates and invertebrates. The sequences between the SH2 domains mediate binding to the amino-terminal domain of the p110 catalytic subunit. SH2 domain-mediated

A. Class I$_A$ PI(3)K p110 catalytic subunits

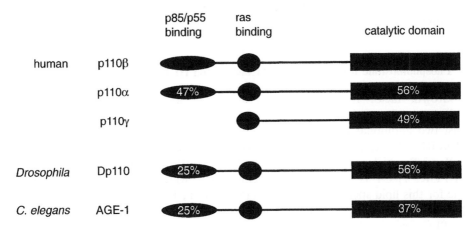

B. p85/p55 adaptor subunits

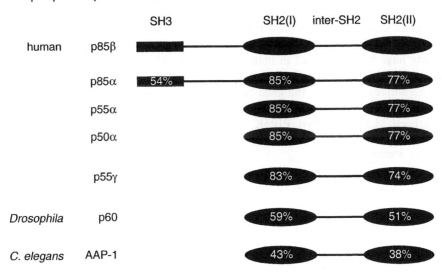

Fig. 1. Class I$_A$ PI3K catalytic and adaptor subunits are conserved between vertebrates and invertebrate species. (A) p110-like catalytic subunits from human, *Drosophila* and *C. elegans* showing positions of intersubunit interaction domains, ras-binding domains and catalytic domains. White numbers show percent identity to human p110β enzyme in intersubunit interaction domains and catalytic domains. The accession numbers for the sequences used for this comparison are: p110β (P42338), p110α (P42336), p110γ (P48736), Dp110 (CAA70291), and AGE-1 (S71792). (B) Comparison of PI3K adaptor subunits with percent identity to p85β shown in white. p85α, p55α, and p50α are splice variants from the same locus and are essentially identical in the SH2 domains. The accession numbers for the sequences shown are: p85β (AAD22671), p85α/p55α/p50α (P27986), p55γ (Q92569), p60 (CAA72030), and AAP-1 (AAF28335).

recruitment of PI3K to activated growth factor receptors is believed to provide a mechanism for recruiting PI3K to the membrane where the enzyme's lipid substrates are located.

3. Phosphoinositide lipid signaling by Class II and III PI3K

Growth factor receptor signaling stimulates a burst of PtdIns-3,4-P_2 and PtdIns-3,4,5-P_3 production, which activates downstream signaling pathways. At least two additional phosphoinositide lipids perform intracellular signaling roles. Class III PI3K enzymes catalyze the production of the singly phosphorylated lipid, PtdIns-3-P, which regulates intracellular vesicle trafficking (Fig. 2). The yeast class III PI3K, Vps34, is essential for protein sorting, which appears to be the general role for this lipid species in eukaryotes (Schu et al., 1993).

Signaling downstream of G-protein coupled receptors can activate phospholipase C (PLC), which cleaves PtdIns-4,5-P_2 to produce 1,2-diacylglycerol (1,2-DAG) and Ins-1,4,5-P_3. Ins-1,4,5-P_3 is a soluble product that diffuses to the ER to activate Ca^{2+} release. The second product of PLC-mediated PtdIns-4,5-P_2 cleavage, 1,2-DAG, activates protein kinase C in the presence of elevated Ca^{2+}. PKC is a serine/threonine protein kinase whose targets include proteins that act in cellular metabolic and growth pathways. PtdIns-4,5-P_2 is produced by the sequential action of PI-4-K and PIP5K and, together with PtdIns-4-P, constitute the majority of cellular phosphoinositide lipid stores (Fig. 2). Proteins with PI-4-K activity have been identified in mammalian cells and are structurally similar to the class I PI3K catalytic subunits (Carpenter and Cantley, 1996). In the yeast, *Saccharomyces cerevisiae*, the gene encoding PI-4-K, *STT4*, is required for viability in staurosporine (Yoshida et al., 1994). Yeast and mammalian PIP5K enzymes have been identified. Two PIP5K enzymes have been studied in *S. cerevisiae*,

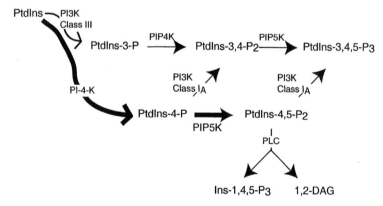

Fig. 2. Pathways for production of phosphoinositide lipids. Large arrows indicate that PtdIns-4-P and PtdIns-4,5-P_2 constitute the majority of cellular phosphoinositide lipids (Fruman et al., 1998). As discussed in the text, in response to activation by growth factor receptors, Class I_A PI3K catalyzes the production of PtdIns-3,4-P_2 and PtdIns-3,4,5-P_3 from PtdIns-4-P and PtdIns-4,5-P_2.

encoded by the genes *FAB1* and *MSS4*. These genes play distinct roles in regulating cellular cytoskeletal dynamics (Desrivieres et al., 1998).

4. PtdIns-3,4,5-P$_3$ activates downstream protein kinases, Akt/PKB and PDK-1

In mammalian cells, PtdIns-3,4,5-P$_3$ production in response to growth factor receptor signaling can couple to several different outputs. Two major targets of PI3K are the serine/threonine protein kinases, Akt/PKB and PDK-1, which mediate the anti-apoptotic effects of PI3K signaling in mammalian cells (Fig. 3) (reviewed by Toker and Cantley, 1997; Datta et al., 1999). Both proteins bind to phospholipids via pleckstrin homology (PH) domains located in the amino terminus of Akt and in the carboxyl terminus of PDK-1. PH domains can mediate phospholipid binding as well as protein–protein interactions. Binding to PtdIns-3,4,5-P$_3$ or PtdIns-3,4-P$_2$ are requisite steps for activating Akt and PDK-1. In addition, Akt must be phosphorylated at two sites, corresponding to Thr 308 and Ser 473 in the human protein. PDK-1 phosphorylates Thr 308, while the enzyme that carries out the second Akt phosphorylation remains elusive.

A number of Akt/PKB substrates have been identified, notably the forkhead transcription factors related to the *C. elegans* protein, DAF-16. Akt/PKB can phosphorylate and inhibit three mammalian FoxA family members, FKHR, FKHR-L1 and AFX, as well as DAF-16 (Brunet et al., 1999; Kops et al., 1999;

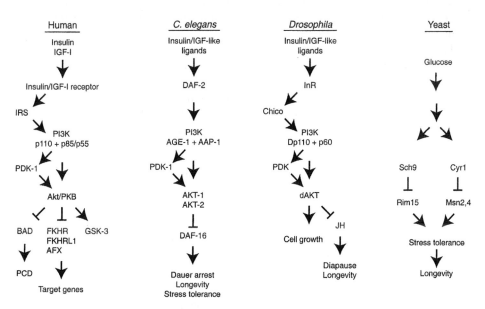

Fig. 3. Components of insulin/insulin-like signaling pathways conserved through evolution. The components of insulin and insulin-like signaling pathways are conserved between humans and invertebrates, including *C. elegans* and *Drosophila*. Glucose sensing in yeast also utilizes a signaling cascade with some similarities to metazoan insulin-like pathways.

Nakae et al., 1999; Takaishi et al., 1999; Cahill et al., 2001; Lee et al., 2001; Lin et al., 2001). 14-3-3 proteins can bind phosphorylated DAF-16, causing cytoplasmic retention and effectively sequestering DAF-16 from its target genes in the nucleus (Cahill et al., 2001). In addition to regulating nuclear entry of these forkhead proteins, Akt can promote cell survival by inhibiting pro-apoptotic factors, such as BAD, and can mediate metabolic outputs by regulating glycogen synthase kinase 3 (GSK-3) activity (Datta et al., 1999).

5. AGE-1/PI3K controls *C. elegans* longevity

In *C. elegans*, mutations in the gene, *age-1*, result in constitutive developmental arrest as dauer larvae (Vowels and Thomas, 1992; Gottlieb and Ruvkun, 1994; Dorman et al., 1995; Malone et al., 1996). Under normal conditions, *C. elegans* larvae proceed through four larval stages, L1–L4, before molting into reproductive adults. When food becomes limiting, or population density increases, animals enter a diapaused state and arrest development as dauer larvae, a long-lived, stress resistant larval form that replaces the L3 larval stage (Riddle et al., 1997). Dauer larvae can survive for several months without food, significantly longer than the two-week lifespan of reproductive adults. When environmental conditions improve, dauer larvae re-enter the reproductive life cycle by molting into L4 larvae. Lifespan is reset after dauer recovery, so that the adult lifespan of recovered dauer larvae is the same as for animals that had never entered the dauer stage. While constitutive dauer arrest results from severe mutations in *age-1*, less severe mutations cause a 2- to 3-fold extension of adult lifespan (Friedman and Johnson, 1988; Johnson, 1990).

The protein encoded by the *age-1* locus is homologous to the mammalian class I_A PI3K p110 catalytic subunits and is the only p110-like protein in the *C. elegans* genome (Morris et al., 1996). The predicted AGE-1 catalytic domain is 37% identical to the catalytic domains of other mammalian p110 PI3K subunits (Fig. 1). This high level of conservation indicates that protein function has been conserved. Nevertheless, the direct demonstration of lipid kinase activity in AGE-1 would validate the bioinformatic predictions.

AGE-1 is also homologous to the mammalian enzymes in the amino-terminal intersubunit interaction domain (Morris et al., 1996). The *C. elegans* genome contains one homolog of the p55-family of class I_A PI3K adaptor subunits. This gene has been designated *aap-1*, and functions genetically in the dauer arrest pathway with *age-1* (Wolkow et al., 2002). Sequence homology between mammalian p110 and AGE-1 in the p110 ras-interaction domain suggests AGE-1 may also be activated by ras binding (Morris et al., 1996). However, this expectation remains to be confirmed, as a ras pathway input into AGE-1/PI3K signaling has not yet been identified.

AGE-1 and AAP-1 act in an insulin-like signaling pathway (Fig. 3). While PI3K signaling carries out diverse functions in mammalian cells, AGE-1/PI3K signaling in *C. elegans* appears to function solely as an effector for insulin-like signaling. In support of this conclusion, *C. elegans* mutants lacking AGE-1/PI3K activity phenocopy animals with mutations in other components of the insulin-like signaling pathway (see below). The *daf-2* gene encodes a protein highly related to the mammalian insulin and IGF-I receptors (Kimura et al., 1997). Mutations inactivating the DAF-2/insulin receptor-like protein cause constitutive arrest at the dauer larval stage and extend adult (Vowels and Thomas, 1992; Kenyon et al., 1993; Gottlieb and Ruvkun, 1994; Dorman et al., 1995). (In order to measure adult lifespan of *daf-2* mutants, animals are raised at the permissive temperature of 15°C through larval development, then shifted to nonpermissive temperature (25°C) as adults.) The phenotypic similarities between animals lacking AGE-1/PI3K and DAF-2/insulin receptor activity supports their grouping as components of the same pathway.

6. AKT-1, AKT-2, and PDK-1 are targets of AGE-1 signaling

Genetic screens for mutations that suppress the dauer arrest phenotype of *age-1* and *daf-2* mutants identified downstream players in the DAF-2/AGE-1 signaling pathway. Dominant, gain-of-function alleles of *akt-1* and *pdk-1*, encoding *C. elegans* homologs of mammalian Akt/PKB and PDK-1, respectively, suppress dauer arrest of *age-1* mutants (Figs. 3 and 4) (Paradis and Ruvkun, 1998; Paradis et al., 1999). In genetics parlance, dominant mutations are those which confer the mutant phenotype in the heterozygous state, in contrast to recessive mutations which are only visible phenotypically when homozygous. In most cases, dominant mutations confer a gain-of-function, or neomorphic, activity to the protein encoded by the mutant gene. Thus, the mutant protein has increased or "gain-of-function" activity that is visible in the heterozygous background.

The *akt-1(mg144)* allele causes threonine to be substituted for alanine (183) and this substitution activates AKT-1 signaling sufficiently to suppress dauer arrest in animals lacking AGE-1/PI3K activity (Fig. 4) (Paradis and Ruvkun, 1998). The *akt-1(mg144)* mutation lies in an unconserved region of the linker between the PH and catalytic domains of AKT-1. The mechanism by which the Ala183Thr substitution activates AKT-1 is not fully understood. It is possible that phospholipid specificity is relaxed in the mutant protein, or that the mutation relieves an inhibitory interaction. Further analysis showed that increased *akt-1* gene dosage can also bypass the requirement for AGE-1/PI3K activity during development (Paradis and Ruvkun, 1998). The AKT-1 kinase activity is required for this effect, as a kinase-dead mutation abrogates the suppression. Interestingly, this finding may point to another source of the activating lipids, $PtdIns-3,4,5-P_3$ or $PtdIns-3,4-P_2$, that is redundant to AGE-1/PI3K.

Akt/PKB activation requires both binding to phospholipids and phosphorylation on two sites, corresponding to Thr 308 and Ser 473 in the human Akt protein. Thr308 phosphorylation is mediated by PDK-1, corresponding to the *C. elegans* gene, *pdk-1*. Mutations that inactivate *C. elegans* PDK-1 cause dauer arrest and extend adult lifespan, the expected outcome of mutations that inactivate signaling through the DAF-2/AGE-1 pathway (Paradis et al., 1999). Furthermore, substituting a valine for ala303 in PDK-1 dominantly activates PDK-1 signaling,

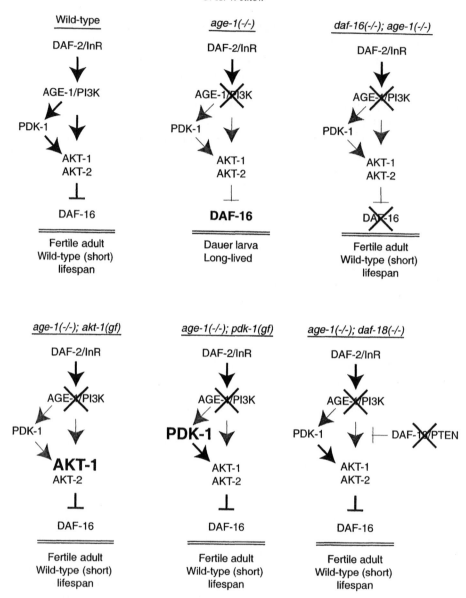

Fig. 4. The genetics of AGE-1/PI3K signaling in *C. elegans*. AGE-1/PI3K signaling controls development and lifespan in *C. elegans*. Each panel illustrates the state of signaling downstream of AGE-1/PI3K in animals of the indicated genotype. Black arrows indicate wild-type signaling and grey arrows show steps where signaling is abrogated due to mutation in upstream components. When a protein is inappropriately activated as a result of mutation, the protein name is shown in large, bold text. The phenotypic outcome of each genetic combination is shown below the pathway illustration. Refer to text for details.

and suppresses dauer arrest in animals lacking AGE-1/PI3K (Fig. 4) (Paradis et al., 1999). Thus, signaling downstream of PI3K is conserved between *C. elegans* and mammals, and relies on the phospholipid-activated serine/threonine kinases, AKT/PKB and PDK-1.

7. Signaling downstream of AGE-1/PI3K converges on DAF-16

The ultimate output of AGE-1/PI(3)K signaling is the forkhead transcription factor, DAF-16. Mutations in *daf-16* suppress all phenotypes of *daf-2/InR* and *age-1/PI3K* mutants (Figs. 3 and 4). Recall that animals lacking AGE-1/PI3K signaling constitutively arrest development as dauer larvae and live longer than wild-type. In this background, the additional loss of DAF-16 activity suppresses these phenotypes, allowing development to proceed normally and restoring shorter wild-type lifespan (Fig. 4) (Kenyon et al., 1993; Gottlieb and Ruvkun, 1994; Dorman et al., 1995; Larsen et al., 1995). Signaling through the AGE-1/PI3K pathway represses DAF-16 activity by AKT-1 and AKT-2-mediated phosphorylation of DAF-16. The DAF-16 forkhead transcription factor contains four sequences that fit the Akt phosphorylation consensus sequence (Lin et al., 1997; Ogg et al., 1997; Paradis and Ruvkun, 1998). In addition, AKT has been shown to phosphorylate DAF-16 *in vitro* and mutation of the Akt consensus sites alters DAF-16 activity *in vivo* (Cahill et al., 2001; Lee et al., 2001; Lin et al., 2001). As mentioned earlier, phosphorylation of DAF-16 and the mammalian orthologs by Akt results in cytoplasmic retention of the DAF-16-like proteins, sequestering the transcription factors away from target gene promoters in the nucleus. In the absence of AGE-1/PI3K signaling, AKT-1 and AKT-2 are inactive; DAF-16 is consequently not phosphorylated, is derepressed, and can activate the expression of target genes promoting dauer arrest and adult longevity (Paradis and Ruvkun, 1998).

Recent efforts have focused on identifying DAF-16 target genes. It has been known for some time that long-lived *age-1* mutants have high levels of catalase and SOD activity (Larsen, 1993). The *C. elegans* genome contains two genes encoding Mn-SOD enzymes, *sod-2* and *sod-3*. The expression of *sod-3* is DAF-16-dependent: *sod-3* mRNA levels are high in long-lived *daf-2* mutants and dauer larvae, and low in *daf-2*; *daf-16* mutants and in wild-type non-dauer animals (Honda and Honda, 1999). The *sod-3* promoter contains one sequence matching the DAF-16 consensus binding site (TTGTTTAC), suggesting that *sod-3* is indeed a DAF-16 target (Honda and Honda, 1999; Furuyama et al., 2000).

Additional genes whose expression is *daf-16*-dependent were identified using differential display PCR (Yu and Larsen, 2001). The genes identified in this study, designated *dao-1* through *dao-8*, were either up- or down-regulated in *daf-2* mutants as compared to wild-type. *dao-1*, *-2*, *-3*, *-4*, *-8*, and *-9* were down-regulated in *daf-2* mutant adults as compared to wild-type or *daf-2*; *daf-16* double mutants. *dao-5*, *-6*, and *-7* were expressed at higher levels in *daf-2* adults than in controls, and therefore constitute potential longevity- or dauer-promoting genes. The expression of *dao-5* and *dao-6* was also upregulated in wild-type and dauer larvae. For the most part, the identity of the *dao* genes does not provide many clues about their function or their relevance to longevity. *dao-5* encodes a protein with homology to xNopp180, involved in transcription and translational control. *dao-6* encodes a novel protein and *dao-7* encodes a putative zinc-finger transcription factor. Perhaps additional insight into the functions of these genes can be gathered by identifying downstream targets of DAO-7.

8. AGE-1/PI3K acts in neurons to control lifespan

How does AGE-1/PI3K signaling control *C. elegans* lifespan? Long-lived *age-1* mutants have additional phenotypes that may provide clues into the basis of this phenotype. Long-lived *age-1* mutants have enhanced resistance to several stresses, including thermal, free radical, and UV stress (Larsen, 1993; Lithgow et al., 1995; Murakami and Johnson, 1996). All of these phenotypes require that DAF-16 be active. Therefore, wild-type DAF-16 activity is required for stress resistance and increased longevity in *age-1* mutants (Figs. 3 and 4). A reasonable model is that DAF-16 directs the expression of target genes, such as *sod-3*, which confer stress resistance and increased longevity. Increasing stress resistance has been shown to be a successful strategy for increasing *C. elegans* lifespan. For example, longevity is increased by feeding animals catalase and SOD mimetics (Melov et al., 2000).

AGE-1/PI3K signaling must occur in neurons to control lifespan. Transgenic animals were constructed in which AGE-1/PI3K signaling was restored to specific cell types, either neurons, muscle or gut, in an otherwise *age-1* mutant background. The cell types where AGE-1/PI3K signaling controlled lifespan were identified by examining the lifespan of the transgenic animals with cell-type restricted AGE-1/PI3K signaling. The long lifespan of *age-1* mutants was only rescued when AGE-1/PI3K signaling was restored to neurons, but not to muscle or intestinal cells (Fig. 5) (Wolkow et al., 2000). Similar results were obtained when the experiment was repeated using *daf-2* mutants with cell-type restricted DAF-2 expression. Thus, AGE-1/PI3K signaling in neuronal cells has a special function for lifespan control. Results from an earlier *daf-2* mosaic analysis were consistent with these findings, showing that the animals are long-lived when *daf-2* activity was lost during development from the AB lineage, which contains cells that give rise to neurons (Fig. 5) (Apfeld and Kenyon, 1998). The AB lineage can be divided into two sub-lineages, the ABa and ABp lineages. When *daf-2* activity was lost from either the ABa or ABp lineages, lifespan was not extended as when *daf-2* activity was lost from the entire AB lineage. This suggests that cells from both parts of the AB lineage control longevity.

Exactly how neuronal AGE-1/PI3K signaling controls lifespan is not understood. One hypothesis is that, in neurons lacking AGE-1/PI3K signaling, derepressed DAF-16 activates expression of target genes that increase neuronal stress resistance, such as *sod-3*. However, the mechanism by which increased neuronal resistance to stress translates into an effect on longevity is unknown. Is longevity controlled by a specific set of lifespan-control neurons, or can any

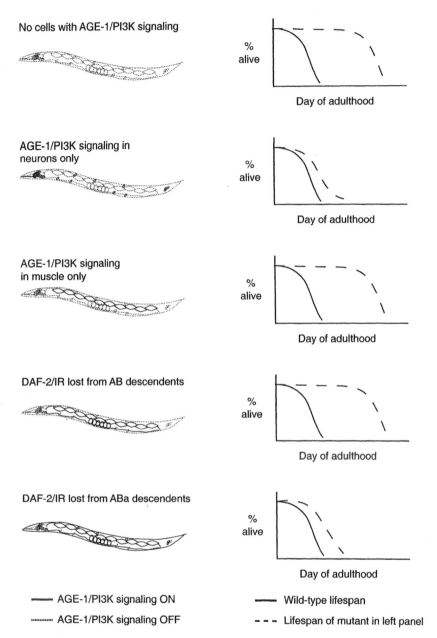

Fig. 5. AGE-1/PI3K signaling in neurons controls *C. elegans* lifespan. Cartoons on the left illustrate which of the animal's cells contain AGE-1/PI3K signaling (outlined in black) and which cells lack AGE-1/ PI3K signaling (gray outlines). The corresponding lifespan curves on the right illustrate the lifespan of animals with AGE-1/PI3K signaling in the indicated cells (dashed lines) as compared with wild-type animals (solid line). The first three panels summarize data from (Wolkow et al., 2000) and the last two panels summarize data from (Apfeld and Kenyon, 1998). Readers should refer to the primary sources for actual lifespans and other relevant controls.

neuron have an effect on lifespan? Current work is focused on answering this question.

9. Upstream pathways that control lifespan via AGE-1 signaling

A number of genes and pathways have been described to extend *C. elegans* lifespan. In many cases, lifespan extension in these experiments depends on the presence of wild-type *daf-16* activity. This observation suggests that these pathways ultimately feed into *daf-16* activity, probably via *daf-2* and *age-1* activity. Genetic analysis of these interactions may reveal new inputs into insulin-like pathways.

In *C. elegans*, the germline sends a lifespan-shortening signal. Wild-type animals lacking a functional germline display a 60% increased lifespan compared to wild-type (Hsin and Kenyon, 1999; Arantes-Oliveira et al., 2002). Animals lacking a germline can be produced by either laser ablating both germline precursor cells (the Z2 and Z3 cells) early in larval development, or by mutations blocking early steps in germline proliferation. In *daf-16* mutants, germline ablation fails to confer extended longevity, showing that the negative influence on lifespan requires *daf-16* activity. Interestingly, longevity of animals lacking germline is enhanced synergistically by disrupting DAF-2/InR signaling. The synergy between defective DAF-2/InR signaling and germline ablation suggests that these pathways extend lifespan by distinct mechanisms. If they worked through the same mechanism, then longevity signals from the germline would not further extend lifespan in animals lacking DAF-2/InR signaling. This evidence suggests that signals from the germline regulate longevity via *daf-16*, but independently of *daf-2*.

Environmental cues also contribute signals regulating *C. elegans* lifespan. The sensory neurons that detect environmental cues must function properly for wild-type lifespan. Mutations disrupting the function of these cells extend lifespan (Apfeld and Kenyon, 1999). As observed with germline signals, longevity cues from the environment may regulate DAF-16/forkhead activity independently from the DAF-2/InR. Clearly the identification and characterization of DAF-2/InR-independent inputs into DAF-16 activity is important. Do these pathways operate via AGE-1/PI3K signaling, or converge upon DAF-16 through an AGE-1-independent pathway? In addition to revealing new mechanisms for lifespan control, the answers to these questions and others will identify new inputs into PI3K signaling.

10. PI3K signaling in *Drosophila melanogaster*

Cell growth in the fruitfly, *Drosophila melanogaster*, is controlled by PI3K signaling. The PI3K p110 catalytic subunit, Dp110, acts with a p55-like adaptor subunit, p60, to transduce signals from the *Drosophila* insulin receptor-like protein, InR (Fig. 3) (Fernandez et al., 1995; Leevers et al., 1996; Weinkove et al., 1997). As with *C. elegans*, the primary function of PI3K signaling in *Drosophila* appears to be as a downstream effector for insulin receptor-like signaling.

PI3K signaling is essential in *Drosophila* and promotes cell growth. The first study revealing this function used a constitutively active version of Dp110 which contained a C-terminal CAAX farnyslation site to target Dp110 to the membrane (Leevers et al., 1996). Transgenic animals overexpressing Dp110-CAAX displayed consistent increases in cell size in the wing and eye. In fact, overexpression of wild-type Dp110 also resulted in increased size in these tissues. In contrast, cell number and size was reduced by the expression of a dominant negative Dp110 mutant.

Null mutations in Dp110 and p60 cause developmental arrest at the third larval stage (Weinkove et al., 1999). In somatic mosaics, clones of *Dp110(−/−)* or *p60(−/−)* cells contain smaller and fewer cells than the neighboring cells containing wild-type PI(3)K signaling (Fig. 6) (Britton et al., 2002). Interestingly, the severity of the phenotypes resulting from loss of Dp110 function was greater than for loss of p60 function, suggesting that alternative, p60-independent pathways for Dp110 activation exist within the cell.

As mentioned above, PI3K is the major transducer of insulin-like signaling in *Drosophila*. The *Drosophila* insulin-like receptor, InR, is essential for growth and development. Some non-null *InR* mutations result in dwarf adults with growth deficits (Tatar et al., 2001). The same is observed in flies with mutations in *chico*, the gene encoding an insulin-receptor substrate (IRS) homolog (Bohni et al., 1999; Poltilove et al., 2000). In *Drosophila* and mammals, insulin receptor activation is transduced to PI3K via the IRS-family of linker proteins. IRS (insulin receptor substrate) proteins contain an amino-terminal PTB-domain for binding specific phosphotyrosines on the activated insulin receptor. Tyrosines in the IRS carboxyl-terminus are substrates for the insulin receptor tyrosine kinase, and, when phosphorylated, provide docking sites for downstream effectors, including PI(3)K. As mentioned earlier, phosphotyrosines within a YXXM motif are the preferred binding sites for the amino-terminal SH2 domain of the PI(3)K adaptor subunits and are the primary targeting mechanisms for recruiting PI(3)K to activated growth factor receptors. The *C. elegans* genome contains an open reading frame with predicted homology to mammalian IRS proteins, encoded by the locus *C54D1.3*, indicating that AGE-1 and AAP-1 may be recruited to DAF-2

Normal sized wild-type cells

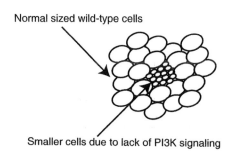

Smaller cells due to lack of PI3K signaling

Fig. 6. PI3K signaling controls cell growth in *Drosophila*. As shown in Britton et al. (2002) using the Gal4/UAS system to inhibit PI3K signaling, cell growth depends on PI3K signaling pathways downstream of the InR/insulin receptor-like protein.

via conserved interactions with the protein encoded by this locus (Wolkow et al., 2002).

11. Control of *Drosophila* lifespan by insulin-like signaling

Insulin signaling is essential for growth and development in *Drosophila*. However, some genetic perturbations which only weaken insulin-like signaling, and therefore do not cause severe growth defects, can extend adult lifespan (Clancy et al., 2001; Tatar et al., 2001). Adult lifespan in mutants with a combination of two *InR* loss-of-function alleles, InR^{E19}/InR^{p5545}, is extended in female flies from 32 days to 60 days and, in male flies less significantly, from 36 to 39 days (Tatar et al., 2001). All other identified *InR* mutations shorten lifespan, suggesting that lifespan may be especially sensitive to the levels or locations of *InR* activity. Mutations in other components of the fly's insulin-like signaling pathway can also increase lifespan. Homozygous *chico¹* mutant female flies live up to 48% longer than wild-type (Clancy et al., 2001). Again, the levels of insulin-like signaling may be an important consideration, as lifespan was reduced compared to wild-type in homozygous *chico¹* males.

InR and *chico* may affect lifespan by controlling hormones that regulate entry into diapause. Levels of juvenile hormone (JH), which controls entry into diapause, were reduced in long-lived InR^{E19}/InR^{p5545} mutant flies (Tatar et al., 2001). Treatment with methoprene, a JH analog, rescued the long lifespan phenotype of these flies. However, JH levels were also reduced in InR^{E19} homozygous mutants, which have slightly shorter lifespan than wild-type. Thus, control of longevity by insulin-like signaling in *Drosophila* is not a straightforward process and these observations could indicate that relative activity levels or cell-type specificity may contribute to the mutant phenotypes observed.

In *C. elegans*, longevity extension as a result of defects in DAF-2/insulin-like signaling are clearly correlated with increased stress resistance (Larsen, 1993; Lithgow et al., 1995; Murakami and Johnson, 1996; Honda and Honda, 1999; Taub et al., 1999). Do long-lived InR^{E19}/InR^{p5545} mutant flies also display enhanced stress resistance? Levels of Cu/Zn SOD activity in long-lived InR^{E19}/InR^{p5545} and *chico¹* mutants is twice that of wild-type animals, suggesting that these genetic mutations confer a state of increased stress resistance and longevity, as observed in *C. elegans* (Clancy et al., 2001; Tatar et al., 2001). It has been shown that *Drosophila* longevity can be extended by increased stress resistance (Parkes et al., 1998; Sun and Tower, 1999). Whether increased stress resistance is required for long lifespan in the *Drosophila* insulin pathway mutants remains to be determined.

PI3K signaling acts as a sensor of nutritional status in *Drosophila* (Britton et al., 2002). A GFP reporter that translocates to the membrane in response to PI3K signaling was used as a cellular marker for detecting PI3K signaling *in vivo*. In these experiments, PI3K activity was found to be decreased in nutritionally-deprived animals. Decreased PI3K activity resulted in smaller cells, showing that PI3K signaling couples nutritional status to cell growth. While cell growth is controlled autonomously by PI3K signaling, there may be downstream hormones (such as JH) whose activities are linked to PI3K signaling through cell growth and other sensors. Additionally, insulin-like signaling in *Drosophila* may be coupled to the expression of target genes via a DAF-16-like transcription factor, although this output has not yet been identified. Further genetic analysis of *Drosophila* insulin-like signaling using the recently reported phenotypes should illuminate the downstream targets of this pathway.

12. An Akt/PKB-like kinase controls lifespan in *Saccharomyces cerevisiae*

Insulin-like signaling pathways provide multicellular organisms with the means to couple growth and differentiation to environmental conditions. For example, when food becomes scarce, *C. elegans* larvae can enter an alternate developmental pathway that provides increased stress resistance and decreased nutritional requirements. Single-cell organisms do not obviously have such complex developmental decisions. It is therefore interesting that mutations disrupting glucose sensing in yeast also extend lifespan (Fabrizio et al., 2001). Sch9 encodes a putative serine/threonine protein kinase that acts in the glucose sensing pathway. The catalytic domain of Sch9 contains nearly 50% homology to the catalytic domains of *C. elegans* AKT-1 and AKT-2 and these enzymes are members of the same protein kinase family. Mutations disrupting *sch9* activity lead to increased stress resistance and extend lifespan (Fabrizio et al., 2001). Similar phenotypes were observed in cells with mutations in *cyr1*, which encodes adenylate cyclase acting in the same pathway. As with the *daf-2* pathway in *C. elegans*, stress resistance and longevity in *sch9* mutants requires downstream transcription factors. Msn2 and Msn4 are required for long lifespan in Cyr1 mutants, and Rim15, which acts independently from Msn2 and Msn4, is required for longevity in both *cyr1* and *sch9* mutants. Thus, lifespan in multicellular and unicellular eukaryotes can be lengthened by perturbations in nutrient sensing signaling pathways coupled to transcriptional outputs.

13. Pathways antagonizing phosphoinositide signaling

The preceeding sections discussed studies showing that invertebrate lifespan is extended by mutations disrupting insulin-like signaling pathways. Using genetic approaches, the proteins known to mediate insulin-like signaling via PI3K have been shown to be required for the shorter, wild-type lifespan. An important question following from these discoveries is the role of proteins that antagonize PI3K signaling.

PTEN is a lipid phosphatase that selectively dephosphorylates the phospholipid products of PI3K (Maehama and Dixon, 1998). PTEN was identified in mammalian cells as the product of a tumor suppressor gene, also referred to as *MMAC1* and *TEP1*, associated with rapid growth of tumors and tumor-derived cell lines (Li and Sun, 1997; Li et al., 1997; Liaw et al., 1997; Steck et al., 1997). In *C. elegans*, the gene *daf-18* encodes a homolog of mammalian PTEN (Ogg and Ruvkun, 1998; Gil et al., 1999; Mihaylova et al., 1999; Rouault et al., 1999). Mutations disrupting DAF-18 activity bypass the requirement for AGE-1/PI3K during development and allow AGE-1 mutants to continue development without arresting as dauer larvae (Fig. 4). As expected, DAF-18 is also required for long lifespan in mutants lacking AGE-1/PI3K (Dorman et al., 1995; Larsen et al., 1995). In a wild-type background, loss of DAF-18 function blocks dauer arrest in the absence of food and is correlated with an incompletely penetrant vulval bursting phenotype (Gil et al., 1999; Ogg and Ruvkun, 1998).

As mentioned above, the loss of DAF-18/PTEN activity bypasses the requirement for normal AGE-1/PI3K signaling. Therefore, in wild-type animals, DAF-18 acts to antagonize activation of AKT-1 and AKT-2 by phospholipids. Consistent with this, AKT-1 and AKT-2 are required for non-dauer development of animals lacking DAF-18/PTEN (Ogg and Ruvkun, 1998). These results predict that, in animals without AGE-1/PI3K activity, DAF-18/PTEN blocks the activation of AKT-1 and AKT-2 by other sources of phospholipids within the cell. In wild-type cells, DAF-18 may insulate the targets of PI3K from activation by phospholipids in the cellular milieu.

14. Final thoughts: outputs of PI3K signaling in vertebrates

The findings discussed above show that PI3K signaling is required for cell growth and developmental decisions in invertebrates. Mutations disrupting PI3K signaling and other components of insulin-like signaling pathways cause the striking phenotype of increased lifespan. In *C. elegans*, the absence of PI3K signaling allows active DAF-16 to promote increased expression of stress resistance genes, which may confer increased longevity. In mammalian cells, however, the absence of PI3K signaling triggers the initiation of programmed cell death via DAF-16-like transcription factors (Datta et al., 1999). Did the outputs of PI3K signaling change so dramatically during mammalian evolution? Are there situations in which PI3K signaling controls vertebrate lifespan pathways as well?

Recent findings suggest that there may be more to learn about the relationship between PI3K signaling and vertebrate longevity. As with invertebrates, PI3K signaling is a major effector for mammalian insulin and IGF-I receptor signaling. Growth hormone (GH) secreted by the pituitary regulates IGF-I secretion. Interestingly, lifespan is extended in dwarf mice that have mutations in the genes directing pituitary development (Brown-Borg et al., 1996; Flurkey et al., 2001). Preliminary evidence that GH receptor knockout mice have extended longevity suggests that the absence of GH in the dwarf mice accounts for extended lifespan (Coschigano et al., 2000). Future experiments investigating the link between GH and IGF-I signaling and vertebrate longevity will help illuminate pathways controlling mammalian longevity. However, regardless of whether a link exists between PI3K signaling and vertebrate longevity, our understanding of longevity control mechanisms has been dramatically improved from studies of invertebrate longevity and PI3K signaling.

References

Apfeld, J., Kenyon, C., 1998. Cell nonautonomy of *C. elegans daf-2* function in the regulation of diapause and life span. Cell 95, 199–210.

Apfeld, J., Kenyon, C., 1999. Regulation of lifespan by sensory perception in *Caenorhabditis elegans*. Nature 402, 804–809.

Arantes-Oliveira, N., Apfeld, J., Dillin, A., Kenyon, C., 2002. Regulation of life-span by germ-line stem cells in *Caenorhabditis elegans*. Science 295, 502–505.

Bohni, R., Riesgo-Escovar, J., Oldham, S., Brogiolo, W., Stocker, H., Andruss, B.F., Beckingham, K., Hafen, E., 1999. Autonomous control of cell and organ size by CHICO, a *Drosophila* homolog of vertebrate IRS1–4. Cell 97, 865–875.

Britton, J., Lockwood, W., Li, L., Cohan, S., Edgar, B., 2002. Drosophila's insulin/PI3-kinase pathway coordinates cellular metabolism with nutritional conditions. Developmental Cell 2, 239–249.

Brown-Borg, H.M., Borg, K., Meliska, C., Bartke, A., 1996. Dwarf mice and the aging process. Nature 384, 33.

Brunet, A., Bonni, A., Zigmond, M.J., Lin, M.Z., Juo, P., Hu, L.S., Anderson, M.J., Arden, K.C., Blenis, J., Greenberg, M.E., 1999. Akt promotes cell survival by phosphorylating and inhibiting a forkhead transcription factor. Cell 96, 857–868.

Cahill, C., Tzivion, G., Nasrin, N., Ogg, S., Dore, J., Ruvkun Alexander-Bridges, M., 2001. Phosphatidylinositol 3-kinase signaling inhibits DAF-16 DNA binding and function via 14-3-3-dependent and 14-3-3-independent pathways. J. Biol. Chem. 276, 13402–13410.

Carpenter, C.L., Cantley, L.C., 1996. Phosphoinositide kinases. Curr. Opin. Cell. Biol. 8, 153–158.

Clancy, D.J., Gems, D., Harshman, L., Oldham, S., Stocker, H., Hafen, E., Leevers, S., Partridge, L., 2001. Extension of life-span by loss of CHICO, a *Drosophila* insulin receptor substrate protein. Science 292, 104–106.

Coschigano, K.T., Clemmons, D., Bellush, L.L., Kopchick, J.J., 2000. Assessment of growth parameters and life span of GHR/BP gene-disrupted mice. Endocrinology 141, 2608–2613.

Datta, S., Brunet, A., Greenberg, M., 1999. Cellular survival: a play in three Akts. Genes and Develop. 13, 2905–2927.

Desrivieres, S., Cooke, F.T., Parker, P.J., Hall, M.N., 1998. MSS4, a phosphatidylinositol-4-phosphate 5-kinase required for organization of the actin cytoskeleton in *Saccharomyces cerevisiae*. J. Biol. Chem. 273, 15787–15793.

Domin, J., Waterfield, M.D., 1997. Using structure to define the function of phosphoinositide 3-kinase family members. FEBS Lett. 410, 91–95.

Dorman, J.B., Albinder, B., Shroyer, T., Kenyon, C., 1995. The *age-1* and *daf-2* genes function in a common pathway to control the lifespan of *Caenorhabditis elegans*. Genetics 141, 1399–1406.

Fabrizio, P., Pozza, F., Pletcher, S., Gendron, C., Longo, V., 2001. Regulation of longevity and stress resistance by Sch9 in yeast. Science 292, 288–290.

Fernandez, R., Tabarini, D., Azpiazu, N., Frasch, M., Schlessinger, J., 1995. The *Drosophila* insulin receptor homolog: a gene essential for embryonic development encodes two receptor isoforms with different signaling potential. EMBO J. 14, 3373–3384.

Flurkey, K., Papaconstantinou, J., Miller, R.A., Harrison, D.E., 2001. Lifespan extension and delayed immune and collagen aging in mutant mice with defects in growth hormone production. Proc. Natl. Acad. Sci. USA 98, 6736–6741.

Friedman, D., Johnson, T., 1988. A mutation in the *age-1* gene in *Caenorhabditis elegans* lengthens life and reduces hermaphrodite fertility. Genetics 118, 75–86.

Fruman, D., Meyers, R., Cantley, L., 1998. Phosphoinositide kinases. Annu. Rev. Biochem. 67, 481–507.

Fruman, D.A., Cantley, L.C., Carpenter, C.L., 1996. Structural organization and alternative splicing of the murine phosphoinositide 3-kinase p85 alpha gene. Genomics 37, 113–121.

Furuyama, T., Nakazawa, T., Nakano, I., Mori, N., 2000. Identification of the differential distribution patterns of mRNAs and consensus binding sequences for mouse DAF-16 homologues. Biochem. J. 349, 629–634.

Gil, E.B., Malone-Link, E., Liu, L.X., Johnson, C.D., Lees, J.A., 1999. Regulation of the insulin-like developmental pathway of *Caenorhabditis elegans* by a homolog of the PTEN tumor suppressor gene. Proc. Natl. Acad. Sci. USA 96, 2925–2930.

Gottlieb, S., Ruvkun, G., 1994. *daf-2, daf-16* and *daf-23*: Genetically interacting genes controlling dauer formation in *Caenorhabditis elegans*. Genetics 137, 107–120.

Honda, Y., Honda, S., 1999. The *daf-2* gene network for longevity regulates oxidative stress resistance and Mn-superoxide dismutase gene expression in *Caenorhabditis elegans*. FASEB J. 13, 1385–1393.

Hsin, H., Kenyon, C., 1999. Signals from the reproductive system regulate the lifespan of *C. elegans*. Nature 399, 362–366.

Johnson, T., 1990. Increased life-span of *age-1* mutants in *Caenorhabditis elegans* and lower gompertz rate of aging. Science 249, 908–912.

Kenyon, C., Chang, J., Gensch, E., Rudner, A., Tabtiang, R., 1993. A *C. elegans* mutant that lives twice as long as wild type. Nature 366, 461–464.

Kimura, K.D., Tissenbaum, H.A., Liu, Y., Ruvkun, G., 1997. *daf-2*, an insulin receptor-like gene that regulates longevity and diapause in *Caenorhabditis elegans*. Science 277, 942–946.

Kops, G.J.P.L., De Ruiter, N.D., De Vries-Smits, A.M.M., Powell, D.R., Bos, J.L., Burgering, B.M.T., 1999. Direct control of the forkhead transcription factor AFX by protein kinase B. Nature 398, 630–634.

Larsen, P.L., 1993. Aging and resistance to oxidative damage in *Caenorhabditis elegans*. Proc. Natl. Acad. Sci. USA 90, 8905–8909.

Larsen, P.L., Albert, P.S., Riddle, D.L., 1995. Genes that regulate both development and longevity in *Caenorhabditis elegans*. Genetics 139, 1567–1583.

Lee, R.Y.N., Hench, J., Ruvkun, G., 2001. Regulation of *C. elegans daf-16* and its human ortholog FKRHL1 by the *daf-2* insulin-like signaling pathway. Curr. Biol. 11, 1950–1957.

Leevers, S.J., Weinkove, D., Macdougall, L.K., Hafen, E., Waterfield, M.D., 1996. The *Drosophila* phosphoinositide 3-kinase Dp110 promotes cell growth. EMBO 15, 6584–6594.

Li, D., Sun, H., 1997. TEP1, encoded by a candidate tumor suppressor locus, is a novel protein tyrosine phosphatase regulated by transforming growth factor beta. Cancer Res. 57, 2124–2129.

Li, J., Yen, C., Liaw, D., Podsypanina, K., Bose, S., Wang, S., Puc, J., Miliaresis, C., Rodgers, L., Mccombie, R., Bigner, S., Giovanella, B., Ittmann, M., Tycko, B., Hibshoosh, H., Wigler, M., Parsons, R., 1997. PTEN, a putative tyrosine phosphatase gene mutated in human brain, breast and prostate cancer. Science 275, 1943–1947.

Liaw, D., March, D., Li, J., Dahia, P., Wang, S., Zheng, Z., Bose, S., Call, K., Tsou, H., Peacocke, M., Eng, C., Parsons, R., 1997. Germline mutations of the PTEN gene in Cowden disease, an inherited breast and thyroid cancer syndrome. Nat. Genet. 16, 64–67.

Lin, K., Dorman, J.B., Rodan, A., Kenyon, C., 1997. *daf-16*: An HNF-3/forkhead family member that can function to double the life-span of *Caenorhabditis elegans*. Science 278, 1319–1322.

Lin, K., Hsin, H., Libina, N., Kenyon, C., 2001. Regulation of the *Caenorhabditis elegans* longevity protein DAF-16 by insulin/IGF-I and germline signaling. Nat. Genet. 28, 139–145.

Lithgow, G., White, T., Melov, S., Johnson, T., 1995. Thermotolerance and extended life-span conferred by single-gene mutations and induced by thermal stress. Proc. Natl. Acad. Sci. 92, 7540–7544.

Maehama, T., Dixon, J., 1998. The tumor suppressor, PTEN/MMAC1, dephosphorylates the lipid second messenger, phosphatidylinositol 3,4,5-trisphosphate. J. Biol. Chem. 273, 13375–13378.

Malone, E.A., Inuoe, T., Thomas, J.H., 1996. Genetic analysis of the roles of *daf-28* and *age-1* in regulating *Caenorhabditis elegans* dauer formation. Genetics 143, 1193–1205.

Melov, S., Ravenscroft, J., Malik, S., Gill, M., Walker, D., Clayton, P., Wallace, D., Malfroy, B., Doctrow, S., Lithgow, G., 2000. Extension of life-span with superoxide dismutase/catalase mimetics. Science 289, 1567–1569.

Mihaylova, V., Borland, C., Manjarrez, L., Stern, M., Sun, H., 1999. The PTEN tumor suppressor homolog in *Caenorhabditis elegans* regulates longevity and dauer formation in an insulin receptor-like signaling pathway. Proc. Natl. Acad. Sci. 96, 7427–7432.

Morris, J.Z., Tissenbaum, H.A., Ruvkun, G., 1996. A phosphatidylinositol-3-OH kinase family member regulating longevity and diapause in *Caenorhabditis elegans*. Nature 382, 536–539.

Murakami, S., Johnson, T., 1996. A genetic pathway conferring life extension and resistance to UV stress in *Caenorhabditis elegans*. Genetics 143, 1207–1218.

Nakae, J., Park, B., Accili, D., 1999. Insulin stimulates phosphorylation of the forkhead transcription factor FKHR on serine 253 through a wortmannin-sensitive pathway. J. Biol. Chem. 274, 15982–15985.

Ogg, S., Paradis, S., Gottlieb, S., Patterson, G.I., Lee, L., Tissenbaum, H.A., Ruvkun, G., 1997. The fork head transcription factor DAF-16 transduces insulin-like metabolic and longevity signals in *C. elegans*. Nature 389, 994–999.

Ogg, S., Ruvkun, G., 1998. The *C. elegans* PTEN homolog, DAF-18, acts in the insulin receptor-like metabolic signaling pathway. Mol. Cell. 2, 887–893.

Paradis, S., Ruvkun, G., 1998. *Caenorhabditis elegans* Akt/PKB transduces insulin receptor-like signals from AGE-1 PI3 kinase to the DAF-16 transcription factor. Genes Dev. 12, 2488–2498.

Paradis, S., Ailion, M., Toker, A., Thomas, J.H., Ruvkun, G., 1999. A PDK1 homolog is necessary and sufficient to transduce AGE-1 PI3 kinase signals that regulate diapause in *Caenorhabditis elegans*. Genes Dev. 13, 1438–1452.

Parkes, T.L., Elia, A.J., Dickinson, D., Hilliker, A.J., Phillips, J.P., Boulianne, G.L., 1998. Extension of *Drosophila* lifespan by overexpression of human SOD1 in motorneurons. Nat. Genet. 19, 171–174.

Poltilove, R., Jacobs, A., Haft, C., Xu, P., Taylor, S., 2000. Characterization of *Drosophila* insulin receptor substrate. J. Biol. Chem. 275, 23346–23354.

Rouault, J.-P., Kuwabara, P., Sinilnikova, O., Duret, L., Thierry-Mieg, D., Billaud, M., 1999. Regulation of dauer larva development in *Caenorhabditis elegans* by DAF-18, a homologue of the tumour suppressor PTEN. Curr. Biol. 9, 329–332.

Schu, P.V., Takegawa, K., Fry, M.J., Stack, J.H., Waterfield, M.D., Emr, S.D., 1993. Phosphatidylinositol 3-kinase encoded by yeast VPS34 gene essential for protein sorting. Science 260, 88–91.

Steck, P., Pershouse, M., Jasser, S., Yung, W., Lin, H., Ligon, A., Langford, L., Baumgard, M., Hattier, T., Davis, T., Frye, C., Hu, R., Swedlund, B., Teng, D., Tavtigian, S., 1997. Identification of a candidate tumour suppressor gene, MMAC1, at chromosome 10q23.3 that is mutated in multiple advanced cancers. Nat. Genet. 15, 356–362.

Sun, J., Tower, J., 1999. Flp recombinase-mediated induction of Cu/Zn-superoxide dismutase transgene expression can extend the life span of adult *Drosophila melanogaster* flies. Mol. Cell. Biol. 19, 216–228.

Takaishi, H., Konishi, H., Matsuzaki, H., Ono, Y., Shirai, Y., Saito, N., Kitamura, T., Ogawa, W., Kasuga, M., Kikkawa, U., Nishizuka, Y., 1999. Regulation of nuclear translocation of forkhead transcription factor AFX by protein kinase B. Proc. Natl. Acad. Sci. 96, 11836–11841.

Tatar, M., Kopelman, A., Epstein, D., Tu, M.-P., Yin, C.-M., Garofalo, R., 2001. A mutant *Drosophila* insulin receptor homolog that extends life-span and impairs neuroendocrine function. Science 292, 107–110.

Toker, A., Cantley, L.C., 1997. Signaling through the lipid products of phosphoinositide-3-OH kinase. Nature 387, 673–676.

Vowels, J.J., Thomas, J.H., 1992. Genetic analysis of chemosensory control of dauer formation in *Caenorhabditis elegans*. Genetics 130, 105–123.

Weinkove, D., Leevers, S.J., Macdougall, L.K., Waterfield, M.D., 1997. p60 is an adaptor for the *Drosophila* phosphoinositide 3-kinase, Dp110. J. Biol. Chem. 272, 14606–14610.

Weinkove, D., Neufeld, T., Twardzik, T., Waterfield, M., Leevers, S., 1999. Regulation of imaginal disc cell size, cell number and organ size by *Drosophila* class Ia phosphoinositide 3-kinase and its adaptor. Curr. Biol. 9.

Wolkow, C.A., Kimura, K., Lee, M., Ruvkun, G., 2000. Regulation of *C. elegans* life-span by insulinlike signaling in the nervous system. Science 290, 147–150.

Wolkow, C.A., Muñoz, M.J., Riddle, D.L., Ruvkun, G., 2002. Insulin receptor substrate and p55 orthologous adaptor proteins function in the *Caenorhabditis elegans daf-2*/insulin-like signaling pathway. J. Biol. Chem. 277, 49591–49597.

Yoshida, S., Ohya, Y., Goebl, M., Nakano, A., Anraku, Y., 1994. A novel gene, STT4, encodes a phosphatidylinositol 4-kinase in the PKC1 protein kinase pathway of *Saccharomyces cerevisiae*. J. Biol. Chem. 269, 11666–11672.

Yu, H., Larsen, P., 2001. DAF-16-dependent and independent expression targets of DAF-2 insulin receptor-like pathway in *Caenorhabditis elegans* include FKBPs. J. Mol. Biol. 314, 1017–1028.

**Advances in
Cell Aging and
Gerontology**

Ceramide-driven stress signals in cancer and aging

Peter P. Ruvolo[1,2,*], Charlene R. Johnson[2], W. David Jarvis[1,2]

[1]*Institute for Molecular Medicine*
[2]*Department of Integrative Biology and Pharmacology*
University of Texas Health Science Center at Houston
2121 W. Holcombe Bivd., Houston, Texas USA.
Correspondence address: Tel.: + 1-713-500-2400; fax: + 1-713-500-2420.
E-mail address: Peter.P.Ruvolo@uth.tmc.edu

Contents

1. Introduction

As each of us age, our prospects for developing diseases such as cancer, diabetes, and arthritis increase. In many ways it is a no win situation where either important cells that comprise the body ultimately succumb to debilitative processes (e.g. osteoporosis, atherosclerosis) or conversely, some cells develop aberrant growth properties resulting in cancer. Either situation can result in death for the organism. For instance, the aging process itself appears to contribute to degeneration of heart tissue. Even in the absence of pathologies that target heart tissue (e.g. hypertension, diabetes, atherosclerosis), the total number of cardiac myocytes that are found in the heart are reduced with age (Olivetti et al., 1991,

1995; Phaneuf and Leeuwenburgh, 2002). On the other hand the incidence of cancer increases with age and thus the development of cancer is dependent on age (Balducci and Beghe, 2001; Sekeres and Stone, 2002). The influence of age on the onset of cancer is most dramatic in patients with Fanconi anemia (Alter, 1996; Fagerlie et al., 2001; Grompe and D'Andrea, 2001). While the onset of leukemia occurs on average at 40 years in the general population, Fanconi anemia patients develop leukemia in their late teens and early 20s (Alter, 1996).

A common goal in biomedical research is to prolong life. To achieve this goal, we will need to understand the cellular processes that regulate normal cellular homeostasis and the mechanisms that are responsible for abnormal cell growth and those responsible for cell death. Interestingly, there is often cross talk between the signal transduction pathways that regulate cell growth and those pathways that regulate cell death. An interesting example of such a cellular regulatory process involves ceramide, a naturally occurring sphingolipid, which is a key biomolecule in the sphingomyelin cycle (Hannun, 1994). Metabolic pathways leading to ceramide formation are depicted in Fig. 1. During normal cell growth, ceramide and phosphatidylcholine are converted to sphingomyelin, an important component in the membranes that make up cells (Hannun, 1994). A byproduct of this reaction is diglyceride, a potent growth and survival agonist by virtue of its

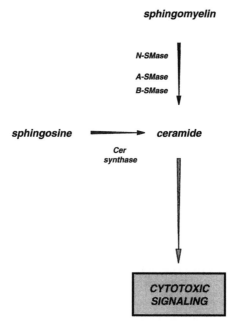

Fig. 1. Metabolic pathways governing the formation of ceramide. Multiple metabolic provisions exist for the formation of ceramide. Ceramide can be produced rapidly from the hydrolysis of sphingomyelin by the actions of acidic (A-SMase), basic (B-SMase), and neutral (N-SMase) isoforms of sphingomyelinase present in different subcellular membrane structures. In addition, ceramide can arise through *de novo* synthesis from fatty acyl CoA and a free sphingoid base by ceramide synthase. Both general types of metabolic pathway have been implicated in the generation of ceramide to initiate a pro-apoptotic signal.

ability to promote cPKC/nPKC-driven signaling pathways (Nishizuka, 1992; May, 1997; McCubrey et al., 2000). Like diglyceride, ceramide is an important second messenger molecule, except that ceramide-mediated signaling usually involves stress pathways (Hannun, 1996; Jarvis et al., 1996b; Smyth et al., 1997; Basu and Kolesnick, 1998; Kolesnick and Kronke, 1998; Hannun and Luberto, 2000). In addition, ceramide is a potent promoter of programmed cell death (a process also known as apoptosis; Kerr et al., 1972). Because at least one enzyme (sphingomyelin synthase) appears to regulate diglyceride and ceramide simultaneously but in opposite directions, it is possible that diglyceride and ceramide have counterbalancing effects on cell growth and cell death (Luberto and Hannun, 1998). In support of a model where diglyceride and ceramide have counterbalancing effects, diglyceride can attenuate the effects of ceramide (Jarvis et al., 1994a,b) and ceramide can antagonize diglyceride by inhibiting cPKCα (Chmura et al., 1996; Lee et al., 1996; Lee et al., 2000). An elegant homeostatic regulatory system is emerging with diglyceride as a pro-growth regulator and ceramide as a negative growth regulator. Since each effector is able to transmodulate the signaling pathways regulated by the other, it is possible that net availability of diglyceride versus ceramide may create a sort of cellular rheostat (Jarvis et al., 1994b; Hannun, 1996; Luberto and Hannun, 1998). It is possible that the dominant second signal molecule produced during growth (*i.e.* diglyceride) or stress (*i.e.* ceramide) conditions could promote signal pathways that determine post-translational modification of potential "survival sensor" molecules as has been have recently proposed for the potent regulator of apoptosis, BCL2 (Ruvolo et al., 2001a). Interestingly, it has been recently found that cardiac myocytes exhibit reduced levels of BCL2 with age and thus may contribute to loss of cardiac myocytes in the aging rat (Phaneuf et al., 2002). It will be interesting to determine the role for ceramide in regulating BCL2 anti-apoptotic potential in cardiac myocytes during aging.

2. Ceramide as a second signal molecule

Acute generation of ceramide has been noted so widely in response to lethal cellular stress that it is considered to represent a nearly universal feature of the apoptotic process (Hannun, 1994). In addition, it is now clear that ceramide-driven signals can impinge upon numerous aspects of cell survival ranging from restriction of cell cycle transit to the induction of differentiation, senescence, and apoptotic cell death. The versatility of ceramide action involves the complex modulation of both proximal and distal signaling elements. The complexities of ceramide metabolism have been reviewed in detail previously (Ariga et al., 1998; Merrill et al., 1998). This section provides a brief overview of known ceramide-regulated kinases and phosphatases, and outlines the apparent roles of their downstream effectors in mediating distinct biological features of ceramide action in mammalian systems. Some of the signaling pathways that are regulated by ceramide are depicted in Fig. 2.

P. P. Ruvolo et al.

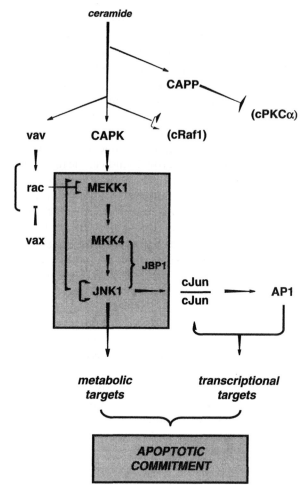

Fig. 2. Overview of ceramide-driven SAPK-JNK cascade activity. Increased intracellular availability of ceramide can engage stress signaling cascades through direct interaction with multiple proximal targets, most notably the ceramide-activated protein kinases (CAPKs) and ceramide-activated protein phosphatases (CAPPs). Among the CAPKs directly implicated in ceramide-driven apoptosis are KSR and aPKCζ These elements, in turn, activate the SAPK sequence MEKK1 MKK4 JNK1/JNK2 (the terminal kinases, JNK1 and 2, appear to require autophosphorylation for maximal activity). JNKs render the transcription factor cJun fully active through phosphorylation and thereby augment AP1-dependent transactivation, a process that is further amplified by increased cJun expression. JNKs additionally influence an array of metabolic targets that lead to apoptotic commitment. Recruitment of the SAPK-JNK cascade may be facilitated by ceramide action in other ways. For example, ceramide enhances the activity of the guanine nucleotide exchange factor vav, which can accelerate activation of small GTP-binding proteins (e.g. rac, ras), potentially resulting in more efficient activation of apical kinases such as MEKK1. Moreover, ceramide influences a number of other signals through the activation of CAPPs (both PP2A and PP1); dephosphorylation of key targets by these enzymes can silence the influence of essential downstream survival elements ranging from MAPK-ERK to PKB-Akt. Thus, this signaling cascade allows for the coordinated activation of pro-apoptotic mechanisms and inactivation of anti-apoptotic mechanisms.

2.1. Ceramide-activated protein kinases

The probable existence of one or more protein kinases subject to the direct influence of ceramide was suggested based upon the obvious signaling potential associated with the sphingomyelin pathway (Kolesnick and Golde, 1994). Several distinct protein kinases subject to selective regulation by ceramide are now known.

Kinase suppressor pf Ras (KSR-CAPK) The best characterized of the ceramide-driven enzymes is a 97-kDa membranal, proline-directed, serine/threonine kinase originally named ceramide-activated protein kinase (CAPK, Joseph et al., 1993; Liu et al., 1994; Mathias et al., 1993). CAPK differs from other members of the proline-directed kinase family in its selective recognition of X-Thr-Leu-Pro-X as a minimal substrate peptide sequence (Joseph et al., 1993). Further molecular characterization of this activity revealed that CAPK is identical with a previously known 100-kDa membrane protein referred to as the kinase suppressor of ras (KSR) (Zhang et al., 1997).

Current information concerning KSR-CAPK signaling and its downstream effectors supports the view that suggests that ceramide signals directly to KSR-CAPK, which in turn results in sequential recruitment of cRaf1 (see further) and the MAPK-ERK cascade in response to cellular stress (Zhang et al., 1997). Thus, the KSR-CAPK → ras/Raf → MAPK-ERK sequence represents at least one pathway through which ceramide can initiate a primary intracellular signaling cascade. At present, there is poor concensus as to the targets(s) for MAPK-ERK outflow in response to KSR activation. Nonetheless, this pathway may function, under some circumstances, through downstream suppression of PKB-Akt activity (Basu et al., 1998); attendant loss of the cytoprotective influence of PKB-Akt ultimately permits dephosphorylation-mediated stimulation of the pro-apoptotic BCL2 family member BAD, a transdominant inhibitory regulator of BCL2. Once relieved of inhibitory phosphorylation, BAD is released from sequestration on 14-3-3 proteins and thereby facilitates apoptotic commitment through blockade of BCL2-dependent cytoprotectivity.

cRaf1 There is also evidence that the well-established serine-threonine protein kinase cRaf1 may be subject to modulation by ceramide. To wit, ceramide has been reported to interact directly with and stimulate the activity of cRaf1 (Huwiler et al., 1996); upon association of cRaf1 with GTP-ras and subsequent translocation to the plasma membrane, cRaf1 can initiate sequential activation of MEK1/MEK2 and ERK1/ERK2, enzymatic elements comprising the primary MAPK module. Consistent with these observations, ceramide can promote, under some circumstances, transient activation of ERK1 (Raines et al., 1993) and multiple downstream of MAPK targets, including phospholipase-A_2 (PLA_2), which reportedly is required for ceramide-dependent apoptotic, and/or inflammatory responses in some settings (Hayakawa et al., 1993). Thus, apart from serving an essential signaling role in normal inflammatory responses, presently it is not clear whether the pathway linking KSR/CAPK to activation of the MAPK cascade (and some downstream elements) also contributes to the induction of apoptosis (Lin et al., 2000).

In this regard, it is noteworthy that ceramide has been reported to stimulate the activity of the guanine nucleotide-exchange factor (GEF) vav, a close homolog of the established GEF sos that is selectively expressed in hematopoietic cells (Gulbins et al., 1994). The functional significance of ceramide-mediated activation of vav is presently uncertain; nonetheless, given the generalized role of GEFs in the activation of various small GTP-binding proteins (e.g., ras, rac), it is tempting to speculate that ceramide may participate in the recruitment of various MAP3Ks such as cRaf1 (or MEKK1) through one or more small G-proteins augmented GDP/GTP exchange.

The atypical protein kinase C isoform-ζ (aPKCζ)   There is abundant evidence to suggest that ceramide can modulate multiple isoforms of PKC. For example, ceramide directly interacts with and stimulates the atypical PKC ζ-isoform (aPKCζ) (Lozano et al., 1994; Muller et al., 1995) as well as cPKCα. Although ectopic expression of aPKCζ reportedly promotes apoptotic cell death in some settings, it remains to be determined whether this represents a physiological effector for ceramide cytotoxicity.

2.2. Ceramide-activated protein phosphatases

Functionally distinct (i.e., non-phosphoryltransferase) targets for ceramide have also been identified. Ceramide prominently stimulates a heterotrimeric cytosolic serine–threonine phosphoprotein phosphatase (designated class-2A; *c*eramide-*a*ctivated *p*rotein *p*hosphatase; CAPP; Dobrowsky and Hannun, 1992; Dobrowsky et al., 1993). As ceramide-related apoptosis is reportedly antagonized by okadaic acid (but not *nor*okadaone) as well as by other inhibitors selective for class-2A phosphatases (Huwiler et al., 1996), a cytotoxic role for CAPP in ceramide action has been inferred (Bielawska et al., 1993). Furthermore, ceramide promotes dephosphorylation of cPKCα in an okadaic acid-sensitive/*nor*-okadaone-insensitive manner, resulting in the loss of phosphorylation on autophosphorylation sites and consequent inactivation of the enzyme (Lee et al., 1996); because multiple isoforms of cPKC/nPKC lie upstream of Raf-1 in various settings and CAPP-mediated PKC and inactivates MAPK. Ceramide is a potent activator of both PP2A (Dobrowsky and Hannun, 1992; Dobrowsky et al., 1993; Ruvolo et al., 1999) and PP1 (Chalfant et al., 1999). The mechanism of ceramide activation of these protein phosphatases is not yet clear. It is known, however, that protein phosphatases play a critical role in ceramide-mediated processes. cPKCα (Lee et al., 1996; Lee et al., 1998) and PKB-Akt (Salinas et al., 2000; Schubert et al., 2000) are inactivated by PP2A in response to ceramide. Ceramide-mediated inactivation of Akt may play a role in diabetes (Teruel et al., 2001). Teruel et al. (2001) have found that ceramide blocks insulin-stimulated glucose uptake in brown adipocyte cells in an Akt-dependent manner. Moreover, PP2A is implicated in this process, in view of evidence that okadaic acid blocks insulin induces dephosphorylation of Akt (Teruel et al., 2001).

Ceramide-mediated activation of PP2A can also directly target cPKCα and Akt substrates, thus allowing for multiple levels of regulation of various stress pathways. For instance, ceramide promotes PP2A dephosphorylation of BCL2 resulting in the loss of BCL2's anti-apoptotic function (Ruvolo et al., 1999). Given that expression of S70E BCL2, a non-phosphorylatable, activated form of BCL2, can protect cells from ceramide-induced apoptosis, one likely mechanism by which ceramide induces cell death is by functionally inactivating BCL2 (Ruvolo et al., 1999). One of the physiologic BCL2 kinases is cPKCα (Ruvolo et al., 1998). Since PP2A inactivates cPKCα (Lee et al., 1996; Lee et al., 1998), ceramide-mediated activation of PP2A can suppress BCL2's anti-apoptotic function indirectly by targeting the BCL2 kinase (Ruvolo, 2001) and/or directly by targeting BCL2 itself (Deng et al., 1998; Ruvolo et al., 1999). Furthermore, PP2A has recently been shown to directly dephosphorylate BAD (Chiang et al., 2001). At least for some members of the BCL2 family, ceramide activation of PP2A promotes anti-survival signaling by targeting BCL2 family members directly while inactivating their physiologic protein kinases (Ruvolo et al., 1999; Ruvolo, 2001).

Ceramide activation of PP1 has not been as well-studied as PP2A. Ceramide has been shown to promote dephosphorylation of pRb in association with growth arrest (Dbaibo et al., 1995). Phosphatidic acid (PtdOH) has been shown to inhibit PP1 but not PP2A (Kishikawa et al., 1999). PtdOH was demonstrated to block ceramide-induced dephosphorylation of pRb and PARP cleavage, thus implicating PP1 in both cell cycle arrest and apoptotic processes, respectively (Kishikawa et al., 1999). A novel apoptosis regulatory pathway has emerged suggesting that ceramide may regulate mRNA alternative splicing of BCL2 family member and caspase genes by a mechanism involving PP1 and serine/arginine rich domain (SR) proteins (Chalfant et al., 2001, 2002). SR proteins are a family of proteins that are required for pre-mRNA processing (Zahler et al., 1992, 1993). Phosphorylation status of SR proteins regulates their cellular localization and function and PP1 plays a major role in this process (Misteli and Spector, 1996; Misteli et al., 1998; Chalfant et al., 2001). Alternative splice products was first shown as a mechanism to regulate apoptosis when it was demonstrated that a dominant negative splice variant (BCL-X$_s$) of the anti-apoptotic BCL2 family member, BCL-X$_L$, existed (Minn et al., 1996). Alternative splicing has also emerged in the regulation of caspase 2 (Jiang et al., 1998; Cote et al., 2001), caspase 9 (Seol and Billiar, 1999), and the pro-apoptotic BCL2 family member, BAX (Zhou et al., 1998; Jin et al., 2001). Chalfant et al. (2002) demonstrated that exogenous C6-ceramide treatment was shown to induce alternative splicing of BCL-X and caspase 9, but not BAX or caspase 2. Furthermore, it was shown in the same study that ceramide promoted the generation of the pro-apoptotic splice variants (i.e. the dominant negative BCL-X$_s$ and caspase 9) at the expense of the anti-apoptotic splice variants (i.e. BCL-X$_L$ and caspase 9b) in a PP1-dependent manner. Fumonisin B1 and myriocin, both inhibitors of the *de novo* sphingomyelin metabolic pathway, blocked ceramide-induced alternative splicing of the apoptotic gene products, suggesting that ceramide generated by the *de novo* pathway rather than ceramide produced by sphingomyelinase is involved (Chalfant et al., 2002).

It is not clear exactly how ceramide activates PP2A or PP1. In the case of PP2A, a possible mechanism appears to involve translocation of PP2A to target membranes (Sontag et al., 1995; McCright et al., 1996a; Ruvolo et al., 2002). It appears that the regulatory B subunits of PP2A are directly involved in this process. Such a finding is not unexpected since the PP2A regulatory B subunits (*a*) are expressed differentially by tissue and temporally during development (Mayer et al., 1991; Ruediger et al., 1991; Mumby and Walter, 1993; McCright and Virshup, 1995; Csortos et al., 1996), (*b*) may determine PP2A substrate specificity (Cegielska et al., 1994) and (*c*) are known to target the PP2A catalytic complex to intracellular sites such as the nucleus or microtubules under certain conditions (Sontag et al., 1995; McCright et al., 1996a). It has recently been shown that ceramide promotes translocation of PP2A to the mitochondrial membranes by a mechanism involving the B56 α-subunit (Ruvolo et al., 2002). The recruitment of PP2A to the mitochondria by ceramide, activates the BCL2 phosphatase and promotes BCL2 dephosphorylation and inactivation (Ruvolo et al., 1999). Interestingly, the regulatory B56 α-subunit is primarily localized in the cytosol under normal conditions in human pre-B ALL cells, however, ceramide promotes B56α movement from the cytosol to mitochondrial membranes (Ruvolo et al., 2002). In addition, ceramide promotes significant mitochondrial translocation of the PP2A subunits that comprise the catalytic subunit (i.e. PP2A/A and PP2A/C) as well, and thus suggests that ceramide promotes translocation of the PP2A catalytic complex to mitochondrial membranes via B56α in the case of BCL2 phosphatase (Ruvolo et al., 2002).

2.3. Downstream effector systems in ceramide action

KSR-CAPK and/or aPKCζ presently represent the most compelling candidate proximal effectors for ceramide-driven cytotoxicity in many systems. As noted above, the proximal target(s) for ceramide that are uniformly essential to the initiation of lethal signaling have evaded definitive identification, and it is therefore to be stressed that ceramide likely elicits apoptosis through different signaling pathways in a highly context-dependent (which is to say, cell type-specific) fashion.

Ceramide and recruitment of the SAPK-JNK cascade Beyond this uncertainty, however, it is clear that the pro-apoptotic influence of ceramide entails the recruitment of several broadly conserved downstream effector mechanisms. Indeed, a critical involvement of both the SAPK-JNK cascade and the mitochondrially gated cytochrome-C/APAF1/caspase-9 system has been widely demonstrated in the apoptotic responses to both natural and synthetic ceramide (Westwick et al., 1995; Jarvis et al., 1997), KSR-CAPK (Basu and Kolesnick, 1998; Basu et al., 1998; Xing et al., 2000) and aPKCζ (Lozano et al., 1994; Muller et al., 1995; Bourbon et al., 2000). Conversely, ceramide can inhibit kinases that are key components of pro-growth signaling processes such as the classical and novel PKC isoforms (e.g. cPKCnPKC; Lee et al., 1996, 1998), or PKB-Akt (Santana et al., 1996; Basu et al., 1998; Scheid and Duronio, 1998; Spiegel and Milstien, 2000). The mechanism of kinase inhibition in each case appears to involve the ability of ceramide to activate one or more species of protein phosphatase. As already discussed, CAPP represents a form of PP2A phosphatase (Dobrowsky and Hannun, 1992; Dobrowsky et al., 1993); in addition, independent findings indicate that ceramide can activate a major form of PP1 as well (Chalfant et al., 1999).

Given that ceramide-regulated enzymes have been implicated as potential tumor suppressor elements (e.g. PP2A; Cohen and Cohen, 1989; Schonthal, 1998), and that ceramide regulatory pathways have been shown to influence pathways controlling known tumor promoters (e.g. BCL2; Ruvolo, 2001), it is highly probable that there is an exact role for ceramide in the regulation of oncogenesis. In fact, data supporting a causal link between ceramide and chemoresistance is emerging, thus supporting such a regulatory role for sphingolipid messengers in the development of neoplasia (Liu et al., 1999, 2000; Senchenkov et al., 2001). For instance, a potential mechanism of drug resistance that has recently been described involves enzymatic modification of ceramide that abrogates the ability of this lipid messenger to act as an effector molecule in programmed cell death pathways (Maurer et al., 2000; Senchenkov et al., 2001). This novel mechanism will be discussed later in this chapter.

Several lines of evidence have demonstrated that ceramide engages, through whatever immediate coupling mechanism, the protein kinase MEKK1; MEKK1 in turn initiates sequential activation of MKK4 (also termed SEK1) and JNK1/JNK2. MKK and JNK together comprise the primary SAPK signaling module. Increased availability of ceramide rapidly engages JNK1 and JNK2, resulting in the activation of multiple downstream SAPK targets; most notable among these c-Jun (which represents an essential component of the various forms of the AP-1 transcription factor complex), and ATF2 (which can serve as a c-Jun dimerization partner for AP1 activation; Sawai et al., 1995; Westwick et al., 1995; Verheij et al., 1996; Jarvis et al., 1996a, 1997, 1999). Induction of the SAPK cascade is also closely associated with multiple ceramide-dependent pro-apoptotic stimuli, ranging from activation of cytotoxic receptor systems to the onset of lethal environmental stresses (Verheij et al., 1996; Jarvis et al., 1997, 1998). Moreover, acute inter-ruption of the SAPK cascade (e.g., through dominant-negative suppression of SEK1 activity) or interference with outflow from JNK1/JNK2 (e.g., through pharmacological inhibition, or dominant-negative quenching of c-Jun transactiva-tion potential) can abolish the apoptotic responses to ceramide or ceramide-dependent lethal stimuli (Sawai et al., 1995; Verheij et al., 1996; Jarvis et al., 1997). Ceramide-mediated induction of cell death thus requires acute activation of the SAPK cascade. In addition, activation of this system is closely associated with, in some instances, reciprocal inactivation of the MAPK-ERK cascade (Jarvis et al., 1997). This notion is consistent with proposals that the two systems are regulated coordinately (Xia et al., 1995; Jarvis et al., 1997).

Ceramide and activation of protein kinase R Ceramide has been shown to activate a novel serine/threonine protein kinase referred to variously as dsRNA-dependent protein kinase, IFNγ-inducible protein kinase, or protein kinase R (PKR); PKR represents a critical physiological regulator of protein synthesis that is ceramide sensitive (Ruvolo et al., 2001b). One mechanism for regulating protein synthesis involves the reversible phosphorylation of the α-subunit of eukaryotic initiation factor-2 (eIF2α) (Merrick, 1992; Samuel, 1993; Sheikh and Fornace, 1999). The phosphorylation of eIF2α results in the inhibition of initiation of protein translation and thus prevents protein synthesis (Hershey, 1993; Rowlands et al., 1988). One of the physiological eIF2α kinases and a key regulator of this process is PKR (Clemens and Elia, 1997; Tan and Katze, 1999; Williams, 1999). Much of what is known about PKR has been derived from studies on its role in the host anti-viral defense mechanism where PKR is activated in response to viral dsRNA (Hovanessian, 1993; Matthews, 1993). PKR, however, has an expanded role in regulating protein synthesis in response to cellular stresses (Williams, 1999). A novel cellular regulator of PKR, RAX (called PACT in human cells; Patel and Sen, 1998), appears to activate PKR in response to stress applications such as IL-3 withdrawal, sodium arsenate treatment, or peroxide treatment (Ito et al., 1999). Importantly, activation of PKR by RAX is independent of dsRNA (Ito et al., 1999). Thus activation of PKR appears to involve two mechanisms: one where PKR is activated by viral dsRNA, the other, where a cellular activator (i.e., RAX) is involved. Evidence suggests that stress stimuli (e.g., IL-3 withdrawal or peroxide treatment) may activate PKR via RAX, presumably by a mechanism whereby RAX associates with and activates PKR (Ito et al., 1999). Inasmuch as the stress treatments that result in PKR activation did not significantly affect total RAX protein expression, a post-translational mechanism appeared to be involved. RAX appears to be activated when it is phosphorylated by an as yet unknown stress-activated kinase (Ito et al., 1999). Because ceramide is produced during almost all apoptotic stress applications (Hannun, 1994) and has been demonstrated to activate stress-activated kinases (Verheij et al., 1996; Jarvis et al., 1997), it is not surprising to find that ceramide induces phosphorylation of RAX with concomitant activation of PKR (Ruvolo et al., 2001b). Consistent with a model where ceramide regulates PKR, ceramide was found to promote eIF2α phosphorylation and inhibit protein synthesis in a dose-dependent manner (Ruvolo et al., 2001b).

Ceramide and recruitment of the intrinsic APAF1/cyt-C/caspase-9 pathway for apopototic commitment As outlined in preceding sections, numerous intracellular signaling systems, recruited both in tandem and in parallel, appear to subserve the cytotoxic influence of ceramide in mammalian systems. The downstream processes that bring about final commitment to apoptosis via the intrinsic mitochondrially gated APAF pathway are yet to be completely characterized. The present section addresses the relationship between ceramide generation and mitochondrial dysfunction.

The major cellular pool of sphingomyelin is found on the outer leaflet of the plasma membrane. Here the generation of ceramide is associated with inhibition of both PKC-induced NFκB activation (Luberto et al., 2000) and platelet-derived growth factor-induced PI3-kinase activity (Zundel et al., 2000). This pool of cellular ceramide, however, has been shown to be insufficient to induce apoptosis. Instead studies targeting bacterial sphingomyelinase to various intracellular compartments have linked ceramide-induced cell death to that generated specifically in the mitochondria (Birbes et al., 2001). This is supported by previous studies in which both C_2 and C_6-ceramide were shown to induce cytochrome c release from isolated mitochondria. This release appeared to be influenced by the redox state of cytochrome c and was prevented by BCL2 expression (Ghafourifar et al., 1999). Likewise, the loss of mitochondrial membrane potential in response to exogenous natural (C_{18}) ceramide can be effectively inhibited by the general caspase inhibitor Z-VAD-fmk (Ghafourifar et al., 1999). Additional evidence supports a direct association between ceramide targeting and the mitochondria. In particular, ceramide has been shown to inhibit complex III of the mitochondrial respiratory chain (Gudz et al., 1997) as well as to induce the generation of reactive oxygen species, both in intact mitochondria (Garcia-Ruiz et al., 1997) and whole cells (France-Lanord et al., 1997; Quillet-Mary et al., 1997).

Accumulating data supports a pivotal role for both the pro- and antiapoptotic BCL2 family members in mediating ceramide-induced cell death. In this regard both BCL2 and BCL-X_L have been shown to inhibit cytochrome c release and subsequent apoptosis (Ghafourifar et al., 1999; Cuvillier et al., 2000; Sawada et al., 2000). BAX, on the other hand, appears to be required for the release of cytochrome c in response to ceramide as antisense BAX has been shown to inhibit cytochrome c release and cell death induced by C_6 ceramide in HL-60 cells (Kim et al., 2001). This action apparently results from the ceramide-induced translocation of BAX from the cytosol to the mitochondria. The redistribution of cellular BAX in combination with an observed decrease in BCL-X_L expression has been suggested to function in ceramide-mediated apoptosis by altering the balance between anti- and pro-apoptotic cellular signaling (Kim et al., 2001).

Recent evidence has also established a role for ceramide in the redistribution of AIF from the mitochondrial intermembrane space to the nucleus (Daugas et al., 2000). As this translocation is coupled to the subsequent induction of peripheral chromatin condensation and large scale DNA fragmentation, it suggests that ceramide can function as a pro-apoptotic second messenger in the absence of cytochrome c release and caspase activation.

3. Regulatory role of ceramide in apoptosis

Apoptosis was first described as a cellular process whereby cells underwent unique morphological changes (e.g. chromatin fragmentation, nuclear condensation, membrane blebbing) in the process of dying (Kerr et al., 1972; Cohen et al., 1994; Zimmerman et al., 2001). Clearly, apoptosis was distinct from other cell death pathways (i.e. necrosis). It would soon be discovered that apoptosis was a critical process for cell differentiation and that aberrations in this process could result in tumorigenesis (Vaux et al., 1988; Wickremasinghe and Hoffbrand, 1999). At the organismal level, apoptotic processes are essential for normal embryonic development and play a key role in the removal of harmful cells by the immune system (Yang and Korsmeyer, 1996; Zimmerman et al., 2001). Yet, the biochemical mechanisms responsible for this process are still not fully understood.

Apoptosis can be initiated by a number of stress signals. While the end result is similar, there are two separate apoptotic pathways (i.e. extrinsic and intrinsic) that are currently known. The extrinsic pathway is triggered by activation of a death receptor like CD95 or the TNF receptor (Chinnaiyan et al., 1996; Muzio et al., 1996; Ashkenazi and Dixit, 1998; Baud and Karin, 2001; Kaufmann and Hentgartner, 2001). These receptors contain "death domains" that are essential to the assembly of the *d*eath-*i*nducing *s*ignal *c*omplex (DISC) when activated (Kaufmann and Hentgartner, 2001; Weber and Vincenz, 2001). In the extrinsic pathway, caspase 8 is activated (Chinnaiyan et al., 1996; Muzio et al., 1996; Ashkenazi and Dixit, 1998). Caspases are a family of cysteine proteases that play a critical role in apoptosis (Cohen, 1997; Thornberry and Lazebnik, 1998; Earnshaw et al., 1999). While the "initiator" caspases (e.g. caspases 8 and 9) can be differentially activated depending on initiation event, the "effector" caspases (e.g. caspases 3) play a role in both intrinsic and extrinsic pathways (Cohen, 1997; Thornberry and Lazebnik, 1998; Earnshaw et al., 1999). Activation of caspase 3 by caspase 8 results in cell death that is BCL2 resistant and thus suggests that this process is independent of the mitochondria (Strasser et al., 1995; Martin et al., 1998). Sphingomyelinase, an enzyme responsible in at least one pathway in the generation of ceramide, is activated by TNF family receptors (Adam et al., 1996; Adam-Klages et al., 1996). Thus ceramide appears to play an important role in TNF-induced cell death (Kolesnick and Kronke, 1998). Still, the role of ceramide in TNF or FAS-induced cell death is not clear. Cells derived from acid sphingomyelinase knockout mice (which are defective in ceramide metabolic pathways involving ceramide) are resistant to various stress stimuli including FAS and TNF-induction depending on cell type (Lin et al., 2000). Still, ceramide appears to be obligate for apoptotic induction by TNF since embryonic fibroblast cells derived from acid sphingomyelinase knockout mice show partial resistance to TNF that can be overcome by natural ceramide (Lozano et al., 2001).

The intrinsic apoptotic pathway is mitochondria-dependent and is triggered by stresses such as chemotherapeutic drug treatment or irradiation that result in mitochondrial damage and release of cytochrome C, which in turn, results in activation of caspase 9 (Green and Reed, 1998; Kaufmann and Hentgartner, 2001). The intrinsic apoptotic pathway is regulated in part by a family of proteins (i.e. BCL2 family) that regulate apoptosis (Korsmeyer, 1992; Yang and Korsmeyer, 1996; Adams and Cory, 1998; Reed, 1998). BCL2 was identified as the cellular oncogene product associated with the t(14,18) translocation commonly seen in B-cell lymphomas (Tsujimoto et al., 1984; Bakhasi et al., 1985; Cleary et al., 1986). While other oncogenes known at the time of BCL2's discovery were found to affect cell proliferation, BCL2's function to prolong cell survival was shown experimentally to be novel (Vaux et al., 1988; Korsmeyer, 1992). BCL2 was demonstrated to be a potent suppressor of the process of programmed cell death because expression of exogenous BCL2 could protect factor-dependent cells from apoptosis in response to growth factor withdrawal (Vaux et al., 1988). However, simple expression of BCL2 is not sufficient to protect cells from apoptosis (Ruvolo et al., 2001a). There are many examples where BCL2 expression alone is not prognostic for outcome in diseases such as heart disease (Olivetti et al., 1997) or acute lymphoblastic leukemia (Coustan-Smith et al., 1996). It has been found that phosphorylation regulates BCL2's anti-apoptotic function (May et al., 1994; Haldar et al., 1995). Phosphorylation can either activate or inactivate BCL2 depending on which sites are phosphorylated (Fig. 3; Ruvolo et al., 2001a). Mono-site phosphorylation at serine 70 is required for BCL2's full and potent anti-apoptotic function (Ito et al., 1997). On the other hand multi-site phosphorylation of BCL2 inactivates the molecule and is associated with cell death (Haldar et al., 1995; Yamamoto et al., 1999). Ceramide has been shown to regulate BCL2 phosphorylation status and function during stress (Ruvolo et al., 1999, 2002). Since expression of S70E BCL2, a non-phosphorylatable, activated form of BCL2, can protect cells from ceramide-induced apoptosis, it is likely that ceramide-induced cell death requires the functional inactivation of BCL2 in cells that express BCL2 (Ruvolo et al., 1999; Ruvolo, 2001). Thus ceramide is critical in regulating the apoptotic machinery in at least some tissues. Similar to BCL2, ceramide has been shown to regulate the phosphorylation status of other BCL2 family members such as BAD (Basu et al., 1998). The ability of ceramide to regulate multiple members of the BCL2 family would likely allow ceramide to be involved in apoptotic regulatory processes in many tissue types. Furthermore, though the pro-apoptotic BCL2 family member BAX is not a phosphoprotein, ceramide has been shown to affect the ability of BAX to kill cells (Pastorino et al., 1999). One mechanism how BAX might kill cells is via the mitochondrial permeability transition (MPT) resulting in the release of cytochrome *c* (Pastorino et al., 1998).

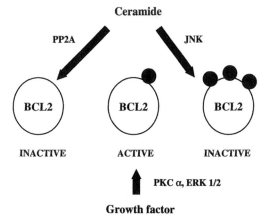

Fig. 3. Model of regulation of BCL2 functional activity by phosphorylation. Monosite phosphorylation of BCL2 as induced by growth factor or other agonists promotes full and potent anti-apoptotic function. The BCL2 survival kinases include PKC α and ERK 1/2. Conversely, ceramide inactivates BCL2 either by dephosphorylation via PP2A or by hyper-phosphorylation of the molecule by stress-kinases such as JNK.

Ceramide has been shown to potentiate the induction of the MPT by BAX (Pastorino et al., 1999). Antisense oligo inhibition of BAX has been shown to block ceramide-induced apoptosis (Kim et al., 2001). Ceramide treatment of cells resulted in a slight increase in BAX expression with translocation of BAX to the mitochondria (Kim et al., 2001). Considering the varied effects of ceramide on BCL2 family members by mechanisms affecting expression, post-translational modification, and sub-cellular localization, it is likely that ceramide acts via multiple mechanisms during apoptosis.

4. Potential roles for ceramide in malignancy and aging

4.1. Ceramide as a regulator of tumor suppressor genes

As mentioned above, PP2A is regulated by ceramide (Dobrowsky and Hannun, 1992; Dobrowsky et al., 1993; Ruvolo et al., 1999). Importantly, a number of studies suggest that PP2A is a potential tumor suppressor (Cohen and Cohen, 1989; Schonthal, 1998). Okadaic acid, a PP2A inhibitor, is a potent tumor promoter (Suganuma et al., 1988; Fujiki and Sugunama, 1993). In addition, viruses with the capability to transform normal cells such as SV40 contain proteins such as the small T antigen and middle T antigen that bind to and inactivate PP2A (Pallas et al., 1990; Yang et al., 1991; Mumby and Walter, 1993; Mateer et al., 1998; Mullane et al., 1998). The small T antigen can bind to the structural A subunit, thus preventing B subunit binding and assembly of the complete ABC PP2A complex (Yang et al., 1991). Interestingly, the middle T antigen and small T antigen differ in their effects on cell signaling though both interact with PP2A (Mullane et al., 1998). The middle T antigen but not the small T antigen activates JNK (Mullane et al., 1998). One potential mechanism how PP2A may act as a tumor suppressor could involve the inactivation of oncogenes such as BCL2. PP2A dephosphorylates and functionally inactivates BCL2 (Ruvolo et al., 1999, 2002).

Inactivation of a PP2A's tumor suppressor properties by mutation or deletion has been postulated to be involved in the development of colon cancer (Wang et al., 1998) and lung cancer (Rasio et al., 1995; Wang et al., 1998; Calin et al., 2000). As mentioned above, one study found 63% of lung cancer tumors examined contained the 11q LOH that contains the PPP2R1B gene (Rasio et al., 1995). More recently, 15% of lung carcinomas have demonstrated mutations in either the α and β isoforms of the A-subunit that give rise to inactive forms of PP2A (Calin et al., 2000). While a recent comparison of cultured versus uncultured (sporadic) human lung cancer cell lines did not analyze the presence of the 11q LOH, it was found that a 3p21 LOH was found in 58% of cultured cell lines and 63% of uncultured cell lines (Wistuba et al., 1999). The gene for the B56 regulatory γ-subunit (PPP2R5C) maps to 3p21 (McCright et al., 1996b). A possible mechanism whereby PP2A's tumor suppressor ability may serve to inhibit aberrant proliferation signaling cascades has emerged since antisense ablation of cPKCα by treatment also reverses tumorigenicity of A549 cells (Wang et al.,

1999). Because ceramide is such an important regulator and potent agonist of PP2A, it would be likely that ceramide would have a role in this process.

Another potential tumor suppressor that ceramide regulates is PKR. As discussed previously, ceramide has been shown to regulate PKR through its cellular activator, RAX (Ruvolo et al., 2001b). Several studies have suggested that suppression of PKR function can lead to malignancy since catalytically inactive PKR has been shown to transform NIH3T3 cells (Koromilas et al., 1992; Meurs et al., 1993; Barber et al., 1995a,b). Consistent with this notion, loss of PKR activity has been associated with tumor growth while WT PKR activity correlates with antitumor activity (Barber et al., 1995a,b). The PKR gene is localized to 2p21–22, a chromosomal region that has been associated with diseases such as large cell lymphoma, myelodysplastic leukemia, and possibly acute myeloid leukemia (Beretta et al., 1996; Basu et al., 1997; Abraham et al., 1998). In addition, PKR has recently been suggested as being dysregulated in breast cancer (Jagus et al., 1999; Savinova et al., 1999). These studies suggest that PKR and its regulatory proteins may play a role in the tumor suppression processes.

4.2. Ceramide as a regulator of oncogenes

BCL2 is the founding member of a family of proteins that regulate programmed cell death pathways (Korsmeyer, 1992; Yang and Korsmeyer, 1996; Reed, 1997, 1998; Adams and Cory, 1998). BCL2 was identified as the cellular oncogene product associated with the t(14,18) translocation commonly seen in B-cell lymphomas (Tsujimoto et al., 1984; Bakhasi et al., 1985; Cleary et al., 1986). While other oncogenes known at the time of BCL2's discovery were found to affect cell proliferation, BCL2's function to prolong cell survival was shown experimentally to be novel. BCL2 was demonstrated to be a potent suppressor of the process of programmed cell death because expression of exogenous BCL2 could protect (IL-3 dependent) cells from apoptosis in response to growth factor withdrawal (Vaux et al., 1988).

Recently, it was discovered that ceramide activates a mitochondrial PP2A responsible for BCL2 dephosphorylation in association with cell death (Ruvolo et al., 1999). A likely mechanism involved in ceramide-induced cell death appears to involve functional inactivation of BCL2, given that expression of S70E BCL2, a non-phosphorylatable (i.e. PP2A resistant) mutant that mimics phosphorylation, protects cells from ceramide-induced apoptosis at doses where WT BCL2 is ineffective (Ruvolo et al., 1999).

4.3. Potential role for ceramide in chemoresistance

As was discussed in the previous sections, ceramide regulates pathways that involve both tumor suppressors (e.g. PP2A and PKR) and tumor promoters (e.g. BCL2). Thus, ceramide can participate in the oncogenic process through direct regulation of the gene products that influence tumorigenesis. It is also clear that ceramide regulates the diverse stress signaling pathways that determine the cell's fate in response to stress stimuli and thus, defects in ceramide metabolism could potentially affect the cellular response to chemotherapy or other anticancer regimens (e.g. irradiation) by rendering the cells more resistant to cell killing. As has been suggested, dysregulation of ceramide metabolic pathways may contribute to multidrug resistance (Senchenkov et al., 2001). Variants of human cells that are TNF α-resistant (Cai et al., 1997) and adriamycin-resistant (Lavie et al., 1996; Michael et al., 1997; Liu et al., 1999, 2000, 2001) have been shown to exhibit defects in ceramide metabolism. One potential mechanism involves the conversion of ceramide to a nontoxic form. For instance, ceramide is converted to glucosylceramide by glucosylceramide synthase (GCS) in adriamycin-resistant cells (Gomez-Munoz et al., 1995). A role for GCS in chemoresistance in adriamycin-resistant cells is supported by studies that show that introduction of GCS into normal MCF-7 cells confers adriamycin-resistance (Liu et al., 1999). Conversely, inhibition of GCS in human adriamycin-resistant cells using an oligonucleotide antisense strategy restores sensitivity to adriamycin (Liu et al., 2000). Liu et al. (2001) have also demonstrated that multidrug resistance to a number of chemotherapeutic agents including anthracyclines and taxanes can be reversed in human breast cancer cells by antisense targeting of GCS.

Defects in ceramide metabolism can also affect the effectiveness of non-chemotherapy anticancer strategies (Senchenkov et al., 2001). An association between resistance to radiation therapy and defective ceramide metabolism has been shown in Burkitt's lymphoma cells (Michael et al., 1997). Thus, dysregulation of ceramide production appears to play a role in chemoresistance, at least in some malignancies.

Ceramide may participate in chemoresistance mechanisms via the regulation of molecules involved in apoptosis such as it has been recently found that ceramide regulation of BCL2 phosphorylation status may contribute to the cellular chemoresistance of human and murine leukemic cell lines (Ruvolo et al., 1999, 2001b). Thus, a recently identified physiologic BCL2 kinase, cPKCα, has been found in retrospective studies to be a prognostic factor for outcome in patients with AML (Kornblau et al., 2000). Consistent with a model where phosphorylated BCL2 promotes BCL2 association with BAX (resulting in loss of BAX's pro-apoptotic function), AML patients with molecular determinants favoring cell survival (a high ratio of [BCL2][cPKCα]/[BAX]) fared much worse than patients with molecular determinants favoring apoptosis (a low ratio of [BCL2] [cPKCα]/ [BAX]). Leukemic cells from patients with favorable/intermediate prognosis cytogenetics displaying a high [BCL2] [cPKCα]/[BAX] ratio had a median remission rate of 65% and median survival of 55 weeks while similar AML patients with leukemic cells exhibiting a low [BCL2] [cPKCα/[BAX] ratio had a median remission rate of 92% and median survival of 141.5 weeks (Kornblau et al., 2000). Interestingly, cPKCα has also been implicated as a prognostic factor in pediatric ALL (Volm et al., 1997). As mentioned earlier, the tumorigenicity of lung carcinoma A549 cells can be reversed by inhibition of cPKCα (Wang et al., 1999). An interesting pattern is emerging where dysregulation of cPKCα may contribute to tumor development and chemoresistance properties (Ruvolo et al., 2001a). Ceramide is likely to have a role in at least some of these processes.

4.4. Ceramide and aging

As interest in aging grows, new data emerges on the implications of aging on expression of gene products involved in cellular processes like the cell cycle, apoptosis, and senescence. It is becoming clear that many of the gene products that are regulated by ceramide show changes in expression as humans and animals age. For instance, the expression of PKR appears to be elevated in aging mice in a broad number of tissues (e.g. brain, heart, lung, liver, pancreas; Ladiges et al., 2000). Cutler and Mattson (2001) have recently postulated that accumulation of ceramide and other sphingolipids during aging contributes to the aging rate and the onset of age-related diseases. Treatment of cells with exogenous ceramide can promote PKR mRNA expression (Ruvolo, Clark, Mitchell, Flagg, Ferguson, and May; unpublished). The exact mechanism how ceramide might promote PKR transcription is as yet unknown. However, if the aging process somehow promotes increased ceramide levels (Cutler and Mattson, 2001), the upregulation of PKR in cells in aging tissues found by Ladiges et al. (2000) may be due to enhanced expression of PKR by ceramide. Since ceramide can also promote inhibition of protein synthesis via PKR (Ruvolo et al., 2001b), aging cells may lack the ability to weather stress as well as younger cells and thus would be more prone to die. Interestingly, hyperactivation of PKR has been associated with Fanconi anemia (Pang et al., 2001). Pang et al. (2001) have shown that activation of PKR results in hematopoietic cells that are excessively apoptotic and hypersensitive to stress in Fanconi anemia cells of the C complementation group, but this phenotype could be reversed using a dominant negative mutant of PKR.

There are other ramifications for potential elevated ceramide levels in cells. If natural ceramides accumulate in aging tissues, ceramide-mediated regulatory pathways may become hypersensitive to stress in these cells. This phenomenon might explain the apparent paradox of why aging heart cells show increased levels of BCL2, but are more prone to death (Olivetti et al., 1997). Despite the presence of increased BCL2 levels, accumulated ceramide would promote the inactive functional state by either the promotion of BCl2 phosphatase activity (i.e. dephosphorylated state; Ruvolo et al., 1999, 2002) or by activation of stress kinases that can lead to inactivation by hyper-phosphorylation of the molecule (Yamamoto et al., 1999).

While the effects of ceramide-mediated signaling pathways have yet to be fully understood in many diseases related to aging (e.g. Alzheimer's disease, arthritis), ceramide-sensitive molecules appear to be important in a number of these diseases. Ceramide appears to promote down-regulation of the tau protein, a molecule that is dysregulated in Alzheimer's disease (Xie and Johnson, 1997). PP2A has been shown to be important in tau regulation (Sontag et al., 1996, 1999). While it is PP2A inactivation resulting in tau hyperphosphorylation that is associated with

Alzheimer's disease (Sontag et al., 1999; Kins et al., 2001), it is possible that ceramide promotes activation of other PP2A isoforms at the expense of the PP2A isoforms that mediate tau phosphorylation. At least in human REH cells, ceramide induces potent migration of PP2A catalytic complex to the mitochondria (Ruvolo et al., 2002). Perhaps the PP2A catalytic complex is recruited away from microtubules (i.e. the site of tau and PP2A association; Sontag et al., 1996, 1999) during the development of Alzheimer's disease. While this notion is purely speculative at this point, it would be a worthy goal to determine if such a mechanism might play a role in diseases associated with PP2A.

References

Abraham, N., Jaramillo, M.L., Duncan, P.I., Methot, N., Icely, P.L., Stojdl, D.F., Barber, G.N., Bell, J.C., 1998. Exp. Cell. Res. 244, 394–404.

Adam, D., Wiegmann, K., Adam-Klages, S., Ruff, A., Kronke, M., 1996. J. Biol. Chem. 271, 1461714622.

Adam-Klages, S., Adam, D., Wiegmann, K., Struve, S., Kolanus, W., Schneider-Mergener, J., Kronke. 1996. Cell 86, 937–947.

Adams, J.M., Cory, S., 1998. Science 281, 1322–1326.

Alter, B.P., 1996. Am. J. Hematol. 53, 99–110.

Ariga, T., Jarvis, W.D., Yu, R.K., 1998. J. Lipid. Res. 39, 1–16.

Ashkenazi, A., Dixit, V.M., 1998. Science 281, 1305–1308.

Bakhasi, A., Jensen, J.P., Goldman, P., Wright, J.J., McBride, O.W., Epstein, A.L., Korsmeyer, S.J., 1985. Cell 41, 889–906.

Balducci, L., Beghe, C., 2001. Crit. Rev. Oncol. Hematol. 37, 137–145.

Barber, G.N., Wambach, M., Thompson, S., Jagus, R., Katze, M.G., 1995a. Mol. Cell. Biol. 15, 3138–3146.

Barber, G.N., Jagus, R., Meurs, E.F., Hovanessian, A.G., Katze, M.G., 1995b. J. Biol. Chem. 270, 17423–17428.

Basu, S., Panayiotidis, P., Hart, S.M., He, L.Z., Man, A., Hoffbr, A.V., Ganeshaguru, K., 1997. Cancer Res. 57, 943–947.

Basu, S., Bayoumy, S., Zhang, Y., Lozano, J., Kolesnick, R., 1998. J. Biol. Chem. 273, 30419–30426.

Basu S., Kolesnick, R., 1998. Oncogene 17, 3277–85.

Baud, V., Karin, M., 2001. Trends Cell. Biol. 11, 372–377.

Beretta, L., Gabbay, M., Berger, R., Hanash, S.M., Sonenberg, N., 1996. Oncogene. 12, 1593–1596.

Bielawska, A., Crane, H.M., Liotta, D., Obeid, L.M., Hannun, Y.A., 1993. J. Biol. Chem. 268, 26226–26232.

Birbes, H., Bawab, S.E., Hannun, Y.A., Obeid, L.M., 2001. FASEB J. 14, 2669–2679.

Bourbon, N.A., Yun, J., Kester, M., 2000. J. Biol. Chem. 275, 35617–35623.

Cai, Z., Bettaieb, A., Mahdani, N.E., Legres, L.G., Stancou, R., Masliah, J., Chouaib, S., 1997. J. Biol. Chem. 272, 6918–6926.

Calin, G.A., di Iasio, M.G., Caprini, E., Vorechovsky, I., Natali, P.G., Sozzi, G., Croce, C.M., Barbanti-Brodano, G., Russo, G., Negrini, M., 2000. Oncogene 19, 1191–1195.

Cegielska, A., Shaffer, S., Derua, R., Goris, J., Virshup, D.M., 1994. Mol. Cell. Biol. 14, 4616–4623.

Chalfant, C.E., Kishikawa, K., Mumby, M.C., Kamibayashi, C., Bielawska, A., Hannun, Y.A., 1999. J. Biol. Chem. 274, 20313–20317.

Chalfant, C.E., Ogretmen, B., Galadari, S., Kroesen, B.J., Pettus, B.J., Hannun, Y.A., 2001. J. Biol. Chem. 276, 44848–44855.

Chalfant, C.E., Rathman, K., Pinkerman, R.L., Wood, R.E., Obeid, L.M., Ogretmen, B., Hannun, Y.A., 2002. J. Biol. Chem. 277, 12587–12595.

Chiang, C.-W., Harris, G., Ellig, C., Masters, S.C., Subramanian, R., Shenolikar, S., Wadzinski, B.E., Yang, E., 2001. Blood 97, 1289–1297.

Chinnaiyan, A.M., O'Rourke, K., Yu, G.L., Lyons, R.H., Garg, M., Duan, D.R., Xing, L., Gentz, R., Ni, J., Dixit, V.M., 1996. Science 274, 990–992.

Chmura, S.J., Nodzenski, E., Weichselbaum, R.R., Quintans, J., 1996. Cancer Res. 56, 2711–2714.

Cleary, M.L., Smith, S.D., Sklar, J., 1986. Cell 47, 19–28.

Clemens, M.J., Elia, A., 1997. J. Interferon Cytokine Res. 17, 503–524.

Cohen, P., Cohen, P.T.W., 1989. J. Biol. Chem. 264, 21435–21438.

Cohen, G.M., Sun, X.M., Fearnhead, H., MacFarlane, M., Brown, D.G., Snowden, R.T., Dinsdale, D., 1994. J. Immunol. 153, 507–516.

Cohen, G.M., 1997. BioChem. J. 326, 1–16.

Cote, J., Dupuis, S., Jiang, Z., Wu, J.Y., 2001. Proc. Natl. Acad. Sci. USA 98, 938–943.

Coustan-Smith, E., Kitanaka, A., Pui, C.H., McNinch, L., Evans, W.E., Raimondi, S.C., Behm, F.G., Arico, M., Campana, D., 1996. Blood 87, 1140–1146.

Csortos, C., Zolnierowicz, S., Bako, E., Durbin, S.D., DePaoli-Roach, A.A., 1996. J. Biol. Chem. 271, 2578–2588.

Cutler, R.G., Mattson, M.P., 2001. Mech. Ageing Dev. 122, 895–908.

Cuvillier, O., Edsall, L., Spiegel, S., 2000. J. Biol. Chem. 275, 15691–15700.

Daugas, E., Susin, S.A., Zamzami, N., Ferri, K.F., Irinopoulou, T., Larochette, N., Prevost, M., Leber, B., Andrews, D., Penninger, J., Kroemer, G., 2000. FASEB J. 14, 729–739.

Dbaibo, G.S., Pushkareva, M.Y., Jayadev, S., Schwarz, J.K., Horowitz, J.M., Obeid, L.M., Hannun, Y.A., 1995. Proc. Natl. Acad. Sci. USA 92, 1347–1351.

Deng X., Ito, T., Carr, B., Mumby, M., May, W.S., 1998. J. Biol. Chem. 273, 34157–34163.

Dobrowskly, R.T., Hannun, Y.A., 1992. J. Biol. Chem. 267, 5048–5051.

Dobrowsky, R.T., Kamibayishi, C., Mumby, M.C., Hannun, Y.A., 1993. J. Biol. Chem. 268, 15523–15530.

Earnshaw, W.C., Martins, L.M., Kaufmann, S.H., 1999. Annu. Rev. BioChem. 68, 383–424.

Fagerlie, S., Lensch, M.W., Pang, Q., Bagby, G.C., 2001. Exp. Hematol. 29, 1371–1381.

France-Lanord, V., Brugg, B., Michel, P.P., Agid, Y., Ruberg, M., 1997. J. Neurochem. 69, 1612–1621.

Fujiki, H., Suganuma, M., 1993. Adv. Cancer Res. 61, 143–194.

Garcia-Ruiz, C., Colell, A., Mari, M., Morales, A., Fernandez-Checa, J.C., 1997. Direct effect of ceramide on the mitochondrial electron transport chain leads to generation of reactive oxygen species. Role of mitochondrial glutathione. J. Biol. Chem. 272, 11369–11377.

Ghafourifar, P., Klein, S.D., Schucht, O., Schenk, U., Pruschy, M., Rocha, S., Richter, C., 1999. J. Biol. Chem. 274, 6080–6084.

Gomez-Munoz, A., Duffy, P.A., Martin, A., O'Brien, L., Byun, H.S., Bittman, R., Brindley, D.N., 1995. Mol. Pharmacol. 47, 833–839.

Green, D.R., Reed, J.C., 1998. Science 281, 1309–1312.

Grompe, M., D., Andrea, A., 2001. Hum. Mol. Genet. 10, 2253–2259.

Gudz, T.I., Tserng, K.Y., Hoppel, C.L., 1997. J. Biol. Chem. 272, 24154–24158.

Gulbins, E., Coggeshall, M., Baier, G., Telford, D., Langlet, C., Baier-Bitterlich, G., Bonnefoy-Berard, N., Burn, P., Wittinghofer, A., Altman, A., 1994. Mol. Cell. Biol. 14, 4749–4758.

Haldar, S., Jena, N., Croce, C.M., 1995. Proc. Natl. Acad. Sci. USA 92, 4507–4511.

Hannun, Y.A., 1994. J. Biol. Chem. 269, 3125–3128.

Hannun, Y.A., 1996. Science, 274, 1855–1899.

Hannun Y.A., Luberto, C., 2000. Trends Cell Biol. 10, 73–80.

Hayakawa, M., Ishida, N., Takeuchi, K., Shibamoto, S., Hori, T., Oku, N., Ito, F., Tsujimoto, M., 1993. J. Biol. Chem. 268, 11290–11295.

Hershey, J.W.B., 1993. Sem. Virol. 4, 201–207.

Hovanessian, A.G., 1993. Sem. Virol. 4, 237–245.

Huwiler, A., Brunner, J., Hummel, R., Vervooddeldonk, M., Dtabels, S., van den Bosch, H., Pfeilshifter, J., 1996. Proc. Natl. Acad. Sci. USA. 93, 6959–6965.

Ito, T., Deng, X., Carr, B.K., May, W.S., 1997. J. Biol. Chem. 272, 11671–11673.

Ito, T., Yang, M., May, W.S., 1999. J. Biol. Chem. 274, 15427–15432.

Jagus, R., Joshi, B., Barber, G.N., 1999. Int. J. BioChem. Cell Biol. 31, 123–138.

Jarvis, W.D., Kolesnick R.N., Fornari F.A., Traylor, R.S., Gewirtz, D.A., Grant, S., 1994a. Proc. Natl. Acad. Sci. 91, 73–77.

Jarvis, W.D., Fornari, F.A., Browning, J.L., Gewirtz, D.A., Kolesnick, R.N., Grant, S., 1994b. J. Biol. Chem. 269, 31685–31692.

Jarvis, W.D., Fornari, F.A., Traylor, R.S., Martin, H.A., Kramer, L.B., Erukulla, R.K., Bittman, R., Grant, S., 1996a. J. Biol. Chem. 271, 8275–8284.

Jarvis, W.D., Grant, S., Kolesnick, R.N., 1996b. Clin. Cancer. Res. 2, 1–6.

Jarvis, W.D., Fornari, F.A., Auer, K.L., Freemerman, A.J., Szabo, E., Birrer, M.J., Johnson, C.R., Barbour, S.E., Dent, P., Grant, S., 1997. Mol. Pharmacol. 52, 935–947.

Jarvis, W.D., Grant, S., 1998. Curr. Opin. Oncol. 10, 552–9.

Jarvis, W.D., Johnson, C.R., Fornari, F.A., Park, J.S., Dent, P., Grant, S., 1999. J. Pharmacol. Exp. Ther. 290,1384–92.

Jiang, Z.H., Zhang, W.J., Rao, Y., Wu, J.Y., 1998. Proc. Natl. Acad. Sci. USA 95, 9155–9160.

Jin, K.L., Graham, S.H., Mao, X.O., He, X., Nagayama, T., Simon, R.P., Greenberg, D.A., 2001. J. NeuroChem. 77, 1508–1519.

Joseph, C.K., Byun, H.S., Bittman, R., Kolesnick, R.N., 1993. J. Biol. Chem. 268, 20002–20006.

Kaufmann, S., Hengartner, M.O., 2001. Trends Cell Biol. 11, 526–534.

Kerr, J.F., Wyllie, A.H., Currie, A.R., 1972. Br. J. Cancer 26, 239–257.

Kim, H.J., Mun, J.Y., Chun, Y.J., Choi, K.H., Kim, M.Y., 2001. FEBS Lett. 505, 264–268.

Kins, S., Crameri, A., Evans, D.R., Hemmings, B.A., Nitsch, R.M., Gotz, J., 2001. J. Biol. Chem. 276, 38193–38200.

Kishikawa, K., Chalfant, C.E., Perry, D.K., Bielawska, A., Hannun Y.A., 1999. J. Biol. Chem. 274, 21335–21341.

Kolesnick, R.N., Golde, D.W., 1994. Cell 77, 325–8.

Kolesnick, R.N., Kronke, M., 1998. Annu. Rev. Physiol. 60, 643–65.

Kornblau, S.M., Vu, H., Ruvolo, P., Estrov, Z., O'Brien, S., Cortes, J., Kantarjian, H., Reeff, M., May, W.S., 2000. Clin. Cancer. Res. 6, 1401–1409.

Koromilas, A.E., Roy, S., Barber, G.N., Katze, M.G., Sonenberg, N., 1992. Science 257, 1685–1689.

Korsmeyer, S.J., 1992. Blood 80, 879–886.

Ladiges, W., Morton, J., Blakely, C., Gale, M., 2000. Mech. Ageing Dev. 114, 123–132.

Lavie, Y., Cao, H., Bursten, S.L., Giuliano, A.E., Cabot, M.C., 1996. J. Biol. Chem. 271, 19530–19536.

Lee, J.Y., Hannun, Y.A., Obeid, L.M., 1996. J. Biol. Chem. 271, 13169–13174.

Lee, J.Y., Hannun, Y.A., Obeid, L.M., 2000. J. Biol. Chem. 275, 29290–29298.

Lin, T., Genestier, L., Pinkoski, M.J., Castro, A., Nicholas, S., Mogil, R., Paris, F., Fuks, Z., Schuchman, E.H., Kolesnick, R.N., Green, D.R., 2000. J. Biol. Chem. 275, 8657–8663.

Liu, J., Mathias, S., Yang, Z., Kolesnick, R.N., 1994. J. Biol. Chem. 269, 3047–3052.

Liu, Y.Y., Han, T.Y., Giuliano, A.E., Cabot, M.C., 1999. J. Biol. Chem. 274, 1140–1146.

Liu, Y.Y., Han, T.Y., Giuliano, A.E., Hansen, N., Cabot, M.C., 2000. J. Biol. Chem. 275, 7138–7143.

Liu, Y.Y., Han, T.Y., Giuliano, A.E., Cabot, M.C., 2001. FASEB J. 15, 719–730.

Lozano, J., Berra, E., Municio, M.M., Diaz-Meco, M.F., Dominquez, I., Sanz, L., Moscat, J., 1994. J. Biol. Chem. 269, 19200–19202.

Lozano, J., Menendez, S., Morales, A., Ehleiter, D., Liao, W.C., Wagman, R., Haimovitz-Friedman, A., Fuks, Z., Kolesnick, R., 2001. J. Biol. Chem. 276, 442–448.

Luberto, C., Hannun, Y.A., 1998. J. Biol. Chem. 273, 14550–14559.

Luberto, C., Yoo, D.S., Suidan, H.S., Bartoli, G.M., Hannun, Y.A., 2000. J. Biol. Chem. 275, 14760–14766.

Mathias, S., Pena, L.A., Kolesnick, R.N., 1998. Biochem. J. 335, 465–80.

Mateer, S.C., Fedorov, S.A., Mumby, M.C., 1998. J. Biol. Chem. 273, 35339–35346.

Matthews, M.B., 1993. Sem. Virol. 4, 247–257.

Martin, D.A., Siegel, R.M., Zheng, L., Lenardo, M.J., 1998. J. Biol. Chem. 273, 4345–4349.

Maurer, B.J., Melton, L., Billups, C., Cabot, M.C., Reynolds, C.P., 2000. J. Natl. Cancer Inst. 92, 1897–1909.

May, W.S., Tyler, P.G., Ito, T., Armstrong, D.K., Qatsha, K.A., Davidson, N.E., 1994. J. Biol. Chem. 269, 26865–26870.

May, W.S., 1997. Adv. Pharmacol. 41, 219–246.

Mayer, R.E., Hendrix, P., Cron, P., Matthies, R., Stone, S.R., Goris, J, Merlevede, W., Hofsteenge, J., Hemmings, B.A., 1991. Biochemistry 30, 3589–3597.

McCright, B., Virshup, D.M., 1995. J. Biol. Chem. 270, 26123–26128.

McCright, B., Rivers, A.M., Audlin, S., Virshup, D.M., 1996a. J. Biol. Chem. 271, 22081–22089.

McCright, B., Brothman, A.R., Virshup, D.M., 1996b. Genomics 36, 168–170.

McCubrey, JA., May, W.S., Duronio, V., Mufson, A., 2000. Leukemia 14, 9–21.

Merrick, W.C., 1992. MicroBiol Rev. 56, 241–315.

Merrill, A.H., Hannun, Y.a., Bell, R.M., 1993. Adv. Lipid Res. 23, 1–26.

Meurs, E.F., Galabru, J., Barber, G.N., Katze, M.G., Hovanessian, A.G., 1993. Proc. Natl. Acad. Sci. USA 90, 232–236.

Minn, A.J., Boise, L.H., Thompson, C.B., 1996. J. Biol. Chem. 271, 6306–6312.

Misteli, T., Spector, D.L., 1996. Mol. Biol. Cell. 7, 1559–1572.

Misteli, T., Caceres, J.F., Clement, J.Q., Krainer, A.R., Wilkinson, M.F., Spector, D.L., 1998. J. Cell. Biol. 143, 297–307.

Mullane, K.P., Ratnofsky, M., Cullere, X., Schaffhausen, B., 1998. Mol. Cell. Biol. 18, 7556–7564.

Muller, G., Ayoub, M., Storz, P., Rennecke, J., Fabbro, P., Pfizenmaier, K., 1995. EMBO J. 14, 1961–1969.

Mumby, M.C., Walter, G., 1993. Physiological Reviews 73, 673–699.

Muzio, M., Chinnaiyan, A.M., Kischkel, F.C., O'Rourke, K., Shevchenko, A., Ni, J., Scaffidi, C., Bretz, J.D., Zhang, M., Gentz, R., Mann, M., Krammer, P.H., Peter, M.E., Dixit, V.M., 1996. Cell 85, 817–827.

Nishizuka, Y., 1992. Science 258, 607–614.

Olivetti, G., Melissari, M., Capasso, J.M., Anversa, P., 1991. Circ. Res. 68, 1560–1568.

Olivetti, G., Abbi, R., Quaini, F., Kajstura, J., Cheng, W., Nitahara, J.A., Quaini, E., Di Loreto, C., Beltrami, C.A., Krajewski, S., Reed, J.C., Anversa, P., 1997. New. Engl. J. Med. 336, 1131–1141.

Pallas, D.C., Shahrik, L.K., Martin, B.L., Jaspers, S., Miller, T.B., Brautigan, D.L., Roberts, T.M., 1990. Cell 60, 167–176.

Pang, Q., Keeble, W., Diaz, J., Christianson, T.A., Fagerlie, S., Rathbun, K., Faulkner, G.R., O'Dwyer, M., Bagby, G.C., 2001. Blood 97, 1644–1652.

Pastorino, J.G., Chen, S.T., Tafani, M., Snyder, J.W., Farber, J.L., 1998. J. Biol. Chem. 273, 7770–7775.

Pastorino, J.G., Tafani, M., Rothman, R.J., Marcinkeviciute, A., Hoek, J.B., Farber, J.L., Marcineviciute, A., 1999. J. Biol. Chem. 274, 31734–31739.

Patel, R.C., Sen, G.C., 1998. EMBO J. 17, 4379–4390.

Phaneuf, S., Leeuwenburgh, C., 2002. Am. J. Physiol. Regul. Integr. Comp. Physiol. 282, R423–430.

Quillet-Mary, A., Jaffrezou, J.P., Mansat, V., Bordier, C., Naval, J., Laurent, G., 1997. J. Biol. Chem. 272, 21388–21395.

Raines, M., Kolesnick, R.N., Golde, D.W., 1993. J. Biol. Chem. 268, 14572–14575.

Rasio, D., Negrini, M., Manenti, G., Dragani, T.A., Croce, C.M., 1995. Cancer Res. 55, 3988–3991.

Reed, J.C., 1997. Nature 387, 773–776.

Reed, J.C., 1998. Oncogene 17, 3225–3236.

Rowlands, A.G., Panniers, R., Henshaw, E.C., 1988. J. Biol. Chem. 263, 5526–5533.

Ruediger, R., Van Wart Hood, J.E., Mumby, M., Walter, G., 1991. Mol. Cell. Bio. 12, 4282–4285.

Ruvolo, P.P., 2001. Leukemia 15, 1153–1160.

Ruvolo, P.P., Deng, X., Carr, B.K., May W.S., 1998. J. Biol. Chem. 273, 25436–25442.

Ruvolo, P.P., Deng, X., Ito, T., Carr, B.K., May, W.S., 1999. J. Biol. Chem. 274, 20296–20300.

Ruvolo, P.P., Deng, X., May, W.S., 2001a. Leukemia 15, 515–522.

Ruvolo, P.P., Gao, F., Blalock, W.L., Deng, X., May, W.S., 2001b. J. Biol. Chem. 276, 11754–11765.

Ruvolo, P.P., Clark, W., Mumby, M., Gao, F., May, W.S., 2002. J. Biol. Chem. 277, 22847–22852.

Salinas, M., Lopez-Valdaliso, R., Martin, D., Alvarez, A., Cuadrado, A., 2000. Mol. Cell. Neurosci. 15, 156–169.

Samuel, C.E., 1993. J. Biol. Chem. 268, 7603–7606.

Santana, P., Pena, L.A., Haimovitz-Friedman, A., Martin, S., Green, D., McLoughlin, M., Cordon-Cardo, C., Schuchman, E.H., Fuks, Z., Kolesnick, R., 1996. Cell 86, 189–199.

Savinova, O., Joshi, B., Jagus, R., 1999. Int. J. Biochem. Cell. Biol. 31, 175–189.

Sawada, M., Nakashima, S., Banno, Y., Yamakawa, H., Takenaka, K., Shinoda, J., Nishimura, Y., Sakai, N., Nozawa, Y., 2000. Oncogene 19, 3508–3520.

Sawai, H., Okazaki, T., Yamamoto, H., Okano, H., Takeda, Y., Teshima, M., Sawada, H., Okuma, H., Ishikura, H., Umehara, H., Domae, N., 1995. J. Biol. Chem. 270, 27326–27331.

Scheid, M.P, Duronio, V., 1998. Proc. Natl. Acad. Sci. USA 95, 7439–7444.

Schonthal, A.H., 1998. Frontiers in Bioscience 3, 1262–1273.

Schubert, K.M., Scheid, M.P., Duronio, V., 2000. J. Biol. Chem. 275, 13330–13335.

Sekeres, M.A., Stone, R.M., 2002. Curr. Opin. Oncol. 14, 24–30.

Senchenkov, A., Litvak, D.A., Cabot, M.C., 2001. J. Natl. Cancer Inst. 93, 347–357.

Seol, D.W., Billiar, T.R., 1999. J. Biol. Chem. 274, 2072–2076.

Sheikh, M.S., Fornace, A.J., 1999. Oncogene 18, 6121–6128.

Smyth, M.J., Obeid, L.M., Hannun, Y.A., 1997. Adv. Pharmacol. 41, 133–154.

Sontag, E., Nunbhakdi-Craig, V., Bloom, G.S., Mumby, M.C., 1995. J. Cell. Biol. 128, 1131–1144.

Sontag, E., Nunbhakdi-Craig, V., Lee, G., Bloom, G.S., Mumby, M.C., 1996. Neuron 17, 1201–1207.

Sontag, E., Nunbhakdi-Craig, V., Lee, G., Brandt, R., Kamibayashi, C., Kuret, J., White, C.L., Mumby, M.C., Bloom, G.S., 1999. J. Biol. Chem. 274, 25490–25498.

Spiegel, S., Milstien, S., 2000. FEBS Lett. 476, 55–57.

Strasser, A., Harris, A.W., Huang, D.C., Krammer, P.H., Cory, S., 1995. EMBO J. 14, 6136–6147.

Suganuma, M., Fujiki, H., Suguri, H., Yoshizawa, S., Hirota, M., Nakayasu, M., Ojika, M., Wakamatsu, K., Yamada, K., 1988. Proc. Natl. Acad. Sci. USA 85, 1768–1771.

Tan, S.L., Katze, M.G., 1999. J. Interferon Cytokine Res. 19, 543–554.

Teruel, T., Hern ez, R., Lorenzo, M., 2001. Diabetes 50, 2563–2571.

Thornberry, N.A., Lazebnik, Y., 1998. Science 281, 1312–1316.

Tsujimoto, Y., Finger, L., Nowell, P.C., Croce, C.M., 1984. Science 226, 1097–1099.

Vaux, D.L., Cory, S., Adams, J.M., 1988. Nature 335, 440–442.

Verheij, M., Bose, R., Lin, X.H., Yao, B., Jarvis, W.D., Grant, S., Birrer, M.J., Szabo, E., Zon, L.I., Kyriakis, J.M., Haimovitz-Friedman, A., Fuks, Z., Kolesnick, R.N., 1996. Nature 380, 75–79.

Volm, M., Zintl, F., Edler, L., Sauerbrey, A., 1997. Medical Pediatric Oncol. 28, 117–126.

Wang, S.S., Esplin, E.D., Li, J.L., Huang, L., Gazdar, A., Minna, J., Evans, G.A., 1998. Science 282, 284–287.

Wang, X.Y., Repasky, E., Liu, H.T., 1999. Exp. Cell. Res. 250, 253–263.

Westwick, J.K., Bielawska, A.E., Dbaibo, G.S., Hannun, Y.A., Brenner, D.A., 1995. J. Biol. Chem. 270, 22689–22692.

Wickremasinghe, R.G., Hoffbrand, A.V., 1999. Blood 93, 3587–3600.

Williams, B.R.G., 1999. Oncogene 18, 6112–6120.

Wistuba, I.I., Bryant, D., Behrens, C., Milchgrub, S., Virmani, A.K., Ashfaq, R., Minna, J.D., Gazdar, A.F., 1999. Clin. Cancer Res. 5, 991–1000.

Weber, C.H., Vincenz, C., 2001. Trends Biochem. Sci. 26, 475–481.

Xia, Z., Dickens, M., Raingeaud, J., Davis, R.J., Greenberg, M.E., 1995. Science 270, 1326–1331.

Xie, H., Johnson, G.V., 1997. J. Neurochem. 69, 1020–1030.

Xing, H.R., Lozano, J., Kolesnick, R., 2000. J. Biol. Chem. 275, 17276–17280.

Yamamoto, K., Ihijo, H., Korsmeyer, S., 1999. Mol. Cell. Biol. 19, 8469–8478.

Yang, E., Korsmeyer, S.J., 1996. Blood 88, 386–401.

Yang, S.I., Lickteig, R.L., Estes, R., Rundell, K., Walter, G., Mumby, M.C., 1991. Mol. Cell. Biol. 11, 1988–1995.

Zahler, A.M., Lane, W.S., Stolk, J.A., Roth, M.B., 1992. Genes Dev. 6, 837–847.
Zahler, A.M., Neugebauer, K.M., Lane, W.S., Roth, M.B., 1993. Science 260, 219–222.
Zhang, Y., Yao, B., Delikat, S., Bayoumy, S., Lin, X.-H., Basu, S., McGinley, M., Chan-Hui, P.-Y., Lichenstein, H., Kolesnick, R.N., 1997. Kinase suppressor of ras is ceramide-activated protein kinase. Cell 89, 63–72.
Zhou, M., Demo, S.D., McClure, T.N., Crea, R., Bitler, C.M., 1998. J. Biol. Chem. 273, 11930–11936.
Zimmermann, K.C., Bonzon, C., Green, D.R., 2001. Pharmacol. Ther. 92, 57–70.
Zundel, W., Swiersz, L.M., Giaccia, A., 2000. Mol. Cell. Biol. 20, 1507–1514.

**Advances in
Cell Aging and
Gerontology**

Sphingolipid metabolism and signaling in atherosclerosis

Subroto Chatterjee and Sergio F. Martin

*Department of Pediatrics, Johns Hopkins University, Baltimore
and Johns Hopkins Singapore
National Heart Centre, Vascular Biology Program, Singapore
E-mail address: schatte2@jhmi.edu*

Contents

Advances in Cell Aging and Gerontology, vol. 12, 71–96

---

1. Introduction

1.1. Overview

Previously atherosclerosis was believed to be an aging phenomenon. Current thinking, however, posits that atherosclerosis is a progressive disorder resulting from a combination of risk factors such as elevated levels of lipid and lipoproteins, cigarette smoking, high level of homocysteine, high sensitivity C-reactive protein (hs-CRP), diabetes, and stress; inflammation also plays a major role in the pathogenesis of this disease (Libby et al., 2002). These factors individually or collectively contribute to the initiation and progression of atherosclerosis (Libby et al., 2002). Although there has been considerable progress in establishing the role of cholesterol and low-density lipoprotein receptors in atherosclerosis, comparatively little is known about the role of sphingolipids. In the past decade, observations emerging from various disciplines point to the fact that sphingolipids are integral components of "lipid rafts" and serve as dynamic bioactive molecules that modulate phenotypic changes such as cell proliferation, cell adhesion, cell migration, and programmed cell death (apoptosis) (Liu et al., 1997; Pena et al., 1997; Chatterjee, 1999, 2000).

In this chapter, we will focus on the enzymes that are involved in the metabolism of the sphingolipids. We will also concentrate on some representative bioactive sphingolipids that have been implicated in mediating signaling events that contribute to phenotypic changes in cells. In particular, we will focus on ceramide (Cer), sphingosine-1-phosphate (S-1-P), lactosylceramide (LacCer), and a disialoganglioside (GD_3). Recently, several review articles have appeared that concentrate on the general aspect of sphingolipid metabolism to which the readers are referred for additional details (Merrill et al., 1997; Usta et al., 2001). Here our focus will be to look at sphingolipids in view of their involvement in atherosclerosis and vascular biology.

1.2. Historical perspectives

Sphingolipids were first discovered by Thudichum in 1886 (Thudichum, 1886). Thudichum characterized a compound, which he called cerebroside, as it was

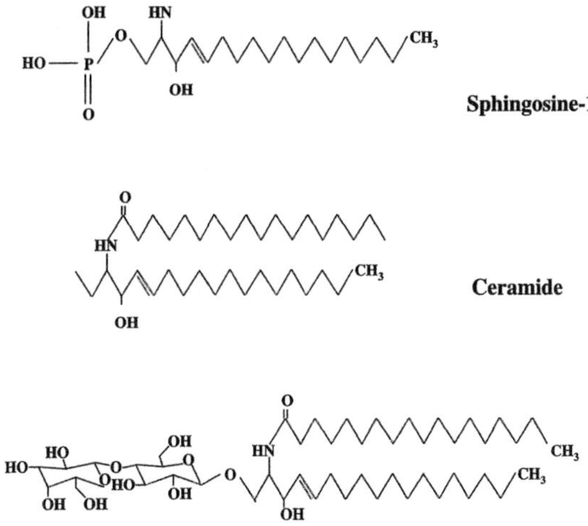

Sphingosine-1-P

Ceramide

Lactosylceramide

Fig. 1. Structure of sphingosine 1-phosphate, ceramide, and lactosylceramide.

isolated from human brain tissue (Thudichum, 1884). A characteristic component of cerebroside and all sphingolipids is an aliphatic amino alcohol called sphingosine (Fig. 1). Thudichum called the compound sphingosine because of its aliphatic chemical nature, i.e. the presence of alcohol, as well as amino-groups, and being soluble in water. More than 100 years of studies on sphingolipids have focused on their structure, biosynthesis, and degradation. In fact, these compounds were nothing more than a curiosity for chemists until it was shown that several of these glycosylated sphingolipids called glycosphingolipids (GSL) were found to accumulate in various tissues leading to metabolic diseases collectively called glycosphingolipidoses. The major sphingolipid in human tissue is sphingomyelin (SM) and it is an independent risk factor for atherosclerosis (Jiang et al., 2000). The other major sphingolipid is Cer, which is a catabolic product of SM and various GSLs. Several of the GSLs are localized on the cell surface, and it has been shown that they serve as cell markers and antigens; e.g. blood group ABO, P, Lewis system, Forsmann antigen, and tumor associated antigen. A major role played by the sphingolipids, and in particular, SM, is that they provide structural rigidity to the cell membrane. In addition, the presence of carbohydrate moieties allow the glycosphingolipids to interact with other carbohydrates projecting outward off of the cell membrane biomolecular leaflet, that in turn leads to signal transduction phenomenon and phenotypic changes. The structure of some sphingolipids involved in atherosclerosis is summarized in Fig. 1.

1.3. Sphingolipids: structure and localization

Sphingosine and dihydro-sphingosine are typical mammalian sphingolipids. The amino group in sphingosine is usually acylated by a fatty acid or a 2-hydroxy fatty acid to form Cer. The most common fatty acids in SM and LacCer are

C16–C18. On the other hand, in glycosphingolipids derived from human atherosclerotic plaques, neutrophils, and brain long chain fatty acids such as C20–C24 are present. Sub-cellular fractionation and employment of freeze fracture techniques have shown that most of the sphingolipids, in particular, SM, is localized on the cell surface. However, other studies have suggested that some SM may be localized on the inner leaflet of the cell membrane. Upon treatment with various cytokines, however, the Cer that is generated due to the hydrolysis of SM via sphingomyelinase action may be localized either on the plasma membrane or in the mitochondria/nucleus. The availability of antibodies against Cer now provides an opportunity to localize this molecule (Vielhaber et al., 2001). On the other hand, although some lacCer is localized on the cell surface, it is predominantly localized in the cytoplasmic vesicles within uroepithelial cells and human neutrophils (Chatterjee et al., 1983; Symington et al., 1987). GD$_3$ is a major ganglioside in human aortic smooth muscle cells, certain carcinomas as well as in breast cancer and is predominantly localized on the cell surface.

GSL cluster to form microdomains on the cell membrane even in the absence of cholesterol. Such microdomains consisting of cholesterol, SM, and GSLs constitute the "lipid rafts" in the plasma leaflet of the cell membrane bilayer. Employing detergent extraction and ultracentrifugation, the "lipid rafts" have been distinguished into "caveolin rich" and "GSL signaling domain" (GSD) fractions. Furthermore, the GSD contained in addition to GSLs c-Src, Fak and Rho which are important players in signal transduction phenomenon (Iwabuchi et al., 1998).

2. Biosynthesis of sphingolipids

2.1. Overview

The biosynthesis of most GSL occurs in the Golgi apparatus, as most if not all biosynthetic enzymes are localized in this organelle. On the other hand, there is controversy as to which organelle, plasma membrane, or Golgi is the major site of SM synthesis. *In vitro*, the activity of SM synthase in plasma membrane is higher than in Golgi and is stimulated by the addition of Cer; the addition of glucosylceramide transferase inhibitor increases SM synthesis. Thus, there is a competition for Cer pool for SM biosynthesis and GSL biosynthesis.

2.2. Enzymes involved in the biosynthesis of sphingolipids

2.2.1. Serine palmitoyl transferase (E.C. 2.3.1.50)

The biosynthesis of sphingolipid begins via the condensation of serine with palmitoyl co-enzyme A in the endoplasmic reticulum, via serine palmitoyltransferase (SPT). This process continues until dihydroceramide is produced through a series of reactions involving 3-ketosphinganine reductase, a NADPH dependent reductase. This enzymatic reaction can be blocked by the use of an inhibitor myriocin (F. 2; T. 1).

2.2.2. *Sphinganine N-acyltransferase (E.C. 2.3.1.24)*

Sphinganine is acylated to dihydroceramide by the enzyme "ceramide synthase" (sphinganine *N*-acyltransferase). A specific inhibitor of ceramide synthase is fumonisin B1. This compound has a very similar structure to sphingosine or sphinganine (which is a substrate for ceramide synthase) and thus blocks ceramide synthase, resulting in a decrease in intracellular Cer level (Bose et al., 1995; Merrill et al., 1996; Garzotto et al., 1998). A fungal metabolite fumonisin B1 is a target for dihydroceramide synthase and recent studies have shown that the chemotherapeutic agent Daunorobicin transiently increases the activity of dihydroceramide synthase. Further studies indicate that dihydroceramide desaturase, a NADPH dependent enzyme, is also important in the biosynthesis of the Cer, although not much is known about this enzyme (English et al., 2000).

2.2.3. *Sphingosine kinase*

In mammalian cells, the phosphorylated form of sphingosine is S-1-P as well as dihydro sphingosine-1-phosphate. A large amount of S-1-P is present in human blood platelets as well as in plasma (Lee et al., 1998). Interluekin 1 (IL-1) and platelet-derived growth factor (PDGF) treatment results in the activation of sphingomyelinase and ceramidase that eventually contribute to the formation of sphingosine and its subsequent phosphorylation by a sphingosine kinase results in the formation of S-1-P (Fatatis and Miller, 1996). A potential role for sphingosine in angiogenesis initiated by hypoxia and/or by vascular endothelial growth factor (VEGF) has generated considerable interest in our understanding of sphingosine 1 kinase (Olivera et al., 1999; Fatatis and Miller, 1996). This enzyme has been recently purified and cloned from human and murine sources and suggests the presence of at least two kinds of mammalian sphingosine kinases; these are called SK1 and SK2, respectively (Kohama et al., 1998; Liu et al., 2000; Pitson et al., 2000). The SK2 protein is larger than SK1 and the preferred substrate is dihydrosphingosine. In contrast, SK1 prefers sphingosine as a substrate (Kohama et al., 1998). Interestingly, these proteins have homology to diacyl glycerol kinase with regard to the ATP binding site of the cystolic domain of the particular enzyme. The SK1 gene is predominantly expressed in the spleen and in the lung, whereas the SK2 gene is expressed in the liver, kidney, and the brain (Murate et al., 2001). A host of biologically active compounds have been shown to modulate the activity of sphingosine-1-kinase. These include: phorbol esters, vitamin D_3, TNF-α, oxidized LDL, PDGF, and nerve growth factor in a variety of cultured cells derived from normal individuals as well as from cancerous cells. In addition, bradykinin and muscarinic receptors were also recently shown to activate sphingosine-1-kinase, leading to the mobilization of calcium (Mazurek et al., 1994; Buehrer et al., 1996; Kleuser et al., 1998; Xia et al., 1998; Auge et al., 1999; Xia et al., 1999; Olivera et al., 31; Blaukat and Dikic, 2001; Edsall et al., 2001). A discussion of the role of S-1-P in angiogenesis and vascular biology will be presented later.

2.2.4. Sphingomyelin synthase

Sphingomyelin synthase (S-Nmase) activity has been shown to be responsible for the synthesis of SM from Cer and phosphocholine derived from phosphatidylcholine. As such, it not only regulates the level of SM production but also generates diacylglycerol. S-Nmase activity is also regulated by TNF-α. More importantly, it has been implicated in cell proliferation, particularly in liver cell regeneration (Miro-Obradors et al., 1993). Since this enzyme has not been purified and cloned, its molecular biology is largely unknown.

2.2.5. Glucosylceramide synthase (E.C. 2.4.1.80)

Glucosylceramide synthase (GlcT-1) catalyzes the transfer of a glucose residue from a nucleotide sugar UDP-glucose to Cer to form glucosylceramide. The enzyme is on the cytosolic side of the *cis*/medial Golgi membrane (Futerman and Pagano, 1991; Jeckel et al., 1992). Mutational and biochemical studies have revealed that histidine-193 of the glucosylceramide synthase is the primary target of the histidine modifying agent diethyl pyrocarbonate, the putative UDP-Glc-binding site in this enzyme (Wu et al., 1999); the enzyme has been cloned. Overexpression of glucosylceramide synthase has been shown to attenuate the adverse effect of Cer upon the use of chemotherapeutic agents such as adriamycin (Liu et al., 1999). Thus, GlcT-1 has been ascribed to protect cells from Cer-induced cell death due to adriamycin treatment. However, this tenet has been challenged recently (Tepper et al., 2000). The activity of GlcT-1 can be inhibited by a glucosylceramide analog 2-decanoylamino-3 morpholino-1-propanol (D-PDMP) (Radin et al., 1993). Kinetic analysis shows that D-PDMP acts by mixed competition against Cer with IC50 and Ki values of 5 and 0.7 M, respectively. It should be noted that PDMP does not inhibit UDP-galactose: N-acylsphingosine galactosyltransferase and beta-glucocerebrosidase. D-PDMP is not a very specific inhibitor of GlcT-1 as it does inhibit the activity of several other GSL glycosyltransferases (Chatterjee et al., 1996a). More recently, new homologs of PDMP have been described that dissociate the depletion of GSLs from secondary metabolic effects such as the accumulation of Cer (Shayman et al., 2000).

N-butyldeoxynojirimycin is known as an inhibitor against both alpha-glucosidase I and GlcCer synthase. Among various N-alkylimino sugars tested, N-butyldeoxygalactonojirimycin possesses no visible cytotoxicity and higher specificity for the inhibition of GlcCer synthesis. Much progress will be expected on determining the mechanism of GlcCer synthesis inhibition by these N-alkylimino sugars and on its medical application for treatment of Gaucher's disease (see below) and in cancer.

2.2.6. Lactosylceramide synthase (E.C. 2.4.1.-)

Lactosylceramide synthase (LacCer synthase) is a *trans*-Golgi localized enzyme that is responsible for the transfer of galactose from UDP-galactose to glucosylceramide to form LacCer. LacCer synthase belongs to the β-14 GalT transferase family. There are at least two LacCer synthases; these are the β-4.

GalT-V, which is constitutively expressed in almost all human embryonic adult tissues (Lo et al., 1998). The other LacCer synthase, GalT-VI, was first purified in human kidney cells (Chatterjee et al., 1992) and subsequently purified and cloned from a rat brain (Nomura et al., 1998). The molecular weight of this enzyme is about 60 kDa and prefers glucosylceramide as a specific substrate. It requires magnesium and detergent such as Triton X-1000 and or cutscum for optimum activity. The expression of β-4 GalT-VI is tissue restrictive, predominantly in human embryonic and adult brain cells, although very little expression is observed in additional adult tissues (Lo et al., 1998). In human endothelial cells, the β-4 GalT-V gene is expressed at high levels, while the β-4 GalT-VI gene is minimally expressed in these cells. On the other hand, in human kidney tissue and human embryonic kidney cells (HEK-293), equal expression of β-4 GalT-V and β-4 GalT-VI genes was found. In human endothelial cells, TNF-α as well as exposure to fluid shear stress was shown to stimulate the activity of LacCer synthase that eventually led to the expression of the intercellular cell adhesion molecule ICAM-1 via a complex signaling cascade involving reactive oxygen species. Since TNF-α and shear stress-induced ICAM-1 expression in HUVECs was abrogated by D-PDMP, this suggests that LacCer synthesis is a target critically involved in this phenotypic change (Bhunia et al., 1998; Yeh et al., 2001). LacCer synthase is also implicated in several proliferative diseases including atherosclerosis, renal cancer, and human and mouse polycystic kidney disease (Chatterjee 1993; Chatterjee et al., 1996b; Chatterjee et al, 1997a). There appears to be a convergence in the activation of LacCer synthase to mediate ox-LDL, PDGF, EGF, and nicotine-induced aortic smooth muscle cell proliferation (Chatterjee, 1998). In human vascular smooth muscle cells, the activation of LacCer synthase by the agonists above produces LacCer. In turn, LacCer recruits an "oxygen sensitive signaling cascade" (see below) that ultimately contributes to the proliferation of aortic smooth muscle cells (Bhunia et al., 1996; Balagopalakrishna et al., 1997; Chatterjee 1997; Chatterjee et al., 1997b). This phenotypic change was completely abrogated by the preincubation of cells with D-PDMP. In contrast, L-PDMP that stimulated the activity of LacCer synthase stimulated the proliferation of human aortic smooth muscle cells (Chatterjee et al., 1997b). Interestingly, the activity of LacCer synthase is higher in human plaques and calcified plaques as well as in human/mouse polycystic kidney tissue (Chatterjee et al., 1996b; Chatterjee et al., 1997a). Collectively, these *in vitro* and *in vivo* studies suggest that LacCer synthase is the target for several cardiovascular and inflammatory risk factors that may contribute to vascular dysfunction.

2.3. Use of inhibitors to study sphingolipid biosynthesis and degradation

Several low molecular weight inhibitors of sphingolipid biosynthesis are known and have become valuable tools for the treatment of sphingolipidoses and other diseases (Kolter and Sandhoff, 1998). Table 1 summarizes a list of enzymes and corresponding inhibitors for use in sphingolipid degradation and

Table 1

List of enzymes and inhibitors corresponding for use in the sphingolipid degradation or/and biosynthesis studies

Enzyme	Inhibitor	Reference
Serine palmitoyltransferase	L-Cycloserine	Chatterjee (1998), Pyne and Pyne (2000)
	Myriocin	Pyne and Pyne (2000)
Dihydroceramide synthase	Fumonisin	Chatterjee (1998), Pyne and Pyne (2000)
Ceramide synthase	Fumonisin B1	Auge et al. (1999)
Sphingomyelinase	Scyphostatin	Arenz et al. (2001)
	Manumycin A	Arenz et al. (2001)
	Ubiquinone, Ubiquinol	Martin et al. (2001), (2002)
Ceramidase	D-MAPP	Auge et al. (1999), Pyne and Pyne (2000)
	N-Oleoylethanolamine	Pyne and Pyne (2000)
Sphingosine kinase	DMS	Auge et al. (1999)
	threo-Dihydrosphingosine	Pyne and Pyne (2000)
Sphingosine-1-phosphate receptor agonists	FTY720	Mandala et al. (2002)
Glucosylceramide synthase	D-PPMP	Pyne and Pyne (2000)
	D-PDMP	Pyne and Pyne (2000), Yeh et al. (2001)
Lactosylceramide synthase	D-PDMP	Chatterjee (2000), (1997), Cho (2001)
	L-PDMP	Chatterjee (1998)

D-MAPP. – D-erythro-2-(N-myristoylamino)-1-phenyl-1-propanol
DMS. – N,N-dimethylsphingosine
D-PDMP. – D-threo-1-phenyl-2-decanoylamino-3-morpholino-1-propanol
L-PDMP. – L-threo-1-phenyl-2-decanoylamino-3-morpholino-1-propanol
D-PPMP. – D-threo-1-phenyl-2-palmitoylamino-3-morpholino-1-propanol

biosynthesis studies. The use of these compounds has been described above and detailed in the references cited in Table 1 (Chatterjee et al., 1997b, 1998, 2000; Auge et al., 1999; Pyne and Pyne, 2000; Martin et al., 2001, 2002; Yeh et al., 2001; Mandala et al., 2002).

3. Alteration of biosynthesis of sphingolipids in atherosclerosis

3.1. Regulation of sphingolipid biosynthesis by lipoproteins

Previous studies have shown that most if not all the plasma sphingo- lipids in particular glycosphingolipids, are associated with various lipoproteins. When plasma was fractionated into very low density protein (LDL) and high density protein (HDL), most of the glycosphingolipid and SM were associated with LDL and HDL (Chatterjee and Kwiterovich, 1976). Very little Cer was found associated with the lipoproteins or lipoprotein-deficient plasma/serum. The distribution of sphingosine and S-1-P in human lipoprotein is not known and is of potential interest as the activation of platelets with the thrombotic phenomenon results in the release of S-1-P into the plasma. How glycosphingolipids are transferred, following synthesis in the liver, to the lipoproteins is not known.

Very few studies have reported on the effects of lipoproteins on sphingolipid metabolism. Some studies have made use of the availability of human skin fibroblasts that express LDL receptors and fibroblasts from familial hypercholesterolemic homozygous patients that do not express LDL receptors. This kind of a model system provides a unique opportunity to address the functional role of lipoproteins and LDL receptors in regulating sphingolipid metabolism via the LDL receptor pathway and in familial hypercholesterolemia. These studies reveal that the addition of human LDL but not HDL or VLDL to normal human fibroblasts did not alter the activity of a serine-paltimoyltransferase (Chatterjee et al., 1986). Although the effects of this lipoprotein on other sphingolipid biosynthetic pathways were not investigated, incorporation studies employing radioactive precursors revealed that LDL did not have a significant effect on the biosyntheses of the lipids in this pathway (Fig. 2). However, LDL inhibited the activity of LacCer synthase in a concentration-dependent manner and with kinetics similar to the inhibition of 3-hydroxy-3-methyl glutaryl-coenzyme A reductase (HMG-CoA), an enzyme important in the feedback regulation of cholesterol biosynthesis. These studies were extended in a number of diploid cells such as human aortic smooth muscle cells, human kidney proximal tubular cells, as well as Chinese hamster ovarian cells (Chatterjee et al., 1986a). Since the binding, internalization, and degradation of LDL were critical in the LDL-mediated downregulation of LacCer synthase activity, it implies an important role for LDL receptors in this process. These findings point to the potential role of apo protein-B in LDL that may have a regulatory effect on LacCer synthase activity. This idea was further supported by the finding that various components in LDL such as serine, fatty acid, phospholipids, and cholesterol failed to alter the activity of LacCer synthase in cultured human skin fibroblasts (Chatterjee, 1990). Moreover, LDL from familial hypercholesterolemic homozygous patients, which contained high levels of glycosphingolipids, including LacCer, had a similar effect on the activity of LacCer synthase as compared to LDL derived from normolipidemic subjects (Chatterjee, 1992). Recently, elevated levels of GlcCer and LacCer have been reported in an apoE KO mouse model of atherosclerosis (Garner et al., 2002) that further validates the previous observations in human atherosclerosis (Chatterjee et al., 1983; Chatterjee, 1999).

3.2. Upregulation of sphingolipid biosynthesis in atherosclerosis

In contrast to studies on normal human derived cells that had a functional LDL receptor, the absence of LDL receptors such as in fibroblasts from patients with homozygous familial hypercholesterolemia resulted in the lack of regulation of LacCer synthase. In fact, in these cells LDL increased the activity of LacCer synthase and contributed to the accumulation of LacCer within cystoplasmic vesicles (Chatterjee et al., 1986). Moreover, modification of the lysine residue in apoprotein B by reductive methylation and/or oxidation of LDL also stimulated the activity of LacCer synthase in human aortic smooth muscle cells and increased

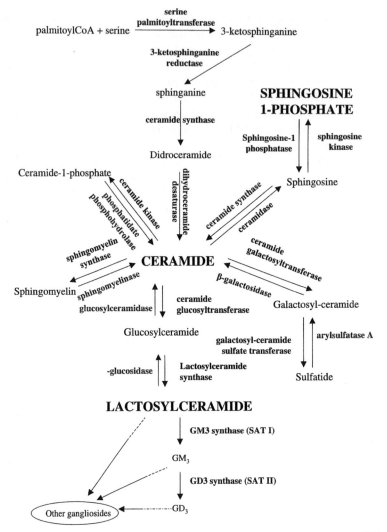

Fig. 2. Pathways for the biosynthesis and degradation of sphingolipids. For convenience all biosynthetic enzymes are indicated in blue and degradation enzymes in red.

LacCer levels. This in turn led to an increase in cell proliferation (Chatterjee et al., 1988; Chatterjee and Ghosh, 1996).

4. Degradation of sphingolipids

4.1. Overview

GSL degradation occurs in the acidic compartments of cells, the endosomes and the lysosomes. Parts of the plasma membrane destined for degradation are endocytosed and transported through the endosomal compartments to reach the lysosome (Sandhoff and Kolter, 1997; Kolter and Sandhoff, 1998). Within the lysosome, hydrolyzing enzymes sequentially cleave off the sugar residues from the nonreducing end; thus, glycosylated GSLs, Cer, and finally sphingosine is produced. Alternatively, the en bloc removal of the oligosaccharide chain in GSL is catalyzed by Cer glycanase to yield Cer and the oligosaccharide chain (Basu et al., 1999). Thus, GSL can leave the lysosome, re-enter the biosynthetic pathway or be further degraded. For GSLs with long carbohydrate chain of more than four sugar residues, the presence of an enzymatically active lysosomal hydrolase is sufficient for degradation *in vivo*. However, degradation of membrane-bound GSLs with short oligosaccharide chains requires the cooperation of an exohydrolase and a protein cofactor, called sphingolipid activator protein (SAP). Several sphingolipid activator proteins are now known including the GM2 activator and the saposins SAP-A, -B, -C, and -D. Inherited deficiencies of either lysosomal hydrolases or activator proteins give rise to GSL storage diseases, the sphingolipidoses (Kolter and Sandhoff, 1998). This is a group of inherited human diseases in which sphingolipids accumulate in one or more organs due to a degradation disorder. The symptoms and course of these diseases vary widely between forms with onset in early childhood and death within the first years of life in some cases and a more chronic course in the other cases. Their pathogenesis is poorly understood and a causal therapy is only available for the non-neuronopathic form of Gaucher's disease. The genes of most proteins involved in sphingolipid degradation have recently been cloned, enabling genotype–phenotype correlation and facilitating the diagnosis of sphingolipidoses. In addition, animal models of many sphingolipidoses have been created by targeted disruption of the respective genes in mice. In the future, these animals will serve as valuable tools for the investigation of the pathogenesis of these diseases and for the study of therapeutic approaches, including substrate deprivation, enzyme replacement, organ transplantation, and gene therapy.

4.2. Enzymes involved in the sphingolipid degradation pathway

4.2.1. Sphingosine-1-phosphatase (E.C. 3.1.3.-)
S-1-P is degraded back to sphingosine by the dephosphorylation mechanism (Mao and Obeid, 2000). Alternatively, it may be broken down by the action of sphingosine-1-phosphate lyase to fatty aldehyde hexadecanal and ethanolamine (Saba et al., 1997).

4.2.2. Ceramidases (E.C. 3.5.1.23)
Ceramidases specifically cleave Cer into sphingosine and fatty acid. At least three kinds of ceramidases have been described; the acid, neutral, and alkaline based upon the pH at which the activity is optimal.

4.2.2.1. Acid ceramidase The deficiency of acid ceramidase, which is predominantly a lysosomal enzyme, results in Farber's disease. The Farber's disease patient has the characteristic of accumulation of Cer in the plasma and has lipogranulomatosis. Although Cer has been implicated in inducing programmed cell death (apoptosis), unpublished data from our laboratory and others have shown that skin fibroblasts from Farber's disease patients undergo apoptosis to the same extent as normal skin fibroblasts. The explanation for this observation is not known but we can speculate that in these cells Cer is stored in a compartment that is not available and does not constitute the signaling pool of Cer that may participate in the apoptotic pathway. The human and mouse acid ceramidase have been cloned and also well characterized. The functional role of the acid ceramidase needs to be investigated further (Mao and Obeid, 2000).

4.2.2.2. Neutral ceramidase   The neutral ceramidase was first purified from bacteria and the gene has been cloned (Saba et al., 1997). Subsequently, the human homolog of the neutral ceramidase has been also cloned. Since the pH optima of the human enzyme is rather broad, it has been termed "non-lysosomal" ceramidase and has been shown to be present in the mitochondria. By virtue of releasing cytochrome *c* that in turn activates caspases involved in the signaling cascade leading to apoptosis the mitochondria plays a critical role in this process. Hence it may be speculated that the mitochondrial ceramidase may play an important role in salvaging the pool of Cer that may be present in mitochondria and/or Cer delivered to mitochondria and may determine the course of apoptosis (Birbes et al., 2001). Because the mitochondrial enzyme shows a very high order of substrate specificity; i.e., D-erythro ceramide as opposed to the L-erythro ceramide, this may suggest a divergence in the catabolic pathway contributing to functional role of these molecules in inducing apoptosis in the mitochondria. In a recent study it was found that the mitochondrial ceramidase may also function as a Cer synthase and therefore again points to the potential role of mitochondria, mitochondrial ceramidase/synthase in regulating the pool of Cer and apoptosis (Saba et al., 1997).

4.2.2.3. Alkaline ceramidase   The dihydroceramidase was found to be present in the endoplasmic reticulum. The two gene products of alkaline ceramidase called YDC1 and YPC1 have been recently cloned from *Sve. cerevisiae* and have been implicated in heat stress response (Mao et al., 2000; Mao et al., 2001). The mammalian alkaline ceramidase activity has been shown to be stimulated by PDGF, presumably producing sphingosine and subsequently S-1-P resulting in mitogenesis.

4.2.3. Sphingomyelinases (SMase; E.C. 3.1.4.12)
At least five major classes of SMases have been identified based upon their pH-optimum, metal ion requirement, and sub-cellular localization. The acid sphingomyelinase was first discovered in the lysosomes and it has been purified and cloned. A closely similar cytosolic acid sphingomyelinase that is dependent on Zn^{2+} and is a product of alternative splicing of the acid sphingomyelinase gene has been implicated in the aggregation of LDL and may have implications in

atherogenesis/atherosclerosis (Taba, 1999). A cell-membrane bound neutral SMase enzyme stimulatable by magnesium has been identified. Some studies suggest that this enzyme is localized on the external leaflet of the plasma membrane, whereas other studies suggest that it is localized on the inner aspect of the plasma membrane (Chatterjee, 1993b; Liu et al., 1997). Another neutral sphingomyelinase, a Mg^{2+} independent sphingomyelinase and an alkaline sphingomyelinase predominantly found in the gastrointestinal mucosa and human bile have been identified (Okazaki et al., 1994; Nilsson and Duan, 1999). The latter enzyme requires taurocholate and a mixture of conjugated bile acids for optimal activity.

4.2.3.1. Acid sphingomyelinase (A-SMase)   The deficiency of A-SMase results in the accumulation of large amounts of SM in the spleen of patients in Niemann-Pick disease, a human autosomal-recessive lysosomal disease (Schuchman and Desnick, 1995). The acid sphingomyelinase gene has been knocked out in mice; fibroblasts derived from the A-SMase KO mice are relatively more resistant to Fas-Apo I-induced apoptosis, whereas A-Smase deficiency has little effect on lymphocyte apoptosis. Moreover, upon exposure to radiation the endothelial cells in the intestine in A-SMase KO mice are remarkably resistant to apoptosis as compared to normal mice, which are not (Saba et al., 1997). Neurons in the brains of A-Smase KO mice exhibit increased resistance to ischemic death in a stroke model (Yu, Z.F., 2000). The availability of the A-SMase KO mice will certainly provide us with more information on whether the A-SMase plays an important role in the generation of Cer that in turn contributes to apoptosis and/ or other cellular phenomenon.

4.2.3.2. Neutral sphingomyelinase   The N-SMase's have been purified and cloned from a variety of sources. The first N-SMase, a bacterial N-SMase that was cloned, is present in the endoplasmic reticulum in mammalian cells and was later shown to be homologous to lyso-PAF phospholipase C (Sawai et al., 1999). Another N-SMase has also been cloned from the human kidney library and is presumably expressed on the surface of mammalian cells (90, 91). The over-expression of this latter N-SMase in human aortic smooth muscle cells results in spontaneous apoptosis irrespective of the presence of tumor necrosis factor-α or oxidized LDL (Chatterjee et al., 1999). However, when exogenous oxidized LDL or TNF-α is added to the N-SMase gene transfected cells, further stimulation of apoptosis was observed as compared to mock gene transfected cells. This is in contrast to the earlier bacterial sphingomyelinase that did not induce apoptosis upon overexpression in human embryonic kidney (HEK-293) cells irrespective of the presence or absence of TNF-α.

4.2.3.3. Physiological role of A-SMase and N-Smase   Employing mutation of the 55 kDa TNF-α receptors (TNF-α-R1), the role of A-SMase and N-SMase in apoptosis has been explored (Fig. 3). For example, the C-terminal cytoplasmic portion of the TNF-α-R1 contains two distinct features that differently associate with N-SMase or A-SMase signaling. For example, an 11 amino acid membrane

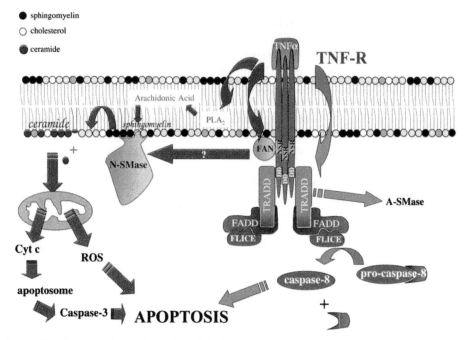

Fig. 3. Signal transduction pathway by which Tumor Necrosis Factor-α recruits ceramide to induce apoptosis.

proximal domain called the N-SMase domain (NSD) that is specifically associated with N-SMase signaling has been identified. An adaptor protein, factor activating N-SMase (FAN), that binds to the TNF-α-R1 proximal membrane motif and in turn promotes N-SMase activity has also been identified. However, overexpression of FAN did not induce apoptosis in these cells. Therefore, the function of FAN remains to be elucidated. On the other hand, a membrane distal region of cytosolic domain of the TNF-α-R1 associates with the acid A-SMase. This death domain represents the 75 amino-acid motif, and deletion and mutation of the death domain abolished TNF-α-R1-induced apoptosis. Subsequently, several death domains indicated in Fig. 3 may be recruited that in turn convert pro-caspase-8 to active caspase-8 leading to apoptosis. The activity of N-SMase has been shown to be stimulated by a variety of compounds including high concentration of oxidized LDL in human aortic smooth muscle cells resulting in apoptosis (Kronke, 1999). The potential role of arachidonic acid in the regulation of N-SMase has also been suggested. In these studies treatment of HL-60 cells and L99 cells with Interferon-gamma (IFN-γ) and TNF-α resulted in the activation of a cytosolic phospholipase A-2 (cPLA2). Phospholipase A2 specifically catalyzes the release of arachidonic acid from the sn-2 position in phosphatidylcholine. Arachidonic acid serves as a precursor to prostaglandins and eicosanoids, a potent activator of the inflammation. Treatment of these cells with arachidonic acid or pharmacological activators of cPLA2 induced N-SMase activity. Interestingly, cells lacking cPLA2 activity failed to activate N-SMase and were resistant to apoptosis induced by TNF-α. On the other hand, restoration of cPLA2 by transfection enabled TNF-α to activate N-SMase and to induce cell death (Jayadev et al., 1994).

A nuclear N-SMase has been reported in liver cells. The activation of this enzyme generated Cer, but it did not induce apoptosis. Rather N-SMase activation in liver cells due to TNF-α treatment results in cell proliferation (Alessenko and Chatterjee, 1995) as well as LDL receptor upregulation via stimulating the maturation of sterol regulatory element binding protein (Lawler et al., 1998) in a cholesterol-independent fashion.

Recent studies have also suggested a potential role for intracellular oxidization in the regulation of N-SMase. Thus, those reactions that lead to N-SMase activation result in the decrease of the cellular level of glutathione, a major anti-oxidant in a variety of cells (Obeid et al., 1993; Liu and Hannun, 1997; Liu et al., 1998). On the other hand, an increase in the endogenous level of glutathione results in a decrease in the N-SMase activity. These studies lead to the suggestion that the N-SMase may serve as a sensor in oxidative stress. However, this study needs to be investigated in copious detail to establish such a relationship.

4.2.4. Glucosylceramide hydrolase and lactosylceramide hydrolase

The cleavage of the β1-4 galactose residue in LacCer is carried out by a β-galatosidase. A rather non-specific β-galatcosidase has been isolated, which not only hydrolyzes LacCer but also several other GSLs having a β-4 galactose residue at the terminal end. On the other hand, glucosylceramide hydrolase, a β-glucosidase has been isolated, purified and cloned and the absence of this β-glucosylceramide hydrolase results in the accumulation of large amounts of glycosylceramide in the spleen of patients in Gaucher's disease, which is an autosomal-recessive neurological disorder that may result in early death. Specific inhibitors for the A-SMase, N-SMase, β-glucosylceramide hydrolase as well as the LacCer hydrolase are not commercially available.

4.3. Alteration in sphingolipid degradation in atherosclerosis

Up to the date research indicates that the activity of LacCer hydrolase in fibroblasts originating in urinary epithelial cells from patients with familial hypercholesterolemia is within the normal range and is not changed upon incubation of cells with LDL (Chatterjee et al., 1986). Whether other SM/GSL degradative enzymes are altered in atherosclerosis is not known.

5. Functional role of selected sphingolipids in atherosclerosis

In this section, we will present the current understanding of the role of Cer, S-1-P, and LacCer in atherosclerosis. While most of the information was generated employing cultured human vascular cells, these findings may well have implications *in vivo* in the vascular wall.

5.1. Ceramide

One of the seminal findings in the sphingolipid field in the past decade has been the demonstration that Cer can serve as an anti-proliferative and pro-apoptotic agent in a variety of mammalian and nonmammalian cultured cells including vascular cells (Obeid et al., 1993). Another important and associated finding was the establishment of structure–function relationships with regard to Cer and apoptosis. The double bond in the D-erythro-Cer molecule was found to be critical in apoptosis, whereas the D-erythro-dihydro ceramide that did not contain the double bond failed to induce apoptosis in cultured mammalian cells. Other functional roles assigned to Cer include cell senescence and differentiation. Exactly how Cer induces apoptosis is not known. However, data from a variety of studies have suggested that Cer generated through the action of neutral sphingomyelinase may well participate in reactions that lead to the activation of caspases, in particular caspase 3 directly, which may in turn execute the cell death process (Fig. 3). Alternatively, Cer generated/delivered in the mitochondria may result in the release of cytochrome c. Such cytochrome c activates caspase-9 and that in turn may lead to apoptosis. In human aortic smooth muscle cells the addition of exogenous membrane permeable forms of ceramide induces apoptosis (Chatterjee et al., 1999).

There may be a dichotomy with regard to apoptosis induced by exogenously supplied Cer from the endogenously generated Cer produced via the action of ox-LDL and TNF-α on N-SMase/A-SMase. In vascular cells of human origin, but not in other cells (Andrieu-Abadie et al., 2001), Cer does not generate ROS in sufficient amounts to induce apoptosis. Rather, Cer may be converted to LacCer and GD$_3$ that produce O_2^- and NO, and peroxynitrite, respectively; peroxynitrite can induce apoptosis (Bhunia et al., 2002). Additional studies have suggested that overexpression of GD$_3$ synthases and the use of GD$_3$ and/or the exogenous supply of LacCer can also induce apoptosis (Maria et al., 1997; Fukumoto et al., 2000). The direct role of mitochondria in LacCer and GD$_3$ induced apoptosis in liver cells as well as mitochondria isolated from liver has been documented (Garcia-Ruiz et al., 2000; Colell et al., 2001). The recruitment of A-SMase initially following Sindbis virus infection to generate Cer and the subsequent activation of N-SMase that provides a sustained increase in Cer results in apoptosis (Jan et al., 2000).

There are several signaling cascades involving Cer and the relevance of the signaling pathways in terms of modern biology and/or vascular biology needs to be examined further. For example, Cer has several targets in the cell. These include Cer activated protein kinases (CAPK), Cer-activated protein phosphatases (CAPP), PKC-, and cathepsin D. These are serine-threonine protein phosphatases: PP1 and PP2a (Chalfant et al., 2000). Potential substrates for CAPP include PKC-α, C2, PKB/Akt and bcl-2, c-jun as well as the retinoblastoma gene product. In contrast, CAPK was shown to be a kinase activator of Ras that may mediate the effect of Cer on the mitogen activated protein kinase. Cer also directly activates PKC- and that in turn has an effect on the SAPK/JNK signaling in stress-induced apoptosis (Verheij et al., 1996; Bourbon et al., 1998). The activation of cathepsin D by Cer but not sphingosine has also been shown (Fig. 3).

5.2. Sphingosine-1-phosphate

Another exciting and seminal finding in the field of sphingolipid biology has been the discovery that S-1-P serves as a proliferative and anti-apoptotic agent (Lee et al., 1998). Even more interesting and exciting finding was that S-1-P induced angiogenesis (Lee et al., 1998). Angiogenesis is a biological phenomenon that involves the formation of new blood vessels from preexisting ones. Angiogenesis is a normal physiological process but it is also implicated in wound healing, embryonic development, and in the pathology in diabetic retinopathy, rheumatoid arthritis, and tumor growth as well as in cardiovascular disease. However, little is known regarding the role of S-1-P in cardiovascular angiogenesis. S-1-P is most probably released from platelets into the plasma due to a thrombotic reaction. How S-1-P associates with lipoproteins or exists freely in plasma is not known. Some of the major players in angiogenesis include VEGF, platelet activation and hypoxia (Fig. 4). All of these activators of angiogenesis may activate ceramidase, which in turn converts Cer to S-1-P and in the presence of sphingosine kinase, sphingosine is converted into S-1-P. S-1-P interactions with endothelial differentiation genes (EDG), a family of G protein

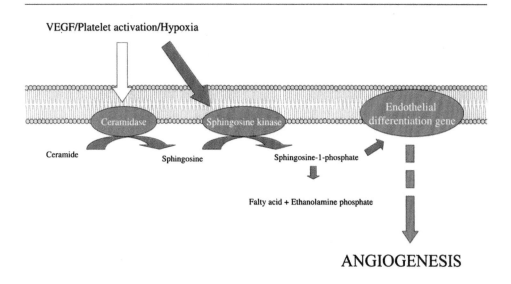

Fig. 4. Signal transduction pathway by which sphingosine-1-phosphate induces angiogenesis.

receptors, have been suggested to activate the expression of the VEGF and induce angiogenesis. S-1-P has also been implicated in endothelial cell proliferation and migration as well as cytoskeletal changes. One of the most important aspects of the activation of these receptors include the participation of pertusis toxin-sensitive Gi protein. Recently, analogs of S-1-P receptors have become available and may be useful in determining the role of S-1-P in the trans-epithelial migration of lymphocytes (Mandala et al., 2002).

5.3. Lactosylceramide

LacCer is another sphingoglycolipid that has been implicated in modulating phenotypic changes in vascular cells such as the human aortic smooth muscle cells, human endothelial cells, and human neutrophil/monocytes.

5.3.1. TNF-α recruits LacCer to induce ICAM-1 expression and cell adhesion in endothelial cells

In human endothelial cells, TNF-α can activate LacCer synthase to generate LacCer. In turn, LacCer activates NADPH oxidase implicated in the generation of reactive oxygen species such as superoxide ($O_2^{\bullet-}$). In human umbilical vein endothelial cells, $O_2^{\bullet-}$ serves as a critical biological sensor that activates a nuclear transcription factor (NF-kB) that ultimately leads to the overexpression of ICAM-1. Interestingly, although TNF-α stimulated the expression of V-CAM-1 and E-selectin in addition to ICAM-1, LacCer specifically stimulated the expression of ICAM-1 and the adhesion of neutrophils/monocytes to endothelial cells (Bhunia et al., 1998). Most importantly, this phenotypic change was abrogated by D-PDMP an inhibitor of LacCer synthase (Fig. 5).

5.3.2. Shear stress recruits LacCer to induce ICAM-1 expression and cell adhesion

In addition to the well-known traditional risk factors such as ox-LDL and TNF-α, hemodynamic factors such as shear stress may also contribute to the initiation and progression of atherosclerosis. One of the underlying mechanisms in this pathogenesis may involve the induction of cell adhesion molecule expression such as ICAM-1 that also is oxidant-sensitive and recruits NF-kB (Yeh et al., 2001). In human umbilical vein endothelial cells, fluid shear stress ($20\,\text{dynes/cm}^2$) activated LacCer synthase to generate LacCer. In turn, LacCer activated the "oxygen-sensitive" signaling cascade above to stimulate the expression of ICAM-1 and the adhesion of monocytes/neutrophils. Interestingly, the shear stress-induced phenotypic change was completely abrogated by D-PDMP and partially by the antioxidant N-acetylcysteine. In contrast, L-PDMP, an activator of LacCer synthase, stimulated LacCer production, O_2^- generation, and ICAM-1 expression. Thus, hemodynamic factors such as fluid shear stress can, independent of other risk factors in atherosclerosis, contribute to "endothelial dysfunction" via involving LacCer and thus contribute to the initiation and progression of atherogenesis.

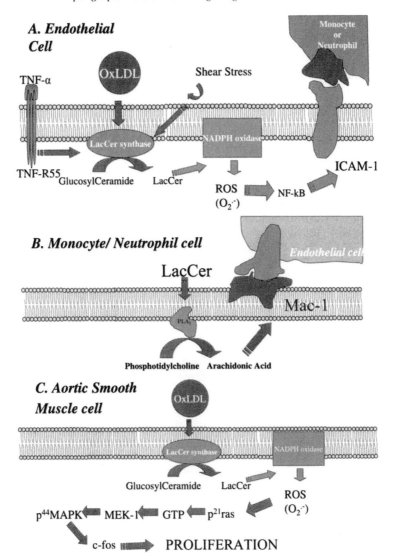

Fig. 5. Signal transduction pathway by which Lactoceramide recruits reactive oxygen species to regulate cell adhesion molecule expression in human endothelial cells (A), in monocyte/neutrophil cells (B), and cell proliferation in human aortic smooth muscle cells (C).

5.3.3. LacCer stimulates the expression of CD11b/CD18 (Mac-1) in human neutrophils/monocytes and adhesion

Neutrophils/monocytes circulating in the blood deploy cell adhesion molecules such as CD11b/CD18 or Mac-1 that serve as ligands that bind to ICAM-1 expressed on the surface of endothelial cells. By way of increasing the expression of Mac-1, LacCer stimulates neutrophil/monocyte adhesion to endothelial cells

(Arai et al., 1998). However, this signaling cascade does not recruit reactive oxygen species. Rather, in human neutrophils LacCer activates phospholipase A2 to generate arachidonic acid from phosphatidylcholine to stimulate Mac-1 expression and the adhesion of neutrophils/monocytes to endothelial cells (Fig. 5).

5.3.4. LacCer stimulates aortic smooth muscle cell proliferation

Although the mechanisms by which LacCer may contribute to cell proliferation *in vivo* is not known, our current understanding, which is based on cell culture studies, is as follows. LDL taken up by endothelial cells may undergo oxidation in the sub-endothelial space. Such minimally modified LDL and/or oxidized LDL can stimulate the activity of LacCer synthase to generate LacCer in aortic smooth muscle cells (Chatterjee, 1992; Balagopalakrishna et al., 1997; Chatterjee, 1997; Chatterjee et al., 1997b; Pyne and Pyne, 2000). Next, LacCer serves as a lipid second messenger that activates an "oxygen sensitive signal transduction pathway." The redox-regulated signaling by LacCer in the proliferation of human aortic smooth muscle involves $p21^{ras}$, GTP loading and activation of kinase cascade including MEK-1 and p44 mitogen-activated protein kinase (Chatterjee, 1991; Bhunia et al., 1996; Chatterjee et al., 1997b). Upon phosphorylation the cytosolic $p^{44}MAPK$ is transported to the nucleus wherein it activates c-fos as well as cyclin expression that eventually leads to the proliferation of aortic smooth muscle cells (Fig. 5). Several interesting features relevant to this phenotypic change are as follows. First, the signaling cascade mentioned above activated in cells incubated with oxidized LDL can be abrogated by the pre-incubation of cells with D-PDMP and was bypassed by pre-incubation of cells with LacCer but not other sphingolipids except GD_3. Second, this signaling cascade was mediated specifically by lactosylceramide and not by its breakdown products such as glucosylceramide, Cer, and/or sphingosine and fatty acids (Chatterjee, 1990). Third, the *in vivo* relevance of this study is evident by the observation that the level of LacCer as well as the activity of LacCer synthase are markedly increased in human atherosclerotic plaques as compared to normal vascular walls (Chatterjee et al., 1997a) and in other hyperproliferative disorders such as human and mouse polycystic kidney disease (Chatterjee et al., 1996b) and human kidney tumor cells (Chatterjee, 1993a). Additional growth factors such as PDGF, EGF, and FGF were all shown to target LacCer synthase to generate LacCer and induce cell proliferation in H-ASMC as well as in other human cells (Chatterjee, 1998).

Cigarette smoking is an independent risk factor in atherosclerosis as well as lung cancer. One of the major components in cigarette smoke that contributes to its deleterious effect is nicotine. Nicotine, like oxidized LDL, can stimulate the activity of LacCer synthase in human aortic smooth muscle cells and contribute to their proliferation. In addition, nicotine stimulates the expression of monocyte chemotatic protein (MCP-1) and this was abrogated by pre-incubation of cells with D-PDMP (Williams and Chatterjee, 2002). Since MCP is implicated in the diapedesis of monocytes, the findings above may have relevance in atherogenesis *in vivo*.

6. Conclusion

Oxidative stress, an excess production of reactive oxygen species, plays a critical role in different pathological conditions such as atherosclerosis, cancer, and arthritis. By virtue of being rapidly diffusible through cell membranes, the reactive oxygen species can impact upon many critical phenotypic changes, including induction of growth, regulation of kinase activity, and the activation of endothelial-derived relaxation factor NO. Moreover, the interplay of reactive oxygen species such as $O_2^{\bullet-}$ and NO produced in response to the treatment of cells with GD3 can generate peroxynitrite that in turn may lead to programmed cell death (Bhunia et al., 2002). Thus, reactive oxygen species can function as intracellular and intercellular second messengers transducing receptor stimulation activation/phosphorylation into a biochemical response. The evidence put forth above using various sphingolipids suggests that S-1-P and LacCer recruit reactive oxygen species as signaling messengers to activate transcription factors and gene expression that eventually contribute to cell proliferation, cell adhesion, and programmed cell death. Therefore, the interplay of the lipid second messengers that up or down-regulate reactive oxygen species production allows a cell to determine which phenotypic change to pursue. Finally, the integration of the responses of the sphingolipid messengers that elicit independent phenotypic changes in cells of the vascular wall may help us understand the role of these molecules in vascular biology and in the pathogenesis in atherosclerosis.

Acknowledgment

We appreciate the skillful assistance of Ms. Ann Koh in the preparation of this review article.

References

Alessenko, A., Chatterjee, S., 1995. Neutral sphingomyelinase: localization in rat liver nuclei and involvement in regeneration/proliferation. Mol. Cell Biochem. 143, 169–174.

Andrieu-Abadie, N., Couaze, V., Salvayre, R., Levade, T., 2001. Ceramide in apoptosis signaling: relationship with oxidative stress. Free Radical Biology & Medicine 31, 717–728.

Arai, Y., Bhunia, A.K., Chatterjee, S., Bulkley, G., 1998. Lactosylceramide stimulates human neutrophils to upregulate Mac-1, adhere to endothelium, and generate reactive oxygen metabolites *in vitro*. Circ. Res. 82, 540–547.

Arenz, C., Gartner, M., Wascholowski, V., Giannis, A., 2001. Synthesis and biochemical investigation of scyphostatin analogues as inhibitors of neutral sphingomyelinase. Bioorganic & Medicinal Chemistry 9, 2901–2904.

Auge, N., Nikolova-Karakashian, M., Carpentier, S., Parthasarathy, S., Negre-Salvayre, A., Salvayre, R., Merrill, A.H. Jr., Levade, T., 1999. Role of sphingosine 1-phosphate in the mitogenesis induced by oxidized low density lipoprotein in smooth muscle cells via activation of sphingomyelinase, ceramidase, and sphingosine kinase. J. Biol. Chem. 274, 21533–21538.

Balagopalakrishna, C., Bhunia, A.K., Snowden, A., Rifkind, J.M., Chatterjee, S., 1997. Minimally modified low density lipoproteins induce aortic smooth muscle cell proliferation via the activation of mitogen activated protein kinase. Mol. Cell. Biochem. 170, 85–89.

Basu, M., Kelly, P., Girzadas, M., Li, Z., Basu, S., 1999. Properties of animal ceramide glycanases. Methods Enzymol. 311, 287–297.

Bhunia, A.K., Schwarzmann, G., Chatterjee, S., 2002. GD3 recruits reactive oxygen species to induce cell proliferation and apoptosis in human aortic smooth muscle cells. J. Biol. Chem. 277, 16396–16402.

Bhunia, A.K., Arai, T., Bulkley, G., Chatterjee, S., 1998. Lactosylceramide stimulates human neutrophils to upregulate Mac-1, adhere to endothelium, and generate reactive oxygen metabolites *in vitro*. J. Biol. Chem. 273, 34349–34357.

Bhunia, A., Han, H., Snowden, A., Chatterjee, S., 1996. Lactosylceramide stimulates Ras-GTP loading, kinases (MEK, Raf), p44 mitogen-activated protein kinase, and c-fos expression in human aortic smooth muscle cells. J. Biol. Chem. 271, 10660–10666.

Birbes, H., El Bawab, S., Hannun, Obeid, L.M., 2001. Selective hydrolysis of a mitochondrial pool of sphingomyelin induces apoptosis. FASEB J. 14, 2669–2679.

Blaukat, A., Dikic, I., 2001. Activation of sphingosine kinase by the bradykinin B2 receptor and its implication in regulation of the ERK/MAP kinase pathway. Biol. Chem. 382, 135–139.

Bose, R., Verheij, M., Haimovitz-Friedman, A., Scotto, K., Fuks, Z., Kolesnick, R., 1995. Ceramide synthase mediates daunorubicin-induced apoptosis: an alternative mechanism for generating death signals. Cell 82, 405–414.

Bourbon, N.A., Yun, J., Kester, M., 1998. Ceramide directly activates protein kinase C zeta to regulate a stress-activated protein kinase signaling complex. J. Biol. Chem. 275, 35617–35623.

Buehrer, B.M., Bardes, E.S., Bell, R.M., 1996. Protein kinase C-dependent regulation of human erythroleukemia (HEL) cell sphingosine kinase activity. Biochim. Biophys. Acta 1303, 233–242.

Chalfant, C.E., Kishikawa, K., Bielawaska, A., Hannun, Y.A., 2000. Analysis of ceramide-activated protein phosphatases. Methods Enzymol. 312, 420–428.

Chatterjee, S. 2002. Sphingomyelinase. In: Encyclopedia Molecular Medicine 5, 2975–2979.

Chatterjee, S., 2000. Assay of lactosylceramide synthase and comments on its potential role in signal transduction. Methods Enzymol. 311, 73–81.

Chatterjee, S., 1999. Neutral sphingomyelinase: past, present and future. Chem. Phys. Lipids 102, 79–96.

Chatterjee, S., Han, H., Rollins, S., Cleveland, T., 1999. Molecular cloning, characterization, and expression of a novel human neutral sphingomyelinase. J. Biol. Chem. 274, 37407–37412.

Chatterjee, S., 1998. Sphingolipids in atherosclerosis and vascular biology. Arterioscl. Thromb. Vas. Biol. Review 1523–1533.

Chatterjee, S., Dey, S., Shi, W.Y., Thomas, K., Hutchins, G.M., 1997a. Accumulation of glycosphingolipids in human atherosclerotic plaque and unaffected aorta tissues. Glycobiology 7, 57–65.

Chatterjee, S., Bhunia, A.K., Han, H., Snowden, A., 1997b. Oxidized low density lipoproteins stimulate galactosyltransferase activity, ras activation, p44 mitogen activated protein kinase and c-fos expression in aortic smooth muscle cells. Glycobiology 7, 703–710.

Chatterjee, S., 1997. Oxidized low density lipoproteins and lactosylceramide both stimulate the expression of proliferating cell nuclear antigen and the proliferation of aortic smooth muscle cells. Indian J. Biochem. Biophys. 34, 56–61

Chatterjee, S., Cleveland, T., Inokuchi, J., Radin, N.S., 1996a. Studies of the action of ceramide-like substances (D- and L-PDMP) on sphingolipid glycosyltransferases and purified lactosylceramide synthase. Glycoconjugate 13, 481–486.

Chatterjee, S., Shi, W.Y., Wilson, P., Mazumdar, A., 1996b. Role of lactosylceramide and MAP kinase in the proliferation of proximal tubular cells in human polycystic kidney disease. J. Lipid Res. 37, 1334–1344.

Chatterjee, S., Ghosh, N., 1996. Oxidized low density lipoprotein stimulates aortic smooth muscle cell proliferation. Glycobiology 6, 303–311.

Chatterjee, S., 1993a. Regulation of synthesis of lactosylceramide in normal and tumor proximal tubular cells. Biochim. Biophys. Acta 1167, 339–344.

Chatterjee, S., 1993b. Neutral sphingomyelinase. Adv. Lipid Res. 26, 25–48.

Chatterjee, S., Ghosh, N., Khurana, S., 1992. Purification of uridine diphosphate-galactose:glucosyl ceramide, beta 1–4 galactosyltransferase from human kidney. J. Biol. Chem. 267, 7148–7153.

Chatterjee, S., 1992. Role of oxidized human plasma low density lipoproteins in atherosclerosis: effects on smooth muscle cell proliferation. J. Mol. and Cell. Biochem. 111, 143–147.

Chatterjee, S., 1991. Lactosylceramide stimulates aortic smooth muscle cell proliferation. Biochem. Biophys. Res. Comm. 181, 554–561.

Chatterjee, S., Ghosh, N., 1990. Phosphatidylcholine stimulates the activity of UDP-Gal beta 1–4 galactosyltransferase in normal human kidney proximal tumour cells. Indian J. Biochem. Biophys. 27, 375–378.

Chatterjee, S., Ghosh, N., Castiglione, E., Kwiterovich, P.O. Jr., 1988. Regulation of glycosphingolipid glycosyltransferase by low density lipoprotein receptors in cultured human proximal tubular cells. J. Biol. Chem. 263, 13017–13022.

Chatterjee, S., Clarke, K.S., Kwiterovich, P.O., Jr., 1986a. Regulation of synthesis of lactosylceramide and long chain bases in normal and familial hypercholesterolemic cultured proximal tubular cells. J. Biol. Chem. 261, 13474–13479.

Chatterjee, S., Clarke, K.S., Kwiterovich, P.O., Jr., 1986b. Uptake and metabolism of lactosylceramide on low density lipoproteins in cultured proximal tubular cells from normal and familial hypercholesterolemic homozygotes. J. Biol. Chem. 261, 13480–13486.

Chatterjee, S., Kwiterovich, P.O. Jr., Gupta, P., Erozan, Y., Alving, C.A., Richard, R., 1983. Localization of urinary lactosylceramide in cytoplasmic vesicles of renal tubular cells in homozygous familial hypercholesterolemia. Proc. Natl. Acad. Sci. U.S.A. 80, 1313–1318.

Chatterjee, S., Kwiterovich, P.O. Jr., 1976. Glycosphingolipids of human plasma lipoproteins. Lipids 11, 462–466.

Colell, A., Garcia-Ruiz, C., Roman, J., Ballesta, A., Fernandez-Checa, J.C., (2001). Ganglioside GD3 enhances apoptosis by suppressing the nuclear factor-kappa B-dependent survival pathway. FASEB J. 15, 1068–1070.

Edsall, L.C., Cuvillier, O., Twitty, S., Spiegel, S., Milstien, S., 2001. Sphingosine kinase expression regulates apoptosis and caspase activation in PC12 cells. J. Neurochem. 76, 1573–1584.

English, D., Welch, Z., Kovala, A.T., Harvey, K., Volpert, O.V., Brindley, D.N., Garcia, J.G., 2000. Sphingosine 1-phosphate released from platelets during clotting accounts for the potent endothelial cell chemotactic activity of blood serum and provides a novel link between hemostasis and angiogenesis. FASEB J. 14, 2255–2265.

Fatatis, A., Miller, R.J., 1996. Sphingosine and sphingosine 1-phosphate differentially modulate platelet-derived growth factor-BB-induced Ca2+ signaling in transformed oligodendrocytes. J. Biol. Chem. 271, 295–301.

Fukumoto, S., Mutoh, T., Hasegawa, T., Miyazaki, H., Okada, M., Goto, G., Furukawa, K., Urano, T., Furukawa, K., 2000. GD3 synthase gene expression in PC12 cells results in the continuous activation of TrkA and ERK1/2 and enhanced proliferation. J. Biol. Chem. 275, 5832–5837.

Futerman, A.H., Pagano, R.E., 1991. Determination of the intracellular sites and topology of glucosylceramide synthesis in rat liver. Biochem. J. 280, 295–302.

Garcia-Ruiz, C., Colell, A., Paris, R., Fernandez-Checa, J., 2000. Direct interaction of GD3 ganglioside with mitochondria generates reactive oxygen species followed by mitochondrial permeability transition, cytochrome c release, and caspase activation. FASEB J. 14, 847–858.

Garzotto, M., White-Jones, M., Jiang, Y., Ehleiter, D., Liao, W.C., Haimovitz-Friedman, A., Fuks, Z., Kolesnick, R., 1998. 12-O-tetradecanoylphorbol-13-acetate-induced apoptosis in LNCaP cells is mediated through ceramide synthase. Cancer Res. 58, 2260–2264.

Garner, B., Priestman, D.A., Stocker, R., Harvey, D.J., Butters, T.D., Platt, F.M., 2002. Increased glycosphingolipid levels in serum and aortae of apolipoprotein E gene knockout mice. J. Lipid Res. 43, 205–214.

Iwabuchi, K., Handa, K., Hakomori, S., 1998. Separation of "glycosphingolipid signaling domain" from caveolin-containing membrane fraction in mouse melanoma B16 cells and its role in cell adhesion coupled with signaling. J. Biol. Chem. 273, 766–33,773.

Jan, J.T., Chatterjee, S., Griffin, D., 2000. Sindbis virus entry into cells triggers apoptosis by activating sphingomyelinase, leading to the release of ceramide. J. Virology 74, 6425–6432.

Jayadev, S., Linardic, C.M., Hannun, Y.A., 1994. Identification of arachidonic acid as a mediator of sphingomyelin hydrolysis in response to tumor necrosis factor alpha. J. Biol. Chem. 269, 5757–5763.

Jeckel, D., Karrenbauer, A., Burger, K.N.J., van Meer, G., Wieland, F., 1992. Glucosylceramide is synthesized at the cytosolic surface of various Golgi subfractions. J. Cell Biol. 117, 259–267.

Jiang, X.-C., Paultre, F., Pearson, T.A., Reed, R.G., Francis, C.K., Lin, M., Berglung, L., Tall, A.R., 2000. Plasma sphingomyelin level as a risk factor for coronary artery disease. Art. Throm. Vasc. Biol. 20, 2614–2618.

Kleuser, B., Cuvillier, O., Spiegel, S., 1998. 1Alpha,25-dihydroxyvitamin D3 inhibits programmed cell death in HL-60 cells by activation of sphingosine kinase. Cancer Res. 58, 1817–1824.

Kohama, T., Olivera, A., Edsall, L., Nagiec, M.M., Dickson, R., Spiegel, S., 1998. Molecular cloning and functional characterization of murine sphingosine kinase. J. Biol. Chem. 273, 23722–23728.

Kolter, T., Sandhoff, K., 1998. Recent advances in the biochemistry of sphingolipidoses. Brain Pathol. 8, 79–100.

Kronke, M., 1999. Involvement of sphingomyelinases in TNF signaling pathways. Chem. Phys. Lipids 102, 157–166.

Lawler, J.G., Yin, M., Diehl, A.M., Roberts, E., Chatterjee, S., 1998. Tumor necrosis factor-alpha stimulates the maturation of sterol regulatory element binding protein-1 in human hepatocytes through the action of neutral sphingomyelinase. J. Biol. Chem. 273, 5053–5059.

Lee, M.J., Van Brocklyn, J.R., Thangada, S., Liu, C.H., Hand, A.R., Menzeleev, R., Spiegel, S., Hla, T., 1998. Sphingosine-1-phosphate as a ligand for the G protein-coupled receptor EDG-1. Science 279, 1552–1555.

Libby, P., Ridker, P.M., Maseri, A., 2002. Inflammation and atherosclerosis. Circulation 105, 1135–1143.

Liu, H., Sugiura, M., Nava, V.E., Edsall, L.C., Kono, K., Poulton, S., Milstien, S., Kohama, T., Spiegel, S., 2000. Molecular cloning and functional characterization of a novel mammalian sphingosine kinase type 2 isoform. J. Biol. Chem. 275, 19513–19520.

Liu, Y.Y., Han, T.Y., Giuliano, A.E., Cabot, M.C., 1999. Expression of glucosylceramide synthase, converting ceramide to glucosylceramide, confers adriamycin resistance in human breast cancer cells. J. Biol. Chem. 274, 1140–1146.

Liu, B., Andrieu-Abadie, N., Levade, T., Zhang, P., Obeid, L.M., Hannun, Y.A., 1998. Glutathione regulation of neutral sphingomyelinase in tumor necrosis factor-alpha-induced cell death. J. Biol. Chem. 273, 11313–11320.

Liu, B., Obeid, L.M., Hannun, Y.A., 1997. Sphingomyelinases in cell regulation. Cell Dev. Biol. 8, 311–322.

Liu, B., Hannun, Y.A., 1997. Inhibition of the neutral magnesium-dependent sphingomyelinase by glutathione. J. Biol. Chem. 272, 16281–16287.

Lo, N.-W., Shaper, J.H., Pevner, J., Shaper, N.L., 1998. The expanding beta 4-galactosyltransferase gene family: messages from the databanks. Glycobiology 8, 517–526.

Mandala, S., Hajdu, R., Bergstrom, J., Quackenbush, E., Xie, J., Milligan, J., Thornton, R., Shei, G.-J., Card, D., Keohane, C.A., Rosenbach, M., Hale, J., Lynch, C.L., Rupprecht, K., Parsons, W., Rosen, H., 2002. Alteration of lymphocyte trafficking by sphingosine-1-phosphate receptor agonists. Science 296, 346–349.

Maria, R.D., Lenti, L., Malisan, d'Agostion, F., Tomassini, B., Zeuner, A., Rippo, M.R., Testi, R., 1997. Requirement for GD3 ganglioside in CD95- and ceramide-induced apoptosis. Science 277, 1652–1655.

Martin, S.F., Gomez-Diaz, C., Bello, R.I., Navas, P., Villalba, J.M., 2002. Inhibition of neutral Mg^{2+}-dependent Sphingomyelinase by ubiquinol-mediated plasma membrane electron transport. Protoplasma (in press).

Martin, S.F., Navarro, F., Forthoffer, N., Navas, P., Villalba, J.M., 2001. Neutral magnesium-dependent sphingomyelinase from liver plasma membrane: purification and inhibition by ubiquinol. J. Bioenerg. Biomembr. 33, 143–153.

Mao, C., Xu, R., Szulc, Z.M., Bielawska, A., Galadari, S.H., Obeid, L.M., 2001. Cloning and characterization of a novel human alkaline ceramidase. A mammalian enzyme that hydrolyzes phytoceramide. J. Biol. Chem. 276, 26577–26588.

Mao, C.G., Obeid, L.M., 2000. Yeast sphingosine-1-phosphatases: assay, expression, deletion, purification, and cellular localization by GFP tagging. Methods Enzymol. 311, 223–233.

Mao, C.G., Xu, R., Bielawska, A., Obeid, L.M., 2000. Cloning and characterization of a Saccharomyces cerevisiae alkaline ceramidase with specificity for dihydroceramide. J. Biol. Chem. 275, 31369–31378.

Mazurek, N., Megidish, T., Hakomori, S., Igarashi, Y., 1994. Regulatory effect of phorbol esters on sphingosine kinase in BALB/C 3T3 fibroblasts (variant A31): demonstration of cell type-specific response – a preliminary note. Biochem. Biophys. Res. Commun. 198, 1–9.

Merrill, A.H.J., Schmelz, E.M., Dillehay, D.L., Spielgel, S., Shayman, J.A., Schroeder, J.J., Riley, R.T., Voss, K.A., Wang, E., 1997. Sphingolipids – the enigmatic lipid class: biochemistry, physiology, and pathophysiology. Toxicol. Appl. Pharmacol. 142, 208–225.

Merrill, A.H. Jr., Wang, E., Vales, T.R., Smith, E.R., Schroeder, J.J., Menaldino, D.S., Alexander, C., Crane, H.M., Xia, J., Liotta, D.C., Meredith, F.I., Riley, R.T., 1996. Fumonisin toxicity and sphingolipid biosynthesis. Adv. Exp. Med. Biol. 392, 297–306.

Miro-Obradors, M.J., Osada, J., Aylagas, H., Sanchez-Vegazo, I., Palacios-Alaiz, E., 1993. Microsomal sphingomyelin accumulation in thioacetamide-injured regenerating rat liver: involvement of sphingomyelin synthase activity. Carcinogenesis 14, 941–946.

Murate, T., Banno, Y., T-Koizumi, K., Watanabe, K., Mori, N., Wada, A., Igarashi, Y., Takagi, A., Kojima, T., Asano, H., Akao, Y., Yoshida, S., Saito, H., Nozawa, Y., 2001. Cell type-specific localization of sphingosine kinase 1a in human tissues. J. Histochem. Cytochem. 49, 845–855.

Nilsson, A., Duan, R.D., 1999. Alkaline sphingomyelinases and ceramidases of the gastrointestinal tract. Chem. Phys. Lipids 102, 97–105.

Nomura, T., Takizawa, M., Aoki, J., Arai, H., Inoue, K., Wakisaka, E., Yoshizuka, N., Imokawa, G., Dohmae, N., Takio, K., Hattori, M., Matsuo, N., 1998. Purification, cDNA cloning, and expression of UDP-Gal: glucosylceramide beta-1,4-galactosyltransferase from rat brain. J. Biol. Chem. 273, 13570–13577.

Obeid, L.M., Linardic, C.M., Karolak, L.A., Hannun, Y.A., 1993. Programmed cell death induced by ceramide. Science 259, 1769–1771.

Okazaki, T., Bielawska, A., Domae, N., Bell, R.M., Hannun, Y.A., 1994. Characteristics and partial purification of a novel cytosolic, magnesium-independent, neutral sphingomyelinase activated in the early signal transduction of 1 alpha,25-dihydroxyvitamin D3-induced HL-60 cell differentiation. J. Biol. Chem. 269, 4070–4077.

Olivera, A., Edsall, L., Poulton, S., Kazlauskas, A., Spiegel, S., 1999. Platelet-derived growth factor-induced activation of sphingosine kinase requires phosphorylation of the PDGF receptor tyrosine residue responsible for binding of PLCgamma. FASEB J. 13, 1593–1600.

Olivera, A., Zhang, H., Carlson, R.O., Mattie, M.E., Schmidt, R.R., Spiegel, S., 1994. Stereospecificity of sphingosine-induced intracellular calcium mobilization and cellular proliferation. J. Biol. Chem. 269, 17924–17930.

Paris, F., Fuks, Z., Kang, A., Capodieci, P., Juan, G., Ehleiter, D., Haimovitz-Friedman, A., Cordon-Cardo, C., Kolesnick, R., 2001. Endothelial apoptosis as the primary lesion initiating intestinal radiation damage in mice. Science 293–297.

Pena, L.A., Fuks, Z., Kolesnick, R., 1997. Stress-induced apoptosis and the sphingomyelin pathway. Biochem. Pharm. 53, 615–621.

Pitson, S.M., D'andrea, R.J., Vandeleur, L., Moretti, P.A., Xia, P., Gamble, J.R., Vadas, M.A., Wattenberg, B.W., 2000. Human sphingosine kinase: purification, molecular cloning and characterization of the native and recombinant enzymes. Biochem. J. 350, 429–441.

Pyne, S., Pyne, N.J., 2000. Sphingosine 1-phosphate signalling in mammalian cells. Biochem. J. 349, 385–402.

Radin, N.S., Shayman, J.A., Inokuchi, J.I., 1993. Metabolic effects of inhibiting glucosylceramide synthesis with PDMP and other substances. Adv. Lipid Res. 26, 183–213.

Saba, J.D., Nara, F., Bielawska, A., Garrett, S., Hannun, Y.A., 1997. The BST1 gene of Saccharomyces cerevisiae is the sphingosine-1-phosphate lyase. J. Biol. Chem. 272, 26087–26090.

Sandhoff, K., Kolter, T., 1997. Biochemistry of glycosphingolipid degradation. Clin. Chim. Acta 266, 51–61.

Sawai, H., Domae, N., Nagan, N., Hannun, Y.A., 1999. Function of the cloned putative neutral sphingomyelinase as lyso-platelet activating factor-phospholipase C. J. Biol. Chem. 274, 38131–38139.

Schuchman, E.H., Desnick, R.J., 1995. In: McGraw-Hill, New York, pp. 2601–2624.

Shayman, J.A., Lee, L., Abe, A., Shu, L., 2000. Inhibitors of glucosylceramide synthase. Methods Enzymol. 311, 373–387.

Symington, F.W., Murray, W.A., Bearman, S.I., Hakomori, S., 1987. Intracellular localization of lactosylceramide, the major human neutrophil glycosphingolipid. J. Biol. Chem. 262, 11356–11363.

Tabas, I., 1999. Secretory sphingomyelinase. Chem. Phys. Lipids 102, 123–130.

Tepper, A.D., Diks, S.H., van Blitterswijk, W.J., Borst, J., 2000. Glucosylceramide synthase does not attenuate the ceramide pool accumulating during apoptosis induced by CD95 or anti-cancer regimens. J. Biol. Chem. 275, 34810–34817.

Thudichum, J.L.W., 1876. In: Reports of the Medical Officer of Privy Council and Local Government Board.

Tomiuk, S., Hofmann, K., Nix, M., Zimmerman, M., Stofell, W., 1998. Cloned mammalian neutral sphingomyelinase: functions in sphingolipid signaling? Proc. Natl. Acad. Sci. USA 95, 3638–3643.

Usta, J., El Bawab, S., Roddy, P., Szulc, Z.M., Hannun, Y.A., Bielawska, A., 2001. Structural requirements of ceramide and sphingosine based inhibitors of mitochondrial ceramidase. Biochemistry 40, 9657–9668.

Verheij, M., Bose, R., Lin, X.H., Yao, B., Jarvis, W.D., Grant, S., Birrer, M.J., Szabo, E., Zon, L.I., Kyriakis, J.M., Haimovitz-Friedman, A., Fuks, Z., Kolesnick, R.N., 1996. Requirement for ceramide-initiated SAPK/JNK signalling in stress-induced apoptosis. Nature 380, 75–79.

Vielhaber, G., Brade, L., Lindner, B., Pfeiffer, S., Wepf, R., Hintze, U., Wittern, K-P., Brade H., 2001. Mouse anti-ceramide antiserum: a specific tool for the detection of endogenous ceramide. Glycobiology 11, 451–457.

Williams, N., Chatterjee, S., 2002. Am. Soc. Biol. Chem. Meeting Abst# 8801.

Wu, K., Marks, D.L., Watanabe, R., Paul, P., Rajan, N., Pagano, R., 1999. Histidine-193 of rat glucosylceramide synthase resides in a UDP-glucose- and inhibitor (D-threo-1-phenyl-2-decanoyla-mino-3-morpholinopropan-1-ol)-binding region: a biochemical and mutational study. Biochem. J. 341, 395–400.

Xia, P., Wang, L., Gamble, J.R., Vadas, M.A., 1999. Activation of sphingosine kinase by tumor necrosis factor-alpha inhibits apoptosis in human endothelial cells. J. Biol. Chem. 274, 34499–34505.

Xia, P., Gamble, J.R., Rye, K.A., Wang, L., Hii, C.S., Cockerill, P., Khew-Goodall, Y., Bert, A.G., Barter, P.J., Vadas, M.A., 1998. Tumor necrosis factor-alpha induces adhesion molecule expression through the sphingosine kinase pathway. Proc. Natl. Acad. Sci. USA 95, 14196–14201.

Yeh, L.H., Kinsey, A.M., Chatterjee, S., Alevriadou, R., 2001. Lactosylceramide mediates shear-induced endothelial superoxide production and intercellular adhesion molecule-1 expression. J. Vasc. Res. 38, 551–559.

Yu, Z.F., Nikolova-Karakashian, M., Zhou, D., Cheng, G., Schuchman, E.H., Mattson, M.P., 2002. Pivotal role for acidic sphingomyelinase in cerebral ishcemia-induced ceramide and cytokine production, and neuronal apoptosis. J. Mol. Neurosci. 15(2), 85–97.

Zhang, Y.H., Yao, B., Delikat, S., Bayoumy, S., Lin, X.H., Basu, S., McGinley, M., Chan-Hui, P.Y., Lichenstein, H., Kolesnick, R., 1997. Kinase suppressor of Ras is ceramide-activated protein kinase. Cell 89, 63–72.

**Advances in
Cell Aging and
Gerontology**

Sphingomyelin and ceramide in brain aging, neuronal plasticity and neurodegenerative disorders

Mark P. Mattson[1,2] and Roy G. Cutler[1]

[1]*Laboratory of Neurosciences, National Institute on Aging Gerontology Research Center,
5600 Nathan Shock Drive, Baltimore, MD 21224, USA.*
[2]*Department of Neuroscience, Johns Hopkins University School of Medicine,
725 N. Wolfe Street, Baltimore, MD 21205, USA.
E-mail address: mattsonm@grc.nia.nih.gov*

Contents

1. Sphingolipid metabolism and ceramide signaling

Sphingolipids are a prominent type of membrane lipid in eukaryotic cells and are particularly abundant in the nervous systems of mammals. Among the different sphingolipids, sphingomyelin has recently received considerable attention because of emerging evidence that it plays important signaling roles in many different cell types (Perry and Hannun, 1998). Sphingomyelin consists of a glycerol backbone with a phosphocholine zwitterionic dihydrophilic headgroup and two long hydrocarbon chains that form a dydrophobic domain. The hydrocarbon

Advances in Cell Aging and Gerontology, vol. 12, 97–116

chains of sphingolipids are longer (> 20 carbons) and contain more saturated bonds than that present in phosphatidylcholine. Sphingolipids play important roles in the biophysical properties of membranes and, in particular, the large disparity in the lengths of the two chains of sphingomyelin allows for interdigitation between the hydrocarbons in the two opposing monolayers of the phospholipid bilayer, thereby providing a means for coupling phase separation with the marked curvature of cell membranes (Levin and Katzen, 1985). Together with their low Tm, which is near body temperature, the unique properties of sphingomyelins allow them to play important roles in the formation of specialized domains in membranes such as lipid rafts. The first step in the synthesis of sphingolipids is catalyzed by serine palmitoyltransferase (SPT), an enzyme that generates 3-ketosphingosine from palmitoyl-CoA (acyl) and serine (Fig. 1). Additional biosynthetic steps result in formation of dihydroceramide and ceramide. Sphingomyelin synthase then catalyzes the formation of sphingomyelin.

In addition to making important contributions to the biophysical properties of cell membranes and to dynamic features of specialized membrane domains such as lipid rafts (Dobrowsky, 2000), sphingolipids are now known to serve important roles in signal transduction. It has been known for more than four decades that membrane receptors for neurotransmitters and hormones that are coupled to certain GTP-binding proteins can activate phospholipase C resulting in the cleavage of phosphatidyl inositol bis-phosphate and the release of diacylglycerol and inositol trisphosphate; diacylglycerol activates protein kinase C, while inositol trisphosphate induces calcium release from intracellular stores (Rebecchi and Pentyala, 2000). In addition, phospholipase-A2 hydrolyzes the acyl group of glycerophospholipids at the sn-2 position resulting in the release of free fatty acids and lysophospholipids, which can be further metabolized to bioactive lipid mediators including platelet activating factor, eicosanoids and lysophosphatidic acid (Farooqui et al., 1997). Similarly, several different intercellular signals and cell stressors can activate one or more enzymes that hydrolyze sphingomyelin resulting in the release of bioactive messenger molecules within the cell. Spingomyelin can be cleaved by two different classes of sphingomyelinases, acid sphingomyelinase (aSMase) and neutral sphingomyelinase (nSMase), resulting in the release of ceramide. Neutral sphingomyelinases are thought to be primarily responsible for hydrolysis of sphingomyelin in the plasma membrane, but may also function in the endoplasmic reticulum Golgi (Tomiuk et al., 2000). Consistent with a major role for nSMases in cell surface receptor-mediated signaling, nSMase colocalizes with sphingomyelin in membrane lipid rafts (Veldman et al., 2001). Two different neutral sphingomyelinases have been identified; one is widely expressed, whereas a second (nSMase2) is expressed mainly by neurons (Hofmann et al., 2000). Acidic sphingomyelinases cleave sphingomyelin located in the endoplasmic reticulum and lysosomes (Monney et al., 1998). At least two different forms of aSMase, arising from a common precursor protein, exist and appear to have distinct subcellular localizations (Ferlinz et al., 1994). Cleavage of sphingomyelin by SMases results in the release of ceramide, a bioactive lipid mediator which, as will be described below, has been shown to exert profound

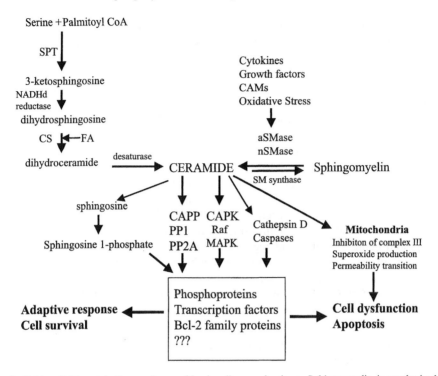

Fig. 1. Sphingolipid metabolism and ceramide signaling mechanisms. Sphingomyelin is synthesized in a series of enzymatic steps beginning with formation of 3-ketosphingosine from serine and palmitoyl CoA catalyzed by the enzyme serine palmitoyl CoA transferase (SPT). Ceramide is an intermediate in sphingomyelin synthesis and can also be rapidly formed by the hydrolysis of sphingomyelin by acidic sphingomyelinase (aSMase) and neutral sphingomyelinase (nSMase). Signals that activate SMases include cytokines, growth factors, cell adhesion molecules (CAMs) and oxyradicals. Ceramide stimulates several different signaling pathways. It can activate kinases such as ceramide-activated protein kinase (CAPK), Raf and mitogen-activated protein kinases (MAPK). Ceramide can also activate protein phosphatases such as ceramide-activated protein phosphatase (CAPP) and protein phosphatase-1 or protein phosphatase-2A (PP2A). In addition, ceramide may activate other enzymes including caspases and cathepsin D. Finally, ceramide can be metabolized to produce sphingosine-1-phosphate, an important intracellular messenger. By affecting the expression and/or function of various proteins ceramide can serve as a mediator of a variety of adaptive cellular responses to physiological signals and cell stress. On the other hand, ceramide may induce cell dysfunction and death, either through some of the same pathways that normally regulate adaptive responses, or through other pathways, such as direct engagement of apoptotic cascades in mitochondria. CS, ceramide synthase; FA, fatty acids.

effects on various cellular processes. Both nSMase and aSMase can be activated by cytokines such as TNF and various types of stresses including oxidative and metabolic insults (Cifone et al., 1994; Wiegmann et al., 1994).

Neutral SMase is associated with plasma membranes and is activated in response to a variety of stimuli including tumor necrosis factor-α (TNFα), Fas ligand and vitamin D (Hannun and Obeid, 1997; Kronke, 1999). Presumably, activation of neutral SMase results in local production of ceramide at that membrane site. The regulation of neutral SMase is not yet fully understood, but may require arachidonic acid and soluble phopsholipase-A2 (Jayadev et al., 1994). Interestingly, neutral SMase activity is also modulated by cellular redox state, with glutathione being a prominent negative regulator of enzyme activity (Liu and Hannun, 1997).

A quite extensive body of literature now suggests that ceramide, liberated as a result of activation of SMase or produced de novo, is an important intracellular messenger that serves multiple physiological roles including the regulation of cell proliferation, differentiation and survival (Billis et al., 1998; Alessenko, 2000). Several different putative effector protein targets of ceramide have been identified (Fig. 1) including a ceramide-activated protein kinase (CAPK), a ceramide-activated protein phosphatase, protein phosphatase-1, cathepsin D and caspases (Dobrowsky et al., 1993; Wolff et al., 1994; Smyth et al., 1996; Zhang et al., 1997; Mathias et al., 1998; Heinrich et al., 2000). In addition, ceramide can be metabolized to produce sphingosine and sphingosine-1-phosphate, the latter compound being an important regulator of cellular calcium homeostasis, proliferation and survival (Spiegel and Milstien, 2000). By activating one or more of these pathways ceramide can affect the function of various cellular proteins via changes in phosphorylation, and can regulate gene expression via the activation or inhibition of transcription factors.

2. Roles of ceramide signaling in neuroplasticity in the developing and adult nervous system

During the development of the nervous system, neural progenitor cells divide and then differentiate into neurons and glial cells. Neurons grow axons and dendrites and form synaptic connections. In addition, programmed cell death occurs largely during a time window in which synaptic connections are being formed. Information is beginning to emerge concerning changes in sphingolipid metabolism during development of the nervous system and the roles of sphingloipid signaling in the processes of neuronal differentiation and survival. Data suggests that the levels, and possibly the membrane localization, of sphingolipids change during development of neurons. For example, levels of ceramide in sphingolipid-rich membrane domains increases during the maturation of cultured cerebellar granule neurons (Prinetti et al., 2001). Studies of the development of glucosylceramide synthase deficient mice, and of embryonic stem cells deficient in this enzyme, suggest that the synthesis of glycosphingolipids, and presumably sphingolipid signaling, is essential for embryonic development and the differentiation of cells in many different tissues (Yamashita et al., 1999). The possible roles of sphingolipid metabolism and ceramide signaling in the regulation of neural stem cells have not been studied. One study did, however, document an increase in levels of sphingomyelin during the differentiation of cultured oligodendrocytes (Nakai et al., 2000).

Ceramide may mediate several biological actions of neurotrophic factors and cytokines on neurons in the developing and adult nervous system. For example, ceramide production is increased in cultured hippocampal neurons exposed to nerve growth factor (NGF) and may mediate the stimulatory effects of NGF on neurite outgrowth (Brann et al., 1999). On the other hand increases in ceramide production were linked to inhibition of neurite outgrowth in cultured rat sympathetic neurons (de Chaves et al., 1997). The effect of ceramide on axon outgrowth of cultured sympathetic neurons is not due to a toxic effect and is associated with decreased cellular uptake of NGF (de Chaves et al., 2001). Low concentrations of ceramide promoted cell survival and stimulated the outgrowth of dendrites in cultured embryonic hippocampal neurons (Mitoma et al., 1998). Thus, ceramide may be an important intracellular messenger that regulates neurite outgrowth and synaptogenesis.

Apoptosis is a form of programmed cell death that occurs in neurons during normal development of the nervous system. Ceramide can induce apoptosis in many different types of cells including neurons. Levels of ceramide in the developing embryonic mouse brain increase during a time period when neuronal differentiation and programmed cell death are occurring (Bieberich et al., 2001). Ceramide induces mitochondrial membrane depolarization and cytochrome c release resulting in caspase activation; it may act directly on mitochondria or may modify proteins that regulate mitochondrial apoptotic changes (Garcia-Ruiz et al., 1997; Ruvolo et al., 1999). Overexpression of sialyltransferase-II in cultured embryonic brain cells decreased the levels of the proapoptotic protein prostate apoptosis response-4 (Par-4) and increased the resistance of the cells to ceramide-induced apoptosis. Par-4 is an important mediator of developmental neuronal death (Chan et al., 1999) and is implicated in the neuronal deaths that occur in Alzheimer's disease (Guo et al., 1998), Parkinson's disease (Duan et al., 1999), stroke (Culmsee et al., 2001a) and amyotrophic lateral sclerosis (Pedersen et al., 2000). Interestingly, although ceramide induces apoptosis, it can also activate the antiapoptotic transcription factor NF-κB (Mattson et al., 1997; Wang et al., 1999), whereas Par-4 inhibits NF-κB (Camandola and Mattson, 2000). Recent studies have shown that ceramide mediates apoptosis of diffentiating neurons (Herget, et al., 2000) suggesting an important role in natural cell death of neurons during development.

The upstream and downstream events of ceramide production in apoptotic cascades are being revealed. In cerebellar granule neurons ceramide-induced death involves activation of p38 MAP kinase (Willaime et al., 2001). The mechanisms by which ceramide induces apoptosis are beginning to be elucidated. It may stimulate the activation of caspases by one or more mechanisms including direct interactions. Ceramide may also suppress anti-apoptotic signaling pathways. For example, it was reported that ceramide can inhibit Akt kinase, a protein that mediates cell survival signaling in many different cell types including neurons (Zhou et al., 1998).

Sphingolipid signaling may also play important roles in synaptic plasticity in the adult nervous system. For example, it was reported that application of the ceramide analog *N*-acetyl-D-sphingosine to hippocampal slices from adult rats results in a long depression of synaptic transmission by a mechanism that may involve protein phosphatase-mediated dephosphorylation of ionotropic glutamate receptors (Yang, 2000). Application of C2-ceramide to rat hippocampal slices resulted in an attenuation of the expression of long-term potentiation (Coogan et al., 1999).

3. Sphingolipid metabolism in normal aging

The majority of studies of membrane lipids during aging have focused on tissues other than the nervous system including liver and muscle. There is a decrease in membrane fluidity during aging, which may result from increased content of cholesterol and decreased content of unsaturated fatty acids (Hegner, 1980). Analyses of the lipid composition of brain cell membranes during aging have revealed increases in cholesterol and sphingomyelin content (Shinitzky, 1987). The age-related increases in sphingomyelin content vary across brain regions (Giusto et al., 1992). Also consistent with age-related alterations in sphingolipid-containing lipid rafts are studies documenting impaired coupling of muscarinic cholinergic receptors to phosphoinositide signaling in cerebral cortical and hippocampal membranes from aged rats (Ayyagari et al., 1998). Ceramide levels increase during senescence of fibroblasts, and exposure of low passage (young) fibroblasts to ceramide induces senescence-like changes including cell cycle arrest and increased expression of beta-galactosidase (Mouton and Venable, 2000). We have found that levels of long-chain ceramides increase in the brains of mice during aging (Fig. 2).

4. Involvement of perturbed sphingolipid metabolism and ceramide production in neurodegenerative disorders

The evidence that alterations in sphingolipid metabolism and ceramide signaling occur during normal aging, suggests a possible role for such alterations in the pathogenesis of age-related disease. Accumulating data from analyses of post-mortem tissues from patients, and experiments in animal and cell culture models, support involvement of perturbed sphingomyelin metabolism and ceramide signaling in several different neurodegenerative disorders. The remainder of this chapter will be devoted to a review of the emerging evidence suggesting that sphingomyelin alterations play important roles in dysfunction and death of neurons in pathological conditions, and that normalizing these alterations may prove effective in preventing and treating one or more of the diseases.

5. Niemann–Pick disease

A deficiency of aSMase causes Niemann–Pick disease in humans, a disorder characterized by progressive accumulation of sphingomyelin in neurons resulting in their dysfunction and degeneration (Elleder, 1989; Kolodny, 2000). Mice deficient in aSMase develop normally for approximately 4 months and then exhibit a progressive ataxia which is associated with degenerative changes in the

Ceramides in hippocampus of 6 month old C57B6/F mouse

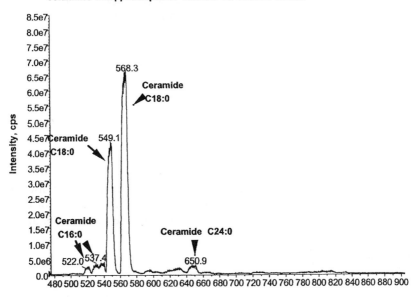

Ceramides in hippocampus of 26 month old C57B6/F mouse

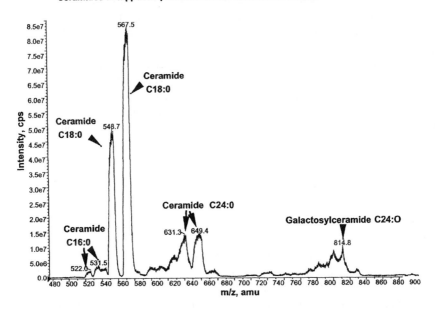

Fig. 2. Levels of long chain ceramides increase in the brain during normal aging. Representative electrospray ionization tandem mass spectrometry chromatograms showing ceramides in a samples from the hippocampus of a 6-month-old mouse and a 26- month-old mouse. Note that levels of C24:0 ceramide and C24:0 galactosylceramide are increased in the sample from the old mouse compared to the young mouse.

cerebellum (Horinouchi et al., 1995). Several aspects of the abnormalities in human Niemann-Pick patients also occur in aSMase knockout mice including: degeneration of cerebellar Purkinje neurons; excessive accumulation of sphingomyelin in the liver, spleen and brain; and motor dysfunction (Otterbach and Stoffel, 1995). In one line of aSMase knockout mice overt degeneration of cerebellar Purkinje cells was evident in two month-old mice; in six month-old mice nearly all Purkinje cells in the anterior lobe had degenerated and by 8 months essentially all Purkinje cells in the cerebellum had died (Sarna et al., 2001).

Cells in numerous tissues of aSMase deficient mice are defective in apoptosis; for example, radiation-induced damage to lung cells is markedly increased and this resistance to cell death is associated with lack of an increase in ceramide levels in lung tissue (Santana et al., 1996). The neurological abnormalities in Niemann-Pick disease may result, in part, from an autoimmune response and/or hyperactivation of microglia because bone marrow transplantation can delay the onset of neurological dysfunction in aSMase deficient mice (Miranda et al., 2000).

6. Ischemic stroke

Stroke is a leading cause of morbidity and mortality in the elderly worldwide. A stroke results from occlusion or rupture of a cerebral blood vessel, and is typically anteceded by atherosclerosis and/or hypertension (Tegos et al., 2000). The degeneration of neurons following a stroke involves metabolic compromise (ATP depletion), oxyradical production, overactivation of glutamate receptors resulting in calcium overload (Fig. 3). These events result in either necrosis or apoptosis depending upon the severity and duration of the ischemia. Typically neurons in the central core of the ischemic infarct die rapidly (minutes to hours) by necrosis, whereas neurons in the surrounding penumbral region undergo apoptotic death that occurs over periods of days to weeks (Mattson et al., 2000). Studies in cell culture and in vivo models of ischemic neuronal injury have shown that pharmacologic and genetic manipulations known to inhibit key steps in apoptosis can protect neurons in the ischemic penumbra. Examples include agents that inhibit caspases or mitochondrial permeability transition pores and overexpression of antiapoptotic genes such as those encoding BCL-2 or inhibitor of apoptosis proteins (Robertson et al., 2000).

A considerable evidence suggests that ceramide production contributes to myocardial ischemia-reperfusion injury (Gottlieb and Engler, 1999). Studies of aSMase-deficient mice have provided strong evidence that ceramide production contributes to neuronal apoptosis following a stroke. Transient occlusion of the middle cerebral artery in wild-type mice resulted in a large increase in aSMase activity, ceramide production and expression of inflammatory cytokines such as tumor necrosis factor and interleukin-1beta, but not in aSMase knockout mice (Yu et al., 2000). The amount of ischemic brain damage was decreased and behavioral outcome was improved in aSMase-deficient mice. The latter findings establish a pivotal role for aSMase in the activation of inflammatory cytokine cascades in response to brain injury, and in the ensuing apoptosis of neurons. In

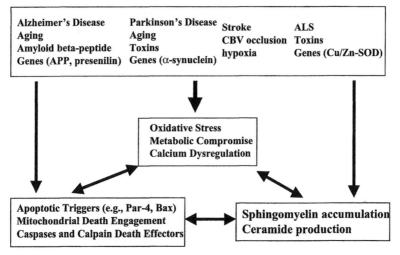

Fig. 3. Working model of the mechanisms responsible for neuronal dysfunction and death in AD, PD, stroke and ALS. Events occurring during normal aging promote neuronal degeneration; these include increased oxidative stress and metabolic compromise. Disease-specific genetic and environmental factors can trigger the neurodegenerative process in specific populations of neurons. For example, mutations in the presenilin and amyloid precursor protein (APP) genes result in increased amyloid production in Alzheimer's disease, environmental toxins may promote degeneration of dopaminergic neurons in Parkinson's disease, occlusion of cerebral blood vessels (CBV) causes a stroke, and mutations in Cu/Zn-superoxide dismutase can cause amyotrophic lateral sclerosis (ALS). Aging and disease-specific factors result in increased oxidative stress and perturbed cellular calcium homeostasis in neurons. The latter events can result in altered sphingomyelin metabolism and increased ceramide production. Dysregulation of sphingolipid signaling may combine with cellular stress to trigger synaptic dysfunction and cell death cascades.

addition, treatment of wild-type mice with D-609, an agent that inhibits Smase, reduced ischemia-induced ceramide and cytokine production and reduced brain damage (Yu et al., 2000). Moreover, the immunosuppressant agent FK506 inhibits both ceramide production and apoptosis signaling in a rat model of focal cerebral ischemia-reperfusion injury (Herr et al., 1999). These findings suggest that drugs that inhibit sphingomyelin hydrolysis may prove beneficial in stroke patients.

Although the majority of data suggests a role for ceramide as a proapoptotic signal, under some conditions ceramide may prevent neuronal death. For example, cultured hippocampal neurons pretreated with low (nanomolar) concentrations of C-6 ceramide exhibited increased resistance to death induced by glutamate, iron and amyloid beta-peptide (Goodman and Mattson, 1996). In a rat model of neonatal cerebral hypoxia-ischemia, intraventricular infusion of C2-ceramide resulted in a decrease in cortical infarct volume (Chen et al., 2001). Similarly, intraventricular infusion of C2-ceramide reduced cortical infarct volume when administered prior to permanent middle cerebral artery occlusion in spontaneously hypertensive adult rats (Furuya et al., 2001). The latter studies suggested that ceramide may induce a preconditioning effect by activating a stress response.

However, the studies of aSMase deficient mice (Yu et al., 2000) strongly suggest endogenous ceramide induction during ischemia/reperfusion.

7. Alzheimer's disease

Alzheimer's disease (AD) is characterized by the degeneration and death of neurons in brain regions involved in learning and memory processes, such as the hippocampus and cerebral cortex. The major histological abnormalities in the brains of AD patients are the extracellular accumulation of insoluble aggregates of a protein called amyloid beta-peptide, and the intracellular aggregation of the microtubule-associated protein tau in degenerating neurons. While most cases of AD have no clear genetic basis, some cases are inherited in an autosomal dominant manner, and three different genes have been identified in which mutations cause AD, namely, presenilin-1, presenilin-2 and amyloid precursor protein (Mattson, 1997). The neurodegenerative process in AD involves oxidative stress, metabolic compromise and disruption of neuronal calcium homeostasis; it is believed that apoptotic and excitotoxic cascades are involved in the death of neurons in AD (Mattson, 1997, 2000). A clue suggesting a possible link of altered sphingomyelin metabolism to AD comes from reports that some patients with Niemann-Pick disease exhibit dementia as a prominent symptom (Hulette et al., 1992). A clue suggesting a possible link of altered sphingomyelin metabolism to AD comes from reports that some patients with Niemann-Pick disease exhibit dementia as a prominent symptom (Hulette et al., 1992). Moreover, neurofibrillary tangles were present in the entorhinal cortex, cingulated cortex and associated regions of the brains of patients with slowly progressing forms of Niemann–Pick disease (Suzuki et al., 1995).

The findings emerging from studies of the pathogenic actions of amyloid beta-peptide add weight to the growing body of evidence implicating sphingolipid-related membrane alterations in AD. Amyloid beta-peptide binds to gangliosides in membranes, and this binding enhances peptide fibril formation (McLaurin and Chakrabartty, 1996; Choo-Smith et al., 1997), suggesting a mechanism whereby amyloid beta-peptide induces membrane damage (Mattson, 1997). Additional studies have shown that amyloid beta-peptide and prion proteins undergo a conformational transition upon interaction with sphingolipids; these amyloidogenic peptides interact with sphingomyelin via V3-like domains (Mahfoud et al., 2002). Thus, sphingomyelin may facilitate the adoption of pathogenic fibril-forming conformations of amyloid peptides. This is particularly interesting because amyloid beta-peptide has been shown to disrupt membrane signal transduction processes present in sphingomyelin-containing lipid rafts. For example, amyloid beta-peptide impaired coupling of muscarinic cholinergic receptors, metabotropic glutamate receptors and thrombin receptors to the GTP-binding protein Gq11 (Kelly et al., 1996; Mattson and Begley, 1996; Blanc et al., 1997). Finally, the detergent-insoluble low density membrane fraction from the brains of presenilin-2 mutant mice contained higher levels of amyloid beta-peptide1-42 and lower levels of sphingomyelin than did the same membrane fraction from wild-type mice (Sawamura et al., 2000). However, the consequences of this alteration for sphingomyelin-mediated signaling is not known. The small amount of preliminary data available, such as those just described, suggest that alterations in the structure and signaling functions of lipid rafts may play important roles in aging and age-related brain disorders including AD.

8. Parkinson's disease

Parkinson's disease (PD) is characterized by progressive degeneration of dopamine-producing neurons in the substantia nigra resulting not only in motor dysfunction, but also affecting other brain regions including limbic and cortical regions that control emotions and learning and memory (Levin and Katzen, 1995). A small number of families with inherited forms of PD have been identified in which mutations in the genes encoding either alpha-synuclein or Parkin are responsible for the disease (Zhang et al., 2000). Interestingly, increasing evidence suggests that environmental toxins including pesticides such as rotenone, may increase risk for PD. Indeed, the most widely employed animal models of PD involve administration of rotenone, MPTP and related mitochondrial toxins which selectively destroy dopaminergic neurons in the substantia nigra. The neurodegenerative process in PD is believed to involve metabolic compromise (particularly inhibition of mitochondrial complex I), oxidative stress and activation of apoptotic death cascades (Duan et al., 1999; Zhang et al., 2000).

More than 30 years ago it was reported that sphingomyelin accumulated in Lewy inclusion bodies, the hallmark lesions in the substantia nigra of PD patients (den Jager, 1969). Surprisingly, however, there is very little information available concerning the possible involvement of alterations in sphingolipid metabolism in the pathogenesis of PD. C2-ceramide has been shown to induce oxyradical production by mitochondrial complex I in differentiated PC12 cells, which appears to be a pivotal event in apoptosis of these cells (France-Lanord et al., 1997). Alpha-synuclein is a synaptic protein that is a major component of Lewy bodies in PD, and some inherited forms of PD are caused by mutations in alpha-synuclein. It was recently reported that wild-type alpha-synuclein can protect cultured neuronal cells from apoptosis induced by ceramide, whereas mutant alpha-synuclein did not protect the neurons (da Costa et al., 2000).

9. Amyotrophic lateral sclerosis

Patients with amyotrophic lateral sclerosis (ALS) manifest progressive degeneration of motor neurons in the spinal cord and brainstem resulting in paralysis and death of the patients by respiratory failure (Haverkamp et al., 1995). While the cause of most cases of ALS is unknown, a small number of families have been identified in which the disease is inherited in an autosomal dominant manner as result of mutations in the antioxidant enzyme Cu/Zn-SOD (Cudkowicz et al., 1997). Transgenic mice expressing the same Cu/Zn-SOD mutations exhibit histopathological and clinical phenotypes remarkably similar to ALS patients (Del Canto and Gurney, 1995). Data obtained from studies of patients, Cu/Zn-SOD mutant mice and cultured neurons suggests that the pathogenic mechanism responsible for motor neuron degeneration involves oxidative stress (Ferrante et al., 1997; Pedersen et al., 1998), overactivation of glutamate receptors (Rothstein et al., 1995; Kruman et al., 1999) and a form of progammed cell death called apoptosis (Pedersen et al., 2000; Sathasivam et al., 2001). Exposure of cultured rat spinal cord neurons to a ceramide analog, promoted cell survival at low concentrations, but induced apoptosis at higher concentrations (Irie and Hirabayashi et al., 1998).

We have discovered quite striking abnormalities in sphingolipid and cholesterol metabolism in the spinal cords of ALS patients and in a Cu/Zn-SOD mutant transgenic ALS mice (Cutler et al., 2002). Mass spectrometry analyses revealed increased levels of sphingomyelin and long-chain ceramides in spinal cord tissue from ALS patients and ALS mice (Cutler et al., 2002). The accumulation of ceramides preceded any evidence of motor dysfunction in the ALS mice suggesting a possible contribution of ceramide to the degeneration of motor neurons. Previous studies have shown that increased levels of oxidative stress occur in spinal cord cells very early in the disease process (Ferrante et al., 1997; Pedersen et al., 1998), while other studies have shown that oxidative stress can induce sphingomyelin metabolism and ceramide production (Fernandez-Ayala et al., 2000). Exposure of cultured motor neurons to oxidative stress increased the accumulation of sphingomyelin, ceramides and cholesterol esters, and inhibition of sphingomyelin synthesis with myriocin prevented accumulation of ceramides and protected motor neurons against death induced by oxidative and excitotoxic insults (Cutler et al., 2002). These findings suggest that altered sphingomyelin metabolism and excessive accumulation of ceramide in motor neuron plays an important role in the pathogenesis of ALS.

10. Conclusions and implications for successful brain aging strategies

Sphingomyelin metabolites, particularly ceramide and sphingosine-1-phosphate, are becoming firmly established as mediators of neuronal plasticity and survival, and perturbations in sphingomyelin metabolism is increasingly recognized for its contributions to the pathogenesis of neurological disorders. Sphingolipid signaling mediates cellular responses to a remarkable array of growth factors, cytokines and neurotransmitters, but is also a prominent mediator of responses oxidative and metabolic stress. If alterations in sphingolipid do indeed play major roles in aging and age-related disease, then one goal of research on this fascinating signaling system is to identify ways to maintain the beneficial functions of sphingolipid metabolism, while preventing pathological alterations in this system. In this regard, pharmacological agents that target specific molecules in sphingomyelin metabolism and ceramide signaling would be valuable. For example, one approach to preventing unwanted cell death-mediated sphingomyelin metabolites is to inhibit sphingomyelin synthesis. One such chemical inhibitor of serine palmitoyl CoA transferase is meryocin (Miyake et al., 1995; Hanada et al., 2000), which has proven effective in protecting motor neurons against death in experimental models of ALS (Cutler et al., 2002). Steps downstream of ceramide production in apoptotic cascades could also be targeted. For example, the tumor suppressor protein p53 is an essential effector of ceramide-mediated apoptosis in some cell types (Dbaibo et al., 1998), and chemical inhibitors of p53 have been demonstrated to be effective in reducing neuronal degeneration in models of stroke and AD in preclinical studies (Culmsee et al., 2001b).

A second approach is to manipulate sphingolipid metabolism through dietary changes. The phospholipid and sphingolipid composition of cell membranes can be modified by diet (Merrill et al., 1997; Cha and Jones, 2000). Interestingly, decreased food intake decreases serine palmitoyl transferase activity and sphingomyelin levels in young rodents (Rotta, et al., 1999). In addition, age related increases in the ratio of sphingomyelin to phospatidylcholine in the cerebral cortex were delayed or prevented in rats maintained on a calorie-restricted diet (Tacconi et al., 1991). The latter findings are interesting because of the extensive literature documenting antiaging effects of caloric restriction in mammals. Moreover, recent studies have shown that dietary restriction can protect neurons against age-related dysfunction and degeneration in experimental models of AD, PD and stroke (Bruce-Keller et al., 1999; Duan and Mattson, 1999; Yu and Mattson, 1999; Zhu et al., 1999). The evidence described above supporting a role for perturbed sphingomyelin metabolism in the pathogenesis of neurodegenerative disorders, suggests a possible role for suppression of patholo-gical sphingomyelin metabolism and ceramide signaling in the neuroprotective effects of dietary restriction. Therefore, one approach for reducing one's risk of age-related neurodegenerative disorders is to reduce calorie intake throughout adult life.

References

Alessenko, A.V., 2000. The role of sphingomyelin cycle metabolites in transduction of signals of cell proliferation, differentiation and death. Membr. Cell Biol. 13, 303–320.

Ayyagari, P.V., Gerber, M., Joseph, J.A., Crews, F.T., 1998. Uncoupling of muscarinic cholinergic phosphoinositide signals in senescent cerebral cortical and hippocampal membranes. Neurochem. Int. 32, 107–115.

Bieberich, E., MacKinnon, S., Silva, J., Yu, R.K., 2001. Regulation of apoptosis during neuronal differentiation by ceramide and b-series complex gangliosides. J. Biol. Chem. 276, 44396–44404.

Billis, W., Fuks, Z., Kolesnick, R., 1998. Signaling in and regulation of ionizing radiation-induced apoptosis in endothelial cells. Recent Prog. Horm. Res. 53, 85–92.

Blanc, E.M., Kelly, J.F., Mark, R.J., Mattson, M.P., 1997. 4-hydroxynonenal, an aldehydic product of lipid peroxidation, impairs signal transduction associated with muscarinic acetylcholine and metabotropic glutamate receptors: possible action on $G\alpha q/11$. J. Neurochem. 69, 570–580.

Brann, A.B., Scott, R., Neuberger, Y., Abulafia, D., Boldin, S., Fainzilber, M., Futerman, A.H., 1999. Ceramide signaling downstream of the p75 neurotrophin receptor mediates the effects of nerve growth factor on outgrowth of cultured hippocampal neurons. J. Neurosci. 19, 8199–8206.

Bruce-Keller, A.J., Umberger, G., McFall, R., Mattson, M.P., 1999. Food restriction reduces brain damage and improves behavioral outcome following excitotoxic and metabolic insults. Ann. Neurol. 45, 8–15.

Camandola, S., Mattson, M.P., 2000. Pro-apoptotic action of PAR-4 involves inhibition of NF-kappaB activity and suppression of BCL-2 expression. J. Neurosci. Res. 61, 134–139.

Cha, M.C., Jones, P.J., 2000. Energy restriction dilutes the changes related to dietary fat type in membrane phospholipid fatty acid composition in rats. Metabolism 49, 977–983.

Chan, S.L., Tammariello, S.P., Estus, S., Mattson, M.P., 1999. Prostate apoptosis response-4 mediates trophic factor withdrawal-induced apoptosis of hippocampal neurons: actions prior to mitochondrial dysfunction and caspase activation. J. Neurochem. 73, 502–512.

Chen, Y., Ginis, I., Hallenbeck, J.M., 2001. The protective effect of ceramide in immature rat brain hypoxia-ischemia involves up-regulation of bcl-2 and reduction of TUNEL-positive cells. J. Cereb. Blood Flow Metab. 21, 34–40.

Choo-Smith, L.P., Garzon-Rodriguez, W., Glabe, C.G., Surewicz, W.K., 1997. Acceleration of amyloid fibril formation by specific binding of Abeta-(1–40) peptide to ganglioside-containing membrane vesicles. J. Biol. Chem. 272, 22987–22990.

Cifone, M.G., De Maria, R., Roncaioli, P., Rippo, M.R., Azuma, M., Lanier, L.L., Santoni, A., Testi, R., 1994. Apoptotic signaling through CD95 (Fas/Apo-1) activates an acidic sphingomyelinase. J. Exp. Med. 180, 1547–1552.

Coogan, A.N., O'Neill, L.A., O'Connor, J.J., 1999. The P38 mitogen-activated protein kinase inhibitor SB203580 antagonizes the inhibitory effects of interleukin-1beta on long-term potentiation in the rat dentate gyrus in vitro. Neuroscience 93, 57–69.

Cudkowicz, M.E., McKenna-Yasek, D., Sapp, P.E., Chin, W., Geller, B., Hayden, D.L., Schoenfeld, D.A., Hosler, B.A., Horvitz, H.R., Brown, R.H., 1997. Epidemiology of mutations in superoxide dismutase in amyotrophic lateral sclerosis. Ann. Neurol. 41, 210–221.

Culmsee, C., Zhu, Y., Krieglstein, J., Mattson, M.P., 2001a. Evidence for the involvement of Par-4 in ischemic neuron cell death. J. Cereb. Blood Flow Metab. 21, 334–343.

Culmsee, C., Zhu, X., Yu, Q.S., Chan, S.L., Camandola, S., Guo, Z., Greig, N.H., Mattson, M.P., 2001b. A synthetic inhibitor of p53 protects neurons against death induced by ischemic and excitotoxic insults, and amyloid beta-peptide. J. Neurochem. 77, 220–228.

Cutler, R.G., Pedersen, W.A., Camandola, S., Rothstein, J.D., Mattson, M.P., 2002. Evidence that accumulation of ceramides and cholesterol esters mediates oxidative stress-induced death of motor neurons in ALS. Ann. Neurol. 52, 448–457.

da Costa, C.A., Ancolio, K., Checler, F., 2000. Wild-type but not Parkinson's disease related ala-53 → Thr mutant alpha -synuclein protects neuronal cells from apoptotic stimuli. J. Biol. Chem. 275, 24065–24069.

Dbaibo, G.S., Pushkareva, M.Y., Rachid, R.A., Alter, N., Smyth, M.J., Obeid, L.M., Hannun, Y.A., 1998. p53-dependent ceramide response to genotoxic stress. J. Clin. Invest. 102, 329–339.

de Chaves, E.I., Bussiere, M., Vance, D.E., Campenot, R.B., Vance, J.E., 1997. Elevation of ceramide within distal neurites inhibits neurite growth in cultured rat sympathetic neurons. J. Biol. Chem. 272, 3028–3035.

de Chaves, E.P., Bussiere, M., MacInnis, B., Vance, D.E., Campenot, R.B., Vance, J.E., 2001. Ceramide inhibits axonal growth and nerve growth factor uptake without compromising the viability of sympathetic neurons. J. Biol. Chem. 276, 36207–36214.

Del Canto, M.C., Gurney, M.E., 1995. Neuropathological changes in two lines of mice carrying a transgene for mutant human Cu,Zn SOD, and in mice overexpressing wild type human SOD: a model of familial amyotrophic lateral sclerosis (FALS). Brain Res. 676, 25–40.

den Jager, W.A., 1969. Sphingomyelin in Lewy inclusion bodies in Parkinson's disease. Arch. Neurol. 21, 615–619.

Dobrowsky, R.T., Kamibayashi, C., Mumby, M.C., Hannun, Y.A., 1993. Ceramide activates heterotrimeric protein phosphatase 2A. J. Biol. Chem. 268, 15523–15530.

Dobrowsky, R.T., 2000. Sphingolipid signalling domains floating on rafts or buried in caves? Cell Signal. 12, 81–90.

Duan, W., Mattson, M.P., 1999. Dietary restriction and 2-deoxyglucose administration improve behavioral outcome and reduce degeneration of dopaminergic neurons in models of Parkinson's disease. J. Neurosci. Res. 57, 195–206.

Duan, W., Zhang, Z., Gash, D.M., Mattson, M.P., 1999. Participation of prostate apoptosis response-4 in degeneration of dopaminergic neurons in models of Parkinson's disease. Ann. Neurol. 46, 587–597.

Elleder, M., 1989. Niemann-Pick disease. Pathol. Res. Pract. 185, 293–328.

Farooqui, A.A., Yang, H.C., Rosenberger, T.A., Horrocks, L.A., 1997. Phospholipase A2 and its role in brain tissue. J. Neurochem. 69, 889–901.

Ferlinz, K., Hurwitz, R., Vielhaber, G., Suzuki, K., Sandhoff, K., 1994. Occurrence of two molecular forms of human acid sphingomyelinase. Biochem. J. 301, 855–862.

Fernandez-Ayala, D.J., Martin, S.F., Barroso, M.P., Gomez-Diaz, C., Villalba, J.M., Rodriguez-Aguilera, J.C., Lopez-Lluch, G., Navas, P., 2000. Coenzyme Q protects cells against serum withdrawal-induced apoptosis by inhibition of ceramide release and caspase-3 activation. Antioxid. Redox Signal. 2, 263–275.

Ferrante, R.J., Browne, S.E., Shinobu, L.A., Bowling, A.C., Baik, M.J., MacGarvey, U., Kowall, N.W., Brown, R.H. Jr., Beal, M.F., 1997. Evidence of increased oxidative damage in both sporadic and familial amyotrophic lateral sclerosis. J. Neurochem. 69, 2064–2074.

France-Lanord, V., Brugg, B., Michel, P.P., Agid, Y., Ruberg, M., 1997. Mitochondrial free radical signal in ceramide-dependent apoptosis: a putative mechanism for neuronal death in Parkinson's disease. J. Neurochem. 69, 1612–1621.

Furuya, K., Ginis, I., Takeda, H., Chen, Y., Hallenbeck, J.M., 2001. Cell permeable exogenous ceramide reduces infarct size in spontaneously hypertensive rats supporting *in vitro* studies that have implicated ceramide in induction of tolerance to ischemia. J. Cereb. Blood Flow Metab. 21, 226–232.

Garcia-Ruiz, C., Colell, A., Mari, M., Morales, A., Fernandez-Checa, J.C., 1997. Direct effect of ceramide on the mitochondrial electron transport chain leads to generation of reactive oxygen species. Role of mitochondrial glutathione. J. Biol. Chem. 272, 11369–11377.

Giusto, N.M., Roque, M.E., Ilincheta de Boschero, M.G., 1992. Effects of aging on the content, composition and synthesis of sphingomyelin in the central nervous system. Lipids 27, 835–839.

Goodman, Y., Mattson, M.P., 1996. Ceramide protects hippocampal neurons against excitotoxic and oxidative insults, and amyloid beta-peptide toxicity. J. Neurochem. 6, 869–872.

Gottlieb, R.A., Engler, R.L., 1999. Apoptosis in myocardial ischemia-reperfusion. Ann. N. Y. Acad. Sci. 874, 412–426.

Guo, Q., Fu, W., Xie, J., Luo, H., Sells, S.F., Geddes, J.W., Bondada, V., Rangnekar, V.M., Mattson, M.P., 1998. Par-4 is a mediator of neuronal degeneration associated with the pathogenesis of Alzheimer disease. Nat. Med. 4, 957–962.

Hanada, K., Nishijima, M., Fujita, T., Kobayashi, S., 2000. Specificity of inhibitors of serine palmitoyltransferase (SPT), a key enzyme in sphingolipid biosynthesis, in intact cells. A novel evaluation system using an SPT-defective mammalian cell mutant. Biochem. Pharmacol. 59, 1211–1216.

Hannun, Y.A., Obeid, L.M., 1997. Ceramide and the eukaryotic stress response. Biochem. Soc. Trans. 25, 1171–1175.

Haverkamp, L.J., Appel, V., Appel, S.H., 1995. Natural history of amyotrophic lateral sclerosis in a database population. Validation of a scoring system and a model for survival prediction. Brain. 118, 707–719.

Hegner, D., 1980. Age-dependence of molecular and functional changes in biological membrane properties. Mech. Ageing Dev. 14, 101–118.

Heinrich, M., Wickel, M., Winoto-Morbach, S., Schneider-Brachert, W., Weber, T., Brunner, J., Saftig, P., Peters, C., Kronke, M., Schutze, S., 2000. Ceramide as an activator lipid of cathepsin D. Adv. Exp. Med. Biol. 477, 305–315.

Herget, T., Esdar, C., Oehrlein, S.A., Heinrich, M., Schutze, S., Maelicke, A., van Echten-Deckert, G., 2000. Production of ceramides causes apoptosis during early neural differentiation *in vitro*. J. Biol. Chem. 275, 30344–30354.

Herr, I., Martin-Villalba, A., Kurz, E., Roncaioli, P., Schenkel, J., Cifone, M.G., Debatin, K.M., 1999. FK506 prevents stroke-induced generation of ceramide and apoptosis signaling. Brain Res. 826, 210–219.

Hofmann, K., Tomiuk, S., Wolff, G., Stoffel, W., 2000. Cloning and characterization of the mammalian brain-specific, Mg2+-dependent neutral sphingomyelinase. Proc. Natl. Acad. Sci. USA 97, 5895–5900.

Hulette, C.M., Earl, N.L., Anthony, D.C., Crain, B.J., 1992. Adult onset Niemann-Pick disease type C presenting with dementia and absent organomegaly. Clin. Neuropathol. 11, 293–297.

Horinouchi, K., Erlich, S., Perl, D.P., Ferlinz, K., Bisgaier, C.L., Sandhoff, K., Desnick, R.J., Stewart, C.L., Schuchman, E.H., 1995. Acid sphingomyelinase deficient mice: a model of types A and B Niemann-Pick disease. Nat. Gen. 10, 288–293.

Irie, F., Hirabayashi, Y., 1998. Application of exogenous ceramide to cultured rat spinal motoneurons promotes survival or death by regulation of apoptosis depending on its concentrations. J. Neurosci. Res. 54, 475–485.

Jayadev, S., Linardic, C.M., Hannun, Y.A., 1994. Identification of arachidonic acid as a mediator of sphingomyelin hydrolysis in response to tumor necrosis factor alpha. J. Biol. Chem. 269, 5757–5763.

Kelly, J., Furukawa, K., Barger, S.W., Mark, R.J., Rengen, M.R., Blanc, E.M., Roth, G.S., Mattson, M.P., 1996. Amyloid β-peptide disrupts carbachol-induced muscarinic cholinergic signal transduction in cortical neurons. Proc. Natl. Acad. Sci. USA 93, 6753–6758.

Kolodny, E.H., 2000. Niemann-Pick disease. Curr. Opin. Hematol. 7, 48–52.

Kronke, M., Involvement of sphingomyelinases in TNF signaling pathways., 1999. Chem. Phys. Lipids 102, 157–166.

Kruman, I.I., Pedersen, W.A., Springer, J.E., Mattson, M.P., 1999. ALS-linked Cu/Zn-SOD mutation increases vulnerability of motor neurons to excitotoxicity by a mechanism involving increased oxidative stress and perturbed calcium homeostasis. Exp. Neurol. 160, 28–39.

Levin, B.E., Katzen, H.L., 1985. Early cognitive changes and nondementing behavioral abnormalities in Parkinson's disease. Adv. Neurol. 65, 85–95.

Liu, B., Hannun, Y.A., 1997. Inhibition of the neutral magnesium-dependent sphingomyelinase by glutathione. J. Biol. Chem. 272, 16281–16287.

Mahfoud, R., Garmy, N., Maresca, M., Yahi, N., Puigserver, A., Fantini, J., 2002. Identification of a common sphingolipid-binding domain in Alzheimer, prion and HIV-1 proteins. J. Biol. Chem. 277, 11292–11296.

Mathias, S., Pena, L.A., Kolesnick, R.N., 1998. Signal transduction of stress via ceramide. Biochem. J. 335, 465–480.

Mattson, M.P., Begley, J.G., 1996. Amyloid βeta-peptide alters thrombin-induced calcium responses in cultured human neural cells. Amyloid 3, 28–40.

Mattson, M.P., 1997. Cellular actions of beta-amyloid precursor protein and its soluble and fibrillogenic derivatives. Physiol. Rev. 77, 1081–1132.

Mattson, M.P., Goodman, Y., Luo, H., Fu, W., Furukawa, K., 1997. Activation of NF-kappaB protects hippocampal neurons against oxidative stress-induced apoptosis: evidence for induction of manganese superoxide dismutase and suppression of peroxynitrite production and protein tyrosine nitration. J. Neurosci. Res. 49, 681–697.

Mattson, M.P., 2000. Apoptosis in neurodegenerative disorders. Nat. Rev. Mol. Cell Biol. 1, 120–129.

Mattson, M.P., Culmsee, C., Yu, Z.F., 2000. Apoptotic and antiapoptotic mechanisms in stroke. Cell Tissue Res. 301, 173–187.

McLaurin, J., Chakrabartty, A., 1996. Membrane disruption by Alzheimer beta-amyloid peptides mediated through specific binding to either phospholipids or gangliosides. Implications for neurotoxicity. J. Biol. Chem. 271, 26482–26489.

Merrill, A.H., Schmelz, E.M., Wang, E., Dillehay, D.L., Rice, L.G., Meredith, F., Riley, R.T., 1997. Importance of sphingolipids and inhibitors of sphingolipid metabolism as components of animal diets. J. Nutr. 127, S830–S833.

Miranda, S.R., Erlich, S., Friedrich, V.L. Jr., Gatt, S., Schuchman, E.H., 2000. Hematopoietic stem cell gene therapy leads to marked visceral organ improvements and a delayed onset of neurological abnormalities in the acid sphingomyelinase deficient mouse model of Niemann-Pick disease. Gene Ther. 7, 1768–1776.

Mitoma, J., 1998. Bipotential roles of ceramide in the growth of hippocampal neurons: promotion of cell survival and dendritic outgrowth in dose- and developmental stage-dependent manners. J. Neurosci. Res. 51, 712–722.

Miyake, Y., Kozutsumi, Y., Nakamura, S., Fujita, T., Kawasaki, T., 1995. Serine palmitoyltransferase is the primary target of a sphingosine-like immunosuppressant, ISP-1/myriocin. Biochem. Biophys. Res. Commun. 211, 396–403.

Monney, L., Olivier, R., Otter, I., Jansen, B., Poirier, G.G., Borner, C., 1998. Role of an acidic compartment in tumor-necrosis-factor-alpha-induced production of ceramide, activation of caspase-3 and apoptosis. Eur. J. Biochem. 251, 295–303.

Mouton, R.E., Venable, M.E., 2000. Ceramide induces expression of the senescence histochemical marker, beta-galactosidase, in human fibroblasts. Mech. Ageing Dev. 113, 169–181.

Nakai, Y., Sakurai, Y., Yamaji, A., Asou, H., Umeda, M., Uyemura, K., Itoh, K., 2000. Lysenin-sphingomyelin binding at the surface of oligodendrocyte lineage cells increases during differentiation *in vitro*. J. Neurosci. Res. 62, 521–529.

Otterbach, B., Stoffel, W., 1995. Acid sphingomyelinase-deficient mice mimic the neurovisceral form of human lysosomal storage disease (Niemann-Pick disease). Cell 81, 1053–1061.

Pedersen, W.A., Fu, W., Keller, J.N., Markesbery, W.R., Appel, S., Smith, R.G., Kasarskis, E., Mattson, M.P., 1998. Protein modification by the lipid peroxidation product 4-hydroxynonenal in the spinal cords of amyotrophic lateral sclerosis patients. Ann. Neurol. 44, 819–824.

Pedersen, W.A., Luo, H., Kruman, I., Kasarskis, E., Mattson, M.P., 2000. The prostate apoptosis response-4 protein participates in motor neuron degeneration in amyotrophic lateral sclerosis. FASEB J. 14, 913–924.

Perry, D.K., Hannun, Y.A., 1998. The role of ceramide in cell signaling. Biochim Biophys Acta. 1436, 233–243.

Prinetti, A., Chigorno, V., Prioni, S., Loberto, N., Marano, N., Tettamanti, G., Sonnino, S., 2001. Changes in the lipid turnover, composition, and organization, as sphingolipid-enriched membrane domains, in rat cerebellar granule cells developing *in vitro*. J. Biol. Chem. 276, 21136–21145.

Rebecchi, M.J., Pentyala, S.N., 2000. Structure, function, and control of phosphoinositide-specific phospholipase C. Physiol. Rev. 80, 1291–1335.

Robertson, G.S., Crocker, S.J., Nicholson, D.W., Schulz, J.B., 2000. Neuroprotection by the inhibition of apoptosis. Brain Pathol. 10, 283–292.

Rothstein, J.D., 1995. Excitotoxicity and neurodegeneration in amyotrophic lateral sclerosis. Clin. Neurosci. 3, 348–359.

Rotta, L.N., Da Silva, C.G., Perry, M.L., Trindade, V.M., 1999. Under-nutrition decreases serine palmitoyltransferase activity in developing rat hypothalamus. Ann. Nutr. Metab. 43, 152–158.

Ruvolo, P.P., Deng, X., Ito, T., Carr, B.K., May, W.S., 1999. Ceramide induces Bcl-2 dephosphorylation via a mechanism involving mitochondrial PP2A. J. Biol. Chem. 274, 20296–20300.

Santana, P., Pena, L.A., Haimovitz-Friedman, A., Martin, S., Green, D., McLoughlin, M., Cordon-Cardo, C., Schuchman, E.H., Fuks, Z., Kolesnick, R., 1996. Acid sphingomyelinase-deficient human lymphoblasts and mice are defective in radiation-induced apoptosis. Cell 86, 189–199.

Sarna, J., Miranda, S.R., Schuchman, E.H., Hawkes, R., 2001. Patterned cerebellar Purkinje cell death in a transgenic mouse model of Niemann Pick type A/B disease. Eur. J. Neurosci. 13, 1873–1880.

Sathasivam, S., Ince, P.G., Shaw, P.J., 2001. Apoptosis in amyotrophic lateral sclerosis: a review of the evidence. Neuropathol. Appl. Neurobiol. 27, 257–274.

Sawamura, N., Morishima-Kawashima, M., Waki, H., Kobayashi, K., Kuramochi, T., Frosch, M.P., Ding, K., Ito, M., Kim, T.W., Tanzi, R.E., Oyama, F., Tabira, T., Ando, S., Ihara, Y., 2000. Mutant presenilin 2 transgenic mice. A large increase in the levels of Abeta 42 is presumably associated with the low density membrane domain that contains decreased levels of glycerophospholipids and sphingomyelin. J. Biol. Chem. 275, 27901–27908.

Shinitzky, M., 1987. Patterns of lipid changes in membranes of the aged brain. Gerontology 33, 149–154.

Smyth, M.J., Perry, D.K., Zhang, J., Poirier, G.G., Hannun, Y.A., Obeid, L.M., 1996. prICE: a downstream target for ceramide induced apoptosis and for the inhibitory action of Bcl-2. Biochem. J. 316, 25–28.

Spiegel, S., Milstien, S., 2000. Sphingosine-1-phosphate: signaling inside and out. FEBS Lett. 476, 55–57.

Suzuki, K., Parker, C.C., Pentchev, P.G., Katz, D., Ghetti, B., D'Agostino, A.N., Carstea, E.D., 1995. Neurofibrillary tangles in Niemann-Pick disease type C. Acta Neuropathol. 89, 227–238.

Tacconi, M.T., Lligona, L., Salmona, M., Pitsikas, N., Algeri, S., 1991. Aging and food restriction: effect on lipids of cerebral cortex. Neurobiol. Aging. 12, 55–59.

Tegos, T.J., Kalodiki, E., Daskalopoulou, S.S., Nicolaides, A.N., 2000. Stroke: epidemiology, clinical picture, and risk factors–Part I of III. Angiology 51, 793–808.

Tomiuk, S., Zumbansen, M., Stoffel, W., 2000. Characterization and subcellular localization of murine and human magnesium-dependent neutral sphingomyelinase. J. Biol. Chem. 275, 5710–5717.

Veldman, R.J., Maestre, N., Aduib, O.M., Medin, J.A., Salvayre, R., Levade, T., 2001. A neutral sphingomyelinase resides in sphingolipid-enriched microdomains and is inhibited by the caveolin-scaffolding domain: potential implications in tumour necrosis factor signalling. Biochem. J. 355, 859–868.

Wang, Y.M., Seibenhener, M.L., Vandenplas, M.L., Wooten, M.W., 1999. Atypical PKC zeta is activated by ceramide, resulting in coactivation of NF-kB/JNK kinase and cell survival. J. Neurosci. Res. 55, 293–302.

Wiegmann, K., Schutze, S., Machleidt, T., Witte, D., Kronke, M., 1994. Functional dichotomy of neutral and acidic sphingomyelinases in tumor necrosis factor signaling. Cell 78, 1005–1015.

Willaime, S., Vanhoutte, P., Caboche, J., Lemaigre-Dubreuil, Y., Mariani, J., Brugg, B., 2001. Ceramide-induced apoptosis in cortical neurons is mediated by an increase in p38 phosphorylation and not by the decrease in ERK phosphorylation. Eur. J. Neurosci. 13, 2037–2046.

Wolff, R.A., Dobrowsky, R.T., Bielawska, A., Obeid, L.M., Hannun, Y.A., 1994. Role of ceramide-activated protein phosphatase in ceramide-mediated signal transduction. J. Biol. Chem. 269, 19605–19609.

Yamashita, T., Wada, R., Sasaki, T., Deng, C., Bierfreund, U., Sandhoff, K., Proia, R.L., 1999. A vital role for glycosphingolipid synthesis during development and differentiation. Proc. Natl. Acad. Sci. USA 96, 9142–9147.

Yang, S.N., 2000. Ceramide-induced sustained depression of synaptic currents mediated by ionotropic glutamate receptors in the hippocampus: an essential role of postsynaptic protein phosphatases. Neuroscience 96, 253–258.

Yoshimura, S., Banno, Y., Nakashima, S., Hayashi, K., Yamakawa, H., Sawada, M., Sakai, N., Nozawa, Y., 1999. Inhibition of neutral sphingomyelinase activation and ceramide formation by glutathione in hypoxic PC12 cell death. J. Neurochem. 73, 675–683.

Yu, Z.F., Mattson, M.P., 1999. Dietary restriction and 2-deoxyglucose administration reduce focal ischemic brain damage and improve behavioral outcome: evidence for a preconditioning mechanism. J. Neurosci. Res. 57, 830–839.

Yu, Z., Nikolova-Karakashian, M., Zhou, D., Cheng, G., Schuchman, E.H., Mattson, M.P., 2000. Pivotal role for acidic sphingomyelinase in cerebral ischemia-induced ceramide and cytokine production, and neuronal apoptosis. J. Mol. Neurosci. 15, 85–98.

Zhang, Y., Yan, B., Delikat, S., Bayoumy, S., Lin, X.H., Basu, S., McGinley, M., Chn-Hui, P.Y., ichenstein, H., Kolesnick, R., 1997. Kinase suppressor of Ras is ceramide-activated protein kinase. Cell 89, 63–72.

Zhang, Y., Dawson, V.L., Dawson, T.M., 2000. Oxidative stress and genetics in the pathogenesis of Parkinson's disease. Neurobiol. Dis. 7, 240–250.

Zhou, H., Summers, S.A., Birnbaum, M.J., Pittman, R.N., 1998. Inhibition of Akt kinase by cell-permeable ceramide and its implications for ceramide induced apoptosis. J. Biol. Chem. 273, 16568–16575.

Zhu, H., Guo, Q., Mattson, M.P., 1999. Dietary restriction protects hippocampal neurons against the death-promoting action of a presenilin-1 mutation. Brain Res. 842, 224–229.

**Advances in
Cell Aging and
Gerontology**

The eicosanoid pathway and
brain aging

Hari Manev and Tolga Uz

*The Psychiatric Institute, Department of Psychiatry,
University of Illinois at Chicago, 1601 West Taylor Street, Chicago, IL 60612, USA
Correspondence address: The Psychiatric Institute, University of Illinois at Chicago,
1601 West Taylor Street, M/C 912, Room 238,
Chicago, Illinois 60612, USA. Tel: +312-413-4558; fax: +312-413-4569.
E-mail address: HManev@psych.uic.edu*

Contents

Abbreviations

AA: arachidonic acid; CNS: central nervous system; COX: cyclooxygenase; $cPLA_2$: cytosolic phospholipase A_2; FLAP: five-lipoxygenase-activating protein; HPETE: hydroperoxyeicosatetraenoic acid; LOX: lipoxygenase; MS: multiple sclerosis; NMDA: N-methyl-D-aspartate; NSAIDs: nonsteroidal anti-inflammatory

Advances in Cell Aging and Gerontology, vol. 12, 117–136
© 2003 Elsevier Science B.V. All Rights Reserved.

drugs; PGHS: prostaglandin endoperoxide H synthase; PPAR: peroxisome proliferator-activated receptor; PUFA: polyunsaturated fatty acids.

1. Introduction

Originally, the term eicosanoid (Greek *eicosa* = twenty) was coined to describe all biologically active oxygenated metabolites of three 20 carbon polyunsaturated fatty acids (PUFA); homo-γ-linolenic acid (20:3n−6), arachidonic acid (20:4n−6), and eicosapentaenoic acid (20:5n−3) (Corey et al., 1980). Typically, the term eicosanoid now refers to biologically active metabolites of arachidonic acid (AA), the most abundant PUFA in the membranes of human cells (Fig. 1). The mammalian eicosanoid pathway encompasses two major enzyme families; the lipoxygenases (LOX) that lead to synthesis of leukotrienes, and the cyclo-oxygenases (COX) that lead to synthesis of prostaglandins and thromboxanes (collectively termed prostanoids) (Funk, 2001). Although insects apparently also possess a functional eicosanoid pathway (Chyb et al., 1999), insect eicosanoids have for the most part been poorly characterized. For example, LOX and COX genes have not been cloned in *Drosophila* (fruit fly), a typical animal model for genomic studies. Generally, it has been believed that AA is not found in terrestrial insects such as flies. Nevertheless, recent data indicate that fly tissue does contain substantial amounts of this PUFA (Aliza et al., 2001).

The LOX family includes proteins encoded by multiple genes. For example, five genes have been identified that express LOX proteins in humans and seven homologs are characterized in mice. These enzymes oxygenate AA in a specific manner (in the first step they add a hydroperoxide group to a select carbon atom in the AA molecule); e.g., 5-LOX oxygenates AA at carbon-5 and produces 5-hydroperoxyeicosatetraenoic acid (5-HPETE) (Brash, 1999). Among various LOX proteins, 5-LOX has been extensively studied and several of its unique characteristics have been pointed out. Its activity appears to be susceptible to modification by a complex protein–protein interaction that requires the presence of an endogenous protein activator. This activator was termed the five-lipo-xygenase-activating protein (FLAP) (Coffey et al., 1994; Steinhilber, 1999). In addition to its well-described enzymatic activity, 5-LOX may have an additional and less understood nonenzymatic function (Lepley and Fitzpatrick, 1994; Provost et al., 1999; Manev et al., 2000b). The expression and the activity of LOX are tissue specific; prominent CNS expression has been demonstrated for 5-LOX (Lammers et al., 1996). In brain cells, the presence of 8-LOX (Jisaka et al., 1997) and 12-LOX (Watanabe et al., 1993; Li et al., 1997) mRNAs have also been reported.

The family of COX enzymes (also termed prostaglandin endoperoxide H synthase; PGHS) encompasses two isoforms referred to as COX-1 and COX-2 (Smith et al., 2000). COX-1 and COX-2 have been extensively studied, in part because they are the targets of widely used nonsteroidal anti-inflammatory drugs (NSAIDs). Selective COX-2 inhibitors are being developed with reduced number of side effects (Wallace, 1999). In the CNS, COX-1 appears to be constitutively expressed whereas COX-2 is inducible and its expression in the brain is highly regulated (O'Banion, 1999; Hoffmann, 2000).

During the process of aging, the brain becomes more susceptible to neuro-degeneration. For example, recovery from a stroke is substantially poorer in very old patients than in younger subjects (Nakayama et al., 1994; Arboix et al., 2000). Similar effects of aging were observed in experimental animals. Thus, the severity of experimental stroke-induced brain injury was greater in 20–24-month-old rats than in 4-month-old rats (Kharlamov et al., 2000) and ischemic brain injury was greater in older than in younger mice (Nagayama et al., 1999). Aging is also

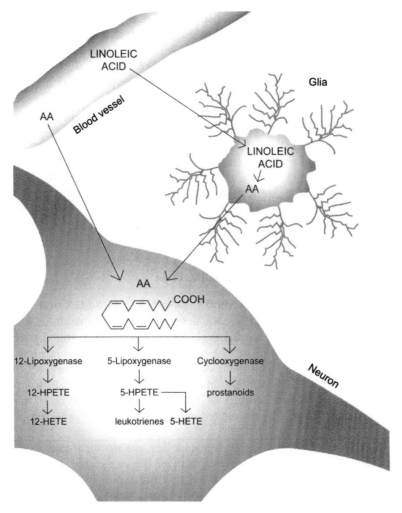

Fig. 1 The CNS eicosanoid pathway encompasses two major enzyme families; the lipoxygenases and the cyclooxygenases. Neurons do not synthesize AA; it is provided to them either by glia (e.g., astrocytes), endothelial cells, or from the periphery (Moore, 2001). Both AA and the essential PUFA, linoleic acid, are capable of crossing the blood-brain barrier (Edmond et al., 1998).

associated with the occurrence and progression of chronic neurodegenerative disorders, such as Alzheimer s disease. Dietary *caloric* restriction with nutritional maintenance can extend the lifespan and may increase the resistance of the nervous system to aging-associated neurodegenerative pathologies (Mattson et al., 2001; Prolla and Mattson, 2001). In contrast, selective dietary *fat* restrictions may retard growth and is generally deleterious to brain development and CNS function (Burr and Burr, 1973; Wainwright, 2001), because essential fatty acids, for example linoleic and alpha-linolenic acids, can be supplied only from dietary sources (Edmond et al., 1998). These fatty acids are precursors for the synthesis of longer-chained PUFA, i.e., AA, which is the main target of enzymes in the eicosanoid pathway. In contrast to most types of mammalian cells (including glial astrocytes), neurons are incapable of elongating or desaturating the precursors of AA; thus, neurons depend on the AA provided to them by cells capable of synthesizing AA from the essential linolenic acid (Moore et al., 1991) (Fig. 1). It has been speculated that the composition of PUFA in brain cells may be altered during aging (Youdim et al., 2000). This might occur in human brain more readily than in experimental animals because animal (e.g., rat) tissues appear to be more active in synthesizing AA than human tissues (Bezard et al., 1994).

Recent data suggest that the eicosanoid pathway may be significantly altered in aging CNS. It has been proposed that aging-associated upregulation of the 5-LOX pathway (Uz et al., 1998; Qu et al., 2000) and COX pathway (Montine et al., 2002) may contribute to aging-related increases in brain vulnerability. Hence, it is conceivable that compounds capable of selectively affecting 5-LOX and COX could become putative neuroprotective therapies for the prevention and treatment of pathologies that affect aging brain.

2. Lipoxygenases and cyclooxygenases in the brain

The expression in the brain of several LOX genes (e.g., 5-LOX and 12-LOX) as well as that of COX has generally been demonstrated using methods for specific mRNA detection. Moreover, biochemical studies have demonstrated that various leukotrienes and prostaglandins can be produced in the mammalian brain, as well as in neurons (Lindgren et al., 1984; Adesuyi et al., 1985; Shimizu et al., 1987; Bishai and Coceani, 1992; Kaufmann et al., 1997). Very recently, it was proposed that in sensory neurons, the product of 12-LOX activity, 12-HPETE, may act as an endogenous regulator of pain-related neuronal vanilloid (capsaicin) receptors (Piomelli, 2001).

In this chapter, we will focus on evidence for the *neuronal* presence of these enzymes, and we will focus on two members of the eicosanoid pathway, 5-LOX and COX-2. However, it has to be stressed that the expression of LOX and COX enzymes in non-neuronal cells may also be important for CNS functioning, particularly in certain pathologies. For example, in conditions associated with the disruption of the blood-brain barrier (e.g., in brain ischemia), cells such as the neutrophils that express the eicosanoid pathway may invade the brain and release eicosanoids (Barone et al., 1992). A similar situation may occur in Alzheimer's disease; circulating monocytes/macrophages, when recruited by chemokines produced by activated microglia and macrophages, could add to the inflammatory destruction of the brain (Fiala et al., 1998). Eicosanoids can also be produced in the walls of blood vessels by resident leukocytes and thereby trigger vasoactive responses (Marceau, 1996). In Alzheimer s disease, β-amyloid peptide could induce vasoconstriction and decrease cerebral blood flow by stimulating 5-LOX and COX-2 pathways (Paris et al., 2000).

2.1. Presence of 5-LOX in the CNS

In the adult mammalian brain, 5-LOX mRNA is expressed in a subpopulation of neurons. *In situ* hybridization and immunohistochemistry demonstrated that 5-LOX and FLAP are expressed in various regions of the rat brain, including the hippocampus, cerebellum, primary olfactory cortex, superficial neocortex, thalamus, hypothalamus, and brainstem (Lammers et al., 1996). The highest levels of expression were observed in the cerebellum and hippocampus; in the latter brain region colocalization was shown for 5-LOX and FLAP in CA1 pyramidal neurons. Moreover, electrophysiological experiments showed that the selective 5-LOX inhibitor MK-886 prevents somatostatin-induced augmentation of the hippocampal K + M-current, suggesting the participation of the 5-LOX pathway in somatostatin-receptor transmembrane signaling (Lammers et al., 1996).

Neuronal localization of 5-LOX protein was also demonstrated immunocytochemically by Ohtsuki et al. (1995), who examined the effects of 5 min of ischemia on brain 5-LOX during reperfusion in a gerbil model of transient forebrain ischemia. In this study, hippocampal neurons exhibited dense 5-LOX immunoreactivity, which was partially redistributed from cytosolic to particulate fractions 3 min during reperfusion. These authors also demonstrated that leukotrienes were generated in neurons and that their content was increased in all forebrain regions during reperfusion. Differences were also observed in the distribution of 5-LOX-like immunoreactivity in various brain areas of adult young (2-month-old) vs. old (24-month-old) male rats (Uz et al., 1998). Greater 5-LOX-like immunoreactivity was found in old vs. young rats, in particular, in the dendrites of pyramidal neurons in limbic structures, including the hippocampus, and in layer V pyramidal cells of the frontoparietal cortex and their apical dendrites.

Mammalian cerebellar granule neurons can be grown *in vitro*; these cultures express 5-LOX and FLAP (Manev and Uz, 1999). In this *in vitro* system, neurons can be studied both in their immature state (i.e., as neuronal precursors) and as mature, differentiated neurons. Using cultures of immature neurons, it was demonstrated that 5-LOX may participate in neurogenesis (Uz et al., 2001a).

The expression of 5-LOX in the brain can be regulated by the rate of neuronal activity. Hence, it was shown that stimulation of receptors for excitatory neurotransmitter glutamate increases 5-LOX expression in the CNS (Manev et al., 1998; 2000a) and that the activation of glutamate receptors also increases the production of leukotrienes in the rat brain (Simmet and Tippler, 1990).

2.1.1. Functional significance of CNS 5-LOX

The 5-LOX metabolites of AA, leukotrienes, may influence CNS functioning on two levels: (a) as intracellular second messengers; and (b) as transcellular mediators (Piomelli, 1994; Lammers et al., 1996; Christie et al., 1999). Nevertheless, our knowledge regarding possible physiological roles of the enzymatic 5-LOX pathway in the brain is limited. Generally, selective pharmacological tools (i.e., agonists/antagonists, inhibitors) are needed for functional characterization of CNS biochemical pathways. Such tools are inadequate for in-vivo investigation of the brain 5-LOX pathway, hence, little is known about the bioavailability of 5-LOX-related drugs and their ability to permeate the blood-brain barrier. When used *in vitro*, LOX inhibitors helped in characterizing, for example, a role for 12-LOX and to a lesser extent for 5-LOX in homosynaptic long-term depression of the rat hippocampus (Normandin et al., 1996).

The basic biology of eicosanoid/5-LOX pathways has been studied in genetically modified mice. These mice studies include cytosolic phospholipase A_2 ($cPLA_2$) gene knockouts that are deficient in leukotriene and prostaglandin synthesis (Sapirstein and Bonventre, 2000) and 5-LOX-deficient transgenic mice that do not synthesize leukotrienes (Chen et al., 1994; Funk and Chen, 2000). Only recently have these experimental animals, i.e., 5-LOX-deficient mice [5-LOX (−); B6129S^{Alox5tm1Fun}] and their controls (B6129SF2/J) been used for studies of CNS functioning (Uz et al., 2002), using the following behavioral tests: elevated plus-maze, marble burying, locomotor activity, rota-road, and the spontaneous alternations in the T-maze. It was observed that in an elevated plus-maze, 5-LOX-deficient mice spent a shorter time in the safe closed arms, a longer time in the anxiogenic open arms, and entered the open arms more frequently. They also covered fewer marbles in the marble-burying anxiety test. No difference was observed between 5-LOX (−) and 5-LOX (+) mice in the other tests (Uz et al., 2002). Thus, it appears that 5-LOX-deficient mice are less prone to anxiety, suggesting a possible role for 5-LOX in affective behaviors. However, it must be stressed that creating congenic 5-LOX-deficient mice by backcrossing into inbred strains would provide additional and better tools to further elucidate the putative role of 5-LOX in CNS functioning (Uz et al., 2002). New strains of 5-LOX-deficient mice would also greatly advance studies of the functional role of 5-LOX in brain aging (Manev et al., 2000b).

In non-neuronal systems, it has been consistently established that inhibitors of 5-LOX reduce cell proliferation. For example, the 5-LOX inhibitors MK-886 and AA-861 reduced proliferation and induced apoptotic cell death in cultures of prostate cancer cells (Ghosh and Myers, 1997; Anderson et al., 1998). In the brain, 5-LOX-regulated cell proliferation might be involved in the growth of brain tumors, which have been shown to secrete leukotrienes. In human brain tumors, 5-LOX is expressed as a multitranscript family encompassing 2.7, 3.1, 4.8, 6.4, and 8.6 kb; the abundance of 5-LOX mRNAs and the expression of the larger transcripts was found to correlate with tumor malignancy (Boado et al., 1992). *In vitro* studies support the notion that 5-LOX might be critical for growth of brain tumors. Thus, 5-LOX inhibitors AA-863 and U-60,257 induced a dose-dependent inhibitory effect on the proliferation of human glioma cells in culture (Gati et al., 1990). Recent data show that LOX pathways (possibly 5-LOX) are also involved in the cell cycle progression of neuroblastoma cells (van Rossum et al., 2002).

A nonpathologic action for 5-LOX in cell proliferation has also been suggested. Primary cultures of rat cerebellar granule neurons express 5-LOX; this expression is higher when neurons are immature and still proliferating (i.e., neuronal precursors) and it decreases in differentiated post-mitotic neurons (Uz et al., 2001a). Treatment of immature neurons with a 5-LOX antisense reduced the expression of 5-LOX proteins and effectively reduced their proliferation. Moreover, [^{3}H]thymidine incorporation (a marker for the cell cycle) was significantly reduced by 5-LOX inhibitors (AA-861, MK-886, and L-655,238); importantly, the anti-proliferative effect of these 5-LOX inhibitors was reversible (Uz et al., 2001a). As such, it appears that the 5-LOX pathway might participate in neurogenesis, including the recently discovered adult neurogenesis (Manev et al., 2001).

2.1.2. 5-LOX and brain pathology

A number of reports indicate the involvement of the 5-LOX pathway in various CNS pathologies. These include developmental diseases such as Sjogren-Larsson syndrome (Willemsen et al., 2001) and neurometabolic diseases characterized by the absence of leukotriene LTC_4 synthesis (Mayatepek, 2000). The 5-LOX pathway has also been associated with chronic neurodegenerative diseases such as multiple sclerosis (MS) (Whitney et al., 2001), which is a debilitating disorder of the CNS characterized by decreased nerve functioning and lesions in the white matter that degenerate the myelin sheath. Recent microarray analysis of gene expression in brain tissue from MS patients and from patients without MS revealed that in the MS brain, 5-LOX was one of the predominantly upregulated genes. Thus, it was suggested that 5-LOX might be a possible treatment target for this devastating neurological disorder (Whitney et al., 2001). Another progressive and fatal neurodegenerative disease in which 5-LOX plays a role is prion disease (transmissible spongiform encephalopathies) (Stewart et al., 2001). Thus, it was demonstrated that the 5-LOX pathway is involved in the neurotoxicity of the prion peptide PrP106-126 and that 5-LOX inhibitors prevent PrP106-126 neurotoxicity.

A direct link of the 5-LOX pathway with stroke was reported by Ohtsuki et al. (1995), who found that neuronal 5-LOX is mobilized and activated during reperfusion of the ischemic brain (e.g., in an experimental model of stroke) and that this activation is accompanied by an increased production of leukotrienes. These findings have recently been confirmed and extended to note that the stroke-triggered increased formation of leukotrienes (LTC_4, LTD_4, and LTE_4) involves an activation of ionotropic glutamate receptors (Ciceri et al., 2001). *In vitro* (Arai et al., 2001) and *in vivo* (Shishido et al., 2001) findings of neuroprotective action of LOX inhibitors against ischemic CNS injury are consistent with the involvement of leukotrienes in the pathology of stroke.

2.2. Presence of COX-2 in the CNS

Since the expression of COX-2 is primarily regulated by mitogenic and growth factors, this enzyme has also been termed a mitogen-inducible COX. When initially cloned from the rat brain, the expression of COX-2 was found to be localized throughout the forebrain in discrete populations of neurons and also found to be enriched in the cortex and hippocampus (Yamagata et al., 1993). Like neuronal 5-LOX expression, the expression of neuronal COX-2 is rapidly and transiently induced by seizures or synaptic activity mediated via glutamate receptors. No expression was detected in glia or vascular endothelial cells. Thus, it was proposed that COX-2 expression could be important in regulating prostaglandin signaling in the brain and that its marked inducibility in neurons by synaptic stimuli indicates a role for the COX-2 pathway in activity-dependent plasticity (Yamagata et al., 1993).

In the human brain, COX-2 immunostaining was found in the cortex and in the hippocampus, including pyramidal neurons (Yasojima et al., 1999; Ho et al., 2001; Yermakova and O'Banion, 2001). Both an increase (Yasojima et al., 1999; Ho et al., 2001) and a decrease (Yermakova and O'Banion, 2001) of neuronal COX-2 content was found in the brains of Alzheimer s patients. Both neuronal and non-neuronal localization of COX-2 was found in the human brain following cerebral ischemia (Iadecola et al., 1999).

In the spinal cord (e.g., in rats), no COX-2 mRNA signals were detected under normal conditions, but a strong expression of these signals was seen bilaterally in non-neuronal cells both within the grey and white matter and along the leptomeninges and blood vessels 6 h after unilateral carrageenan injection into the hind paw, but not after peripheral nerve injury (Ichitani et al., 1997). These results suggest that COX-2 expressed in non-neuronal cells contributes to prostaglandin production in and around the spinal cord under peripheral inflammatory processes.

2.2.1. Functional significance of CNS COX-2

Prostaglandins, the products of COX-2 activity, could affect neuronal function by modulating neurotransmitter release; this has been demonstrated for the sensory neurons (Nicol et al., 1992). In the CNS, the COX-2 pathway has been associated with glutamate receptor-mediated neurotransmission. COX-2 behaves as an immediate-early gene; its neuronal expression is dramatically and transiently induced in response to NMDA (N-methyl-D-aspartate) glutamate receptor activation. Moreover, in models of acute excitotoxic neuronal injury, elevated and sustained levels of COX-2 appear to promote neuronal apoptosis, suggesting that upregulated COX-2 activity is injurious to neurons (Iadecola et al., 2001).

Constitutively elevated neuronal COX-2 was recently achieved in transgenic mice that overexpressed COX-2 in neurons and produced elevated levels of prostaglandins in the brain (Andreasson et al., 2001). Behavioral studies using COX-2 transgenic mice showed that these animals developed an age-dependent deficit in spatial memory at 12 and 20 months but not at 7 months, and a deficit in aversive behavior at 20 months of age. These behavioral changes were associated with a parallel age-dependent increase in neuronal apoptosis occurring at 14 and 22 months but not at 8 months (Andreasson et al., 2001). These findings suggest that neuronal COX-2 significantly influences CNS function and that, under pathologic conditions, the COX-2 pathway could contribute to the mechanisms of age-related brain disorders by promoting memory dysfunction and neuronal death in an age-dependent manner.

Similar to the findings indicating a role for 5-LOX in neurogenesis (Manev et al., 2001; Uz et al., 2001a), it has also been suggested that COX-2 could be required for adult neurogenesis in the mammalian CNS (Kumihashi et al., 2001). Transient global ischemia causes neurogenesis in the dentate gyrus of adult rodents and also induces COX-2. In experiments with gerbils, adult animals were chronically treated with acetylsalicylic acid, a nonselective COX inhibitor, and the proliferation of cells in response to ischemia was studied in the dentate gyrus. Acetylsalicylic acid significantly reduced the number of proliferating cells, suggesting that COX, probably COX-2, and prostaglandins play an important role in adult neurogenesis, e.g., the proliferation of neural cells after ischemia (Kumihashi et al., 2001).

2.2.2. COX-2 and brain pathology

Recently, several excellent reviews have summarized a possible role for the COX-2 pathway in CNS pathologies (Kaufmann et al., 1997; O'Banion, 1999; Hoffmann, 2000). Thus, COX-2 has been associated with the pathobiologic mechanisms of brain ischemia, pain and hyperalgesia, fever, seizures, and Alzheimer's disease.

A possible link between COX-2 and Alzheimer's disease has attracted considerable attention. This interest was initiated by epidemiological observations and subsequent epidemiological studies that have demonstrated a reduced prevalence of Alzheimer's disease among users of NSAIDs (McGeer et al., 1996; Stewart et al., 1997). An early small clinical trial indicated that the NSAID indomethacin protected mild-to-moderately impaired Alzheimer's disease patients from the degree of cognitive decline exhibited by a well-matched, placebo-treated group (Rogers et al., 1993). Although the beneficial effect of NSAIDs in Alzheimer s patients has been attributed to the inhibitory action of these drugs on COX and is often used as an argument for the presence of an inflammatory component in Alzheimer's pathology, recent findings suggest that a subset of NSAIDs might influence Alzheimer's pathology via a mechanism unrelated to COX inhibition (Weggen et al., 2001). Thus, Weggen et al. (2001) found that $A\beta42$ peptide production (typically associated with Alzheimer's disease) is reduced by NSAIDs, indomethacin, ibuprofen, and sulindac sulphide in a variety of cultured cells and that this effect was not mediated by COX inhibition. These authors also found that short-term administration of ibuprofen to mice that produce mutant β-amyloid precursor protein lowered their brain levels of $A\beta42$. These new results indicate that an alternative mechanism unrelated to COX-2 might be operative in the beneficial actions of NSAIDs in Alzheimer's disease;

these findings also point to possible shortcomings when currently available pharma-cological tools are used to characterize the CNS function of the COX-2 pathway.

3. Regulation of LOX and COX expression

3.1. Hormones and 5-LOX

The cell type-specific expression of 5-LOX appears to be determined by DNA methylation of the 5-LOX promoter (see below). Natural mutations occur in the human 5-LOX promoter (In et al., 1997); they consist of the deletion of one or two, or the addition of one Sp1-binding site and these mutations have been implicated in the response of asthma patients to 5-LOX inhibitors (Drazen et al., 1999). Moreover, the sequence of the 5-LOX promoter indicates that this gene could be susceptible to hormonal regulation in tissues in which 5-LOX is expressed (Hoshiko et al., 1990). Studies with human monocytes (Riddick et al., 1997) and human mast cells (Colamorea et al., 1999) have pointed to an apparent contradiction; the typical "anti-inflammatory" hormones, i.e., glucocorticoids, stimulate the expression of a "pro-inflammatory" enzyme, 5-LOX. In the CNS, glucocorticoids stimulate 5-LOX expression both *in vivo* (Uz et al., 1999) and *in vitro* (Uz et al., 2001b), which appears to involve the glucocorticoid receptor (Uz et al., 2001b). It has been proposed that the glucocorticoid-triggered upregulation of neuronal 5-LOX expression may contribute to the well-known neurotoxic action of these hormones (Uz et al., 1999).

Another hormone-mediated mechanism that affects 5-LOX expression includes an orphan nuclear receptor termed ROR/RZR; it was shown that the pineal hormone melatonin suppresses 5-LOX expression via this receptor (Steinhilber et al., 1995). Although further research is needed to confirm these findings, *in vivo* studies have shown that melatonin deficiency increases 5-LOX expression in the brain (Uz et al., 1997). Moreover, the pineal expression of 5-LOX mRNA conforms to a circadian rhythm in a manner opposite to the rhythm of melatonin synthesis (Uz and Manev, 1998). Since the function of the pineal gland declines during aging, it remains to be elucidated whether this aging-associated melatonin deficiency influences the level of 5-LOX expression.

3.2. Hormones and COX-2

Initial studies on the regulation of neuronal COX-2 expression indicated that in contrast to 5-LOX the expression of this gene is suppressed by glucocorticoids (Yamagata et al., 1993). However, more detailed studies revealed that the inhibitory action of glucocorticoids on neuronal COX-2 expression is not universal. Thus, whereas the glucocorticoid dexamethasone suppressed induced COX-2 expression in the cortex it was ineffective in altering COX-2 expression in the hippocampus (Koistinaho et al., 1999). Hence, it appears that glucocorticoids are not effective agents to inhibit the pathological activation of COX-2 in the hippocampus. In addition, considering their stimulatory effect on the expression of hippocampal 5-LOX (Uz et al., 1999), the use of these hormones for treatment of neurodegenerative pathologies should be reevaluated.

3.3. DNA methylation and regulation of gene expression

Recent studies point to an important role for DNA methylation in the epigenetic regulation of gene expression in the CNS (Tucker, 2001). The DNA of animal cells is subject to covalent modification by methylation at the 5 carbon position of cytosine residues at CpG dinucleotides. DNA methylation is catalyzed by DNA methyltransferase (Dnmt 1) and is crucial for modulating gene expression, and consequently, neuronal functioning DNA methylation of promoter-containing CpG islands is associated with transcriptional inactivation, i.e., gene silencing. Typically, the same gene is not methylated in the tissue where it is expressed and it is methylated in tissues or cells where it is not expressed. Thus, DNA methylation may be responsible for neuron-specific gene expression. Alterations in DNA methylation can be induced pharmacologically, which has generally been confined to cancer treatment (Szyf, 2001).

Under certain experimental conditions, a selective hypomethylation in post-mitotic neurons may cause significant functional impairment and cell death (Fan et al., 2001). There is evidence for global and gene-specific hypomethylation in aging cells and tissues (Mays-Hoopes, 1989). On the other hand, hypermethylation of certain gene promoters has also been found to occur during aging and has been linked to neoplasia (Issa et al., 1994). The causes of these changes, which have been observed during aging, both hypo- and hypermethylation, are not clear and their relevance for aging-associated CNS alterations is yet to be evaluated. However, it has been speculated that the incidence of Alzheimer's disease might be linked to a genetic imprinting mechanism involving DNA methylation (Farrer et al., 1991).

3.3.1. DNA methylation and 5-LOX

5-LOX transcription requires the transcription factors Egr-1 and Sp1, which functionally interact with the human 5-LOX promoter (Hoshiko et al., 1990). In addition, the 5-LOX promoter is characterized by the presence of repeated G+C-rich elements and the lack of TATAA or CCAAT boxes. These characteristics point to similarities between the 5-LOX promoter and so-called housekeeping genes. However, whereas the promoters of G+C-rich housekeeping genes are usually unmethylated, the methylation of the G+C-rich 5-LOX promoter is cell-specific and is associated with the suppression of 5-LOX mRNA expression (Uhl et al., 2001).

Uhl et al. (2001) analyzed the methylation status of the 5-LOX promoter in human myeloid cell lines and found that the 5-LOX promoter is methylated in 5-LOX-negative cell lines and unmethylated in 5-LOX-positive cells. Moreover, treatment of 5-LOX-negative cells with the demethylating agent 5-aza-2′-deoxycytidine triggered the expression of 5-LOX primary transcripts and mature mRNA. Demethylating agents are also effective in stimulating 5-LOX expression in neural cell cultures (Manev and Uz, unpublished observation). These results suggest that epigentic regulation of the 5-LOX gene via DNA methylation of its promoter may be responsible for the previously observed cell differentiation- and tissue-specific expression of 5-LOX (Steinhilber, 1999).

3.3.2. DNA methylation and COX-2

COX-2 expression also appears to be regulated by DNA methylation, particularly in cancerous cells (Song et al., 2001). Song et al. (2001) characterized the methylation of the COX-2 CpG island (spanning from −590 to +186 with respect to the transcription initiation site) in human gastric carcinoma cell lines and found that three gastric cell types (SNU-601, -620, and -719) without COX-2 expression demonstrated hypermethylation at the COX-2 CpG island. Treatment with demethylating agents effectively reactivated the expression of COX-2. In addition, COX-2 promoter activity was completely blocked by *in vitro* methylation of all CpG sites in the COX-2 promoter region. These results indicate that transcriptional repression of COX-2 is caused by hypermethylation of the COX-2 CpG island in gastric carcinoma cell lines (Song et al., 2001). The role of DNA methylation in neuronal COX-2 expression has not yet been characterized.

4. 5-LOX and COX-2 in aging brain

Although aging-associated neuropathology often goes hand in hand with pathological alterations of cerebral vasculature, and the eicosanoid pathway is important for the functioning of the vasculatory system and for inflammatory processes, neuronal expression of 5-LOX and COX-2 appears to be operative in neurodegeneration as well. Ultimately, neurodegenerative diseases involve neuronal death; oxidative stress induces age-dependent neuronal apoptosis, which is one of the mechanisms that produce age-dependent vulnerability to neurotoxins (Kim and Chan, 2001). Oxidative stress also favors lipid peroxidation in biomembranes, which can proceed enzymatically, e.g., via 5-LOX and COX-2 pathways. The capacity of CNS neurons to compromise their own viability through endogenously generated toxins (e.g., glutamate excitotoxicity) has been well recognized (Schwarcz et al., 1984; Manev et al., 1990). A similar noninflammatory pathophysiological mechanism might be based on neuronal 5-LOX and COX-2, and could mediate aging-associated increased brain vulnerability.

4.1. Aging-associated alterations in the CNS eicosanoid pathway

Microarray studies of gene expression during *in vitro* replicative senescence revealed a high degree of cell-specific patterns of expression; in some cells, senescence was accompanied by an upregulation of inflammation-associated genes (Shelton et al., 1999).

Both 5-LOX and COX-2 pathways are upregulated in the brain of aging mammals and it has been proposed that these increases might be associated with aging-related brain pathologies. The rationale for this hypothesis can be summarized by the following evidence: (a) the aging CNS is increasingly vulnerable to neurodegeneration; (b) 5-LOX and COX are upregulated in the aging CNS; and (c) the inhibition of 5-LOX and/or COX-2 protects CNS neurons from degeneration.

5-LOX activity of splenocytes was found to be increased in 24-month-old vs. 4-month-old mice (Hayek et al., 1994) and 5-LOX expression was increased during aging in peripheral blood vessels (Ibe et al., 1997). In the brain, 5-LOX expression was found to be greater in old than in young rats (Uz et al., 1998; Qu et al., 2000). In rats, kainate (a glutamate receptor agonist) increased hippocampal 5-LOX expression (Manev et al., 1998) and activity (Simmet and Tippler, 1990); and it was found that the use of a 5-LOX inhibitor was neuroprotective. Moreover, hippocampal excitotoxic injury induced by kainate was greater in old rats (Uz et al., 1998). In earlier studies it was shown that two LOX inhibitors nordihydroquaiaretic acid and AA861, can protect cultured hippocampal neurons against death induced by amyloid beta-peptide (Goodman et al, 1994). Considering the prominent effect of aging on DNA methylation, it is possible that the upregulation of 5-LOX expression in the aging brain is in part mediated via aging-associated alterations in the methylation state of the 5-LOX promoter.

COX activity was also found to be greater in the brain of 24-month-old rats than 6-month-old rats (Baek et al., 2001). Since, unlike the aging-increased brain content of 5-LOX, COX-2 mRNA and protein levels showed little corresponding age-related changes, it has been suggested that the increase in COX activity with age is due to activation of COX catalytic reactions, presumably by reactive oxygen species (Baek et al., 2001). Neuroprotection was demonstrated for a number of COX inhibitors in several models of neurodegeneration both *in vitro* and *in vivo* (O'Banion, 1999; Hoffmann, 2000) and by experiments in transgenic mice. Namely, COX-2 deficient ("knockout") mice exhibited a reduced severity of brain injury in a model of stroke and in response to glutamate receptor-mediated excitotoxicity (Iadecola et al., 2001). On the other hand, transgenic mice over-expressing COX-2 in neurons developed age-dependent cognitive deficits and neuronal apoptosis, suggesting that neuronal COX-2 may contribute to the pathophysiology of age-related diseases such as Alzheimer's (Andreasson et al., 2001).

4.2. Eicosanoid pathway as a putative CNS therapeutic target

The evidence for the presence and activity of 5-LOX and COX-2 in the brain including neurons, and the evidence of aging-associated alterations of the brain eicosanoid pathway that we review here indicate that the CNS eicosanoid pathway is amenable to pharmacological alterations and that it could be targeted for therapy of disorders of aging brain. Drugs to be considered for such a therapy can be grouped as drugs effective on the 5-LOX pathway and drugs acting on the COX-2 pathway. 5-LOX-related drugs include 5-LOX inhibitors (e.g., zileuton) and leukotriene receptor antagonists (e.g., CysLT$_1$ antagonists montelukast and zafirlukast); COX-2-related drugs include COX-2 inhibitors (e.g., rofecoxib). Whereas this pharmacological approach appears to be feasible in an *in vitro* setting, the possibility these compounds could be used in the CNS is uncertain due to insufficient information about the blood–brain barrier permeability of these drugs.

Recent data indicate that leukotrienes and prostaglandins bind the peroxisome proliferator-activated receptors (PPAR). These proteins act as transcription factors; three types, PPAR alpha, beta, and gamma, are expressed in the mammalian CNS (Cullingford et al., 1998). Functionally, the role of these receptors has not been fully characterized. Activation of PPAR-gamma was recently shown to be effective in inhibiting glutamate toxicity in primary cultures of cerebellar granule neurons (Uryu et al., 2002). To apply specific leukotriene or PPAR receptor pharmacological treatments to the CNS, the type and precise localization of these receptors in the brain cells would have to be better characterized.

In the CNS, the expression of 5-LOX and/or COX-2 can be altered by hormones (e.g., glucocorticoids), activation of neurotransmitter receptors (e.g., glutamate receptors), during neuronal proliferation and maturation, and in aging. Thus, it appears that the pharmacological targeting of the *expression* of these genes could be a therapeutic option. This could be achieved via indirect pathways (e.g., targeting the signaling systems that influence 5-LOX or COX-2 expression) or more specifically by affecting the 5-LOX or COX-2 promoters via epigenetic mechanisms that involve DNA methylation. If further research confirms the possibility of altering the neuronal expression of 5-LOX or COX-2 via alterations in DNA methylation, it would be apparent that novel pharmacology directed toward DNA methylation might be relevant for treatment of aging brain-related pathologies.

Acknowledgment

Support by the National Institute on Aging grant RO1-AG15347 is kindly acknowledged.

References

Adesuyi, S.A., Cockrell, C.S., Gamache, D.A., Ellis, E.F., 1985. Lipoxygenase metabolism of arachidonic acid in brain. J. Neurochem. 45, 770–776.

Aliza, A.R.N., Bedick, J.C., Rana, R.L., Tunaz, H., Hoback, W.W., Syanley, D.W., 2001. Arachidonic and eicosapentaenoic acids in tissue of the firefly, *Photinus pyralis* (Insecta: Coleoptera). Comp. Biochem. Physiol. 128 Part A, 251–257.

Anderson, K.M., Seed, T., Vos, M., Mulshine, J., Meng, J., Alrefai, W., Ou, D., Harris, J.E., 1998. 5-Lipoxygenase inhibitors reduce PC-3 cell proliferation and initiate nonnecrotic cell death. Prostate 37, 161–173.

Andreasson, K.I., Savonenko, A., Vidensky, S., Goeliner, J.J., Zhang, Y., Shaffer, A., Kaufmann, W.E., Worley, P.F., Isakson, P., Markowska, A.L., 2001. Age-dependent cognitive deficits and neuronal apoptosis in cyclooxygenase-2 transgenic mice. J. Neurosci. 21, 8198–8209.

Arai, K., Nishiyama, A., Matsuki, N., Ikegaya, Y., 2001. Neuroprotective effects of lipoxygenase inhibitors against ischemic injury in rat hippocampal slice cultures. Brain Res. 904, 167–172.

Arboix, A., Garcia-Eroles, L., Massons, J., Oliveres, M., Targa, C., 2000. Acute stroke in very old people: clinical features and predictors of in-hospital mortality. J. Am. Geriatr. Soc. 48, 36–41.

Baek, B.S., Kim, J.W., Lee, J.H., Kwon, H.J., Kim, N.D., Kang, H.S., Yoo, M.A., Yu, B.P., Chung, H.Y., Gerontol, J., 2001. Age-related increase of brain cyclooxygenase activity and dietary modulation of oxidative status. A. Biol. Sci. Med. Sci. 56, B426–431.

Barone, F.C., Schmidt, D.B., Hillegass, L.M., Price, W.J., White, R.F., Feuerstein, G.Z., Clark, R.K., Lee, E.V., Griswold, D.E., Sarau, H.M., 1992. Reperfusion increases neutrophils and leukotriene B4 receptor binding in rat focal ischemia. Stroke 23, 1337–1347.

Bezard, J., Blond, J.P., Bernard, A., Clouet, P., 1994. The metabolism and availability of essential fatty acids in animal and human tissues. Reprod. Nutr. Dev. 34, 539–568.

Bishai, I., Coceani, F., 1992. Eicosanoid formation in the rat cerebral cortex. Contribution of neurons and glia. Mol. Chem. Neuropathol. 17, 219–238.

Boado, R.J., Pardridge, W.M., Vinters, H.V., Black, K.L., 1992. Differential expression of arachidonate 5-lipoxygenase transcripts in human brain tumors: evidence for the expression of multitranscript family. Proc. Natl. Acad. Sci. USA 89, 9044–9048.

Brash, A.R., 1999. Lipoxygenases: occurrence, functions, catalysis and acquisition of substrate. J. Biol. Chem. 274, 23679–23680.

Burr, G.O., Burr, M.M., 1973. A new deficiency disease produced by the rigid exclusion of fat from the diet. Nutr. Rev. 31, 248–249.

Chen, X.S., Sheller, J.R., Johnson, E.N., Funk, C.D., 1994. Role of leukotrienes revealed by target disruption of the 5-lipoxygenase gene. Nature 372, 179–182.

Christie, M.J., Vaughan, C.W., Ingram, S.L., 1999. Opioids, NSAIDs and 5-lipoxygenase inhibitors act synergistically in brain via arachidonic acid metabolism. Inflamm. Res. 48, 1–4.

Chyb, S., Raghu, P., Hardie, R.C., 1999. Polyunsaturated fatty acids activate the *Drosophila* light-sensitive channels TRP and TRPL. Nature 397, 255–259.

Ciceri, P., Rabuffetti, M., Monopoli, A., Nicosia, S., 2001. Production of leukotrienes in a model of focal cerebral ischemia in the rat. Br. J. Pharmacol. 133, 1323–1329.

Coffey, M.J., Wilcoxen, S.E., Peters-Golden, M., 1994. Increases in 5-lipoxygenase activating protein expression account for enhanced capacity for 5-lipoxygenase metabolism that accompanies differentiation of peripheral blood monocytes into alveolar macrophages. Am. J. Respir. Cell. Mol. Biol. 11, 153–158.

Colamorea, T., DiPaola, R., Macchia, F., Guerrese, M.C., Tursi, A., Butterfield, J.H., Caiaffa, M.F., Haeggstrom, J.Z., Macchia, L., 1999. 5-Lipoxygenase upregulation by dexamethasone in human mast cells. Biochem. Biophys. Res. Commun. 265, 617–624.

Corey, E.J., Albright, J.O., Barton, A.E., Hashimoto, S., 1980. Chemical and enzymic syntheses of 5-HPETE, a key biological precursor of slow-reacting substance of anaphylaxis (SRS) and 5-HETE. J. Am. Chem. Soc. 102, 1435–1436.

Cullingford, T.E., Bhakoo, K., Peuchen, S., Dolphin, C.T., Patel, R., Clark, J.B., 1998. Distribution of mRNAs encoding the peroxisome proliferator-activated receptor alpha, beta, and gamma, and the retionid X receptor alpha, beta, and gamma in rat central nervous system. J. Neurochem. 70, 1366–1375.

Drazen, J.M., Yandava, C.N., Dube, L., Szczerbach, N., Hippensteel, R., Pillari, A., Israel, E., Schork, N., Silverman, E.S., Katz, D.A., Drajsek, J., 1999. Pharmacogenetic association between ALOX5 promoter genotype and the response to anti-asthma treatment. Nat. Genet. 22, 168–170.

Edmond, J., Higa, T.A., Korsak, R.A., Bergner, E.A., Lee, W.-N.P., 1998. Fatty acid transport and utilization for the developing brain. J. Neurochem. 70, 1227–1234.

Fan, G., Beard, C., Chen, R.Z., Csankovszki, G., Sun, Y., Siniaia, M., Biniszkiewicz, D., Bates, B., Lee, P.P., Kuhn, R., Trumpp, A., Poon, C., Wilson, C.B., Jaenisch, R., 2001. DNA hypomethylation perturbs the function and survival of CNS neurons in postnatal animals. J. Neurosci. 21, 788–797.

Farrer, L.A., Cupples, L.A., Connor, L., Wolf, P.A., Growdon, J.H., 1991. Association of decreased paternal age and late-onset Alzheimer's disease. An example of genetic imprinting? Arch. Neurol. 48, 599–604.

Fiala, M., Zhang, L., Gan, X., Sherry, B., Taub, D., Graves, M.C., Hama, S., Way, D., Weinand, M., Witte, M., Lorton, D., Kuo, Y.M., Roher, A.E., 1998. Amyloid-beta induces chemokine secretion and monocyte migration across a human blood-brain barrier model. Mol. Med. 4, 480–489.

Funk, C.D., 2001. Prostaglandins and leukotrienes: advances in eicosanoid biology. Science 294, 1871–1875.

Funk, C.D., Chen, X.S., 2000. 5-Lipoxygenase and leukotrienes. Transgenic mouse and nuclear targeting studies. Am. J. Respir. Crit. Care Med. 161, S120–124.

Gati, I., Bergstrom, M., Csoka, K., Muhr, C., Carlsson, J., 1990. Effects of the 5-lipoxygenase inhibitors AA-863 and U-60,257 on human glioma cell lines. Prostaglandins Leukot. Essent. Fatty Acids 40, 117–124.

Ghosh, J., Myers, C.E., 1997. Arachidonic acid stimulates prostate cancer cell growth: critical role of 5-lipoxygenase. Biochem. Biophys. Res. Commun. 235, 418–423.

Goodman, Y., Steiner, M.R., Steiner, S.M., Mattson, M.P., 1994. Nordihydroquaiaretic acid protects hippocampal neurons against amyloid beta-peptide toxicity, and attenuates free radical and calcium accumulation. Brain Res. 654, 171–176.

Hayek, M.G., Meydani, S.N., Meydani, M., Blumberg, J.B., 1994. Age differences in eicosanoid production of mouse splenocytes: effects on mitogen-induced T-cell proliferation. J. Gerontol. 49, B197–B207.

Ho, L., Purohit, D., Haroutunian, V., Luterman, J.D., Willis, F., Naslund, J., Buxbaum, J.D., Mohs, R.C., Aisen, P.S., Pasinetti, G.M., 2001. Neuronal cyclooxygenase 2 expression in the hippocampal formation as a function of the clinical progression of Alzheimer disease. Arch. Neurol. 58, 487–492.

Hoffmann, C., 2000. COX-2 in brain and spinal cord – implications for therapeutic use. Curr. Med. Chem. 7, 1113–1120.

Hoshiko, S., Radmark, O., Samuelsson, B., 1990. Characterization of the human 5-lipoxygenase gene promoter. Proc. Natl. Acad. Sci. USA 87, 9073–9077.

Iadecola, C., Forster, C., Nogawa, S., Clark, H.B., Ross, M.E., 1999. Cyclooxygenase-2 immunoreactivity in the human brain following cerebral ischemia. Acta Neuropathol. (Berl.) 98, 9–14.

Iadecola, C., Niwa, K., Nogawa, S., Zhao, X., Nagayama, M., Araki, E., Morham, S., Ross, M.A., 2001. Reduced susceptibility to ischemic brain injury and N-methyl-D-aspartate-mediated neurotoxicity in cyclooxygenase-2-deficient mice. Proc. Natl. Acad. Sci. USA 98, 1294–1299.

Ibe, B., Anderson, J., Raj, J., 1997. Leukotriene synthesis by isolated perinatal ovine intrapulmonary vessels correlates with age-related changes in 5-lipoxygenase protein. Biochem. Mol. Med. 61, 63–71.

Ichitani, Y., Shi, T., Haeggstrom, J.Z., Samuelsson, B., Hokfelt, T., 1997. Increased levels of cyclooxygenase-2 mRNA in the rat spinal cord after peripheral inflammation: an in situ hybridization study. Neuroreport 8, 2949–2952.

In, K.H., Asano, K., Beier, D., Grobholz, J., Finn, P.W., Silverman, E.K., Silverman, E.S., Collins, T., Fischer, A.R., Keith, T.P., Serino, K., Kim, S.W., Desanctis, G.T., Yandava, C., Pillari, A., Rubin, P., Kemp, J., Israel, E., Busse, W., Ledford, D., Murray, J.J., Segal, A., Tinkleman, D., Drazen, J.M., 1997. Naturally occurring mutations in the human 5-lipoxygenase gene promoter that modify transcription factor binding and reporter gene transcription. J. Clin. Invest. 99, 1130–1137.

Issa, J.-P., Ottaviano, Y.L., Celano, P., Hamilton, S.R., Davidson, N.E., Baylin, S.B., 1994. Methylation of the oestrogen receptor CpG islands links ageing and neoplasia in human colon. Nat. Genet. 7, 536–540.

Jisaka, M., Kim, R.B., Boeglin, W.E., Nanney, L.B., Brash, A.R., 1997. Molecular cloning and functional expression of a phorbol ester-inducible 8S-lipoxygenase from mouse skin. J. Biol. Chem. 272, 24410–24416.

Kaufman, W.E., Andreasson, K.I., Isakson, P.C., Worley, P.F., 1997. Cyclooxygenases and the central nervous system. Prostaglandins 54, 601–624.

Kharlamov, A., Kharlamov, E., Armstrong, D.M., 2000. Age-dependent increase in infarct volume following photochemically induced cerebral infarction: putative role of astroglia. J. Gerontol. A. Biol. Sci. Med. Sci. 55, B135–B141.

Kim, G.W., Chan, P.H., 2001. Oxidative stress and neuronal DNA fragmentation mediate age-dependent vulnerability to the mitochondrial toxin, 3-nitropropionic acid, in the mouse striatum. Neurobiol. Dis. 8, 114–126.

Koistinaho, J., Koponen, S., Chan, P.H., 1999. Expression of cyclooxygenase-2 mRNA after global ischemia is regulated by AMPA receptors and glucocorticoids. Stroke 30, 1900–1905.

Kumihashi, K., Uchida, K., Miyazaki, H., Kobayashi, J., Tsushima, T., Machida, T., 2001. Acetylsalicylic acid reduces ischemia-induced proliferation of dentate cells in gerbils. Neuroreport 12, 915–917.

Lammers, C.H., Schweitzer, P., Facchinetti, P., Arrang, J.M., Madamba, S.G., Siggins, G.R., Piomelli, D., 1996. Arachidonate 5-lipoxygenase and its activating protein: prominent hippocampal expression and role in somatostatin signaling. J. Neurochem. 66, 147–152.

Lepley, R.A., Fitzpatrick, F.A., 1994. 5-Lipoxygenase contains a functional Src homology 3-binding motif that interacts with the Src homology 3 domain of Grb2 and cytoskeletal proteins. J. Biol. Chem. 269, 24163–24168.

Li, Y., Maher, P., Schubert, D., 1997. A role for 12-lipoxygenase in nerve cell death caused by glutathione depletion. Neuron 19, 453–463.

Lindgren, J.A., Hokfelt, T., Dahlen, S.E., Patrono, C., Samuelsson, B., 1984. Leukotrienes in the rat central nervous system. Proc. Natl. Acad. Sci. USA 81, 6212–6216.

Manev, H., Uz, T., 1999. Primary cultures of rat cerebellar granule cells as a model to study neuronal 5-lipoxygenase and FLAP gene expression. Ann. NY Acad. Sci. 890, 183–190.

Manev, H., Costa, E., Wroblewski, J.T., Guidotti, A., 1990. Abusive stimulation of excitatory amino acid receptors: a strategy to limit neurotoxicity. FASEB J. 4, 2789–2797.

Manev, H., Uz, T., Qu, T., 1998. Early upregulation of hippocampal 5-lipoxygenase following systemic administration of kainate to rats. Restor. Neurol. Neurosci. 12, 81–85.

Manev, H., Uz, T., Qu, T., 2000a. 5-Lipoxygenase and cyclooxygenase mRNA expression in rat hippocampus: early response to glutamate receptor activation by kainate. Exp. Gerontol. 35, 1201–1209.

Manev, H., Uz, T., Sugaya, K., Qu, T., 2000b. Putative role of neuronal 5-lipoxygenase in an aging brain. FASEB J. 14, 1464–1469.

Manev, H., Uz, T., Manev, R., Zhang, Z., 2001. Neurogenesis and neuroprotection in the adult brain. A putative role for 5-lipoxygenase? Ann. N.Y. Acad. Sci. 939, 45–51.

Marceau, F., 1996. Evidence for vascular tone regulation by resident or infiltrating leukocytes. Biochem. Pharmacol. 52, 1481–1488.

Mattson, M.P., Duan, W., Lee, J., Guo, Z., 2001. Suppression of brain aging and neurodegenerative disorders by dietary restriction and environmental enrichment: molecular mechanisms. Mech. Ageing Dev. 122, 757–778.

Mayatepek, E., 2000. Leukotriene C4 synthesis deficiency: a member of a probably underdiagnosed new group of neurometabolic diseases. Eur. J. Pediatr. 159, 811–818.

Mays-Hoopes, L.L., 1989. DNA methylation in aging and cancer. J. Gerontol. 44, 35–36.

McGeer, P.L., Schulzer, M., McGeer, E.G., 1996. Arthritis and anti-inflammatory agents as possible protective factors for Alzheimer's disease: a review of 17 epidemiologic studies. Neurology 47, 425–432.

Montine, K.S., Montine, T.J., Morrow, J.D., Frei, B., Milatovic, D., Eckenstein, F., Quinn, J.F., 2002. Mouse cerebral prostaglandins, but not oxidative damage, change with age and are responsive to indomethacin treatment. Brain Res. 930, 75–82.

Moore, S.A., 2001. Polyunsaturated fatty acid synthesis and release by brain-derived cells *in vitro*. J. Mol. Neurosci. 16, 195–200.

Moore, S.A., Yoder, E., Murphy, S., Dutton, G.R., Spector, A.A., 1991. Astrocytes, not neurons, produce docosahexaenoic acid (22:6 omega-3) and arachidonic acid (20:4 omega-6). J. Neurochem. 56, 518–524.

Nagayama, M., Aber, T., Nagayama, T., Ross, M.E., Iadecola, C., 1999. Age-dependent increase in ischemic brain injury in wild-type mice and in mice lacking the inducible nitric oxide synthase gene. J. Cereb. Blood Flow Metab. 19, 661–666.

Nakayama, H., Jorgensen, H.S., Raaschou, H.O., Olsen, T.S., 1994. The influence of age on stroke outcome. The Copenhagen Stroke Study. Stroke 25, 808–813.

Nicol, G.D., Klingberg, D.K., Vasko, M.R., 1992. Prostaglandin E2 increases calcium conductance and stimulates release of substance P in avian sensory neurons. J. Neurosci. 12, 1917–1927.

Normandin, M., Gagne, J., Bernard, J., Elie, R., Miceli, D., Baudry, M., Massicotte, G., 1996. Involvement of the 12-lipoxygenase pathway of arachidonic acid metabolism in homosynaptic long-term depression of the rat hippocampus. Brain Res. 730, 40–46.

O'Banion, M.K., 1999. Cyclooxygenase-2: molecular biology, pharmacology, and neurobiology. Crit. Rev. Neurobiol. 13, 45–82.

Ohtsuki, T., Matsumoto, M., Hayashi, Y., Yamamoto, K., Kitagawa, K., Ogawa, S., Yamamoto, S., Kamada, T., 1995. Reperfusion induces 5-lipoxygenase translocation and leukotriene C4 production in ischemic brain. Am. J. Physiol. 268, H1249–1257.

Paris, D., Town, T., Parker, T., Humphrey, J., Mullan, M., 2000. A beta vasoactivity: an inflammatory reaction. Ann. NY Acad. Sci. 903, 97–109.

Piomelli, D., 1994. Eicosanoids in synaptic transmission. Crit. Rev. Neurobiol. 8, 65–83.

Piomelli, D., 2001. The ligand that came from within. Trends. Pharmacol. Sci. 22, 17–19.

Prolla, T.A., Mattson, M.P., 2001. Molecular mechanisms of brain aging and neurodegenerative disorders: lessons from dietary restriction. Trends Neurosci. 24, S21–31.

Provost, P., Samuelsson, B., Radmark, O., 1999. Interaction of 5-lipoxygenase with cellular proteins. Proc. Natl. Acad. Sci. USA 96, 1881–1885.

Qu, T., Uz, T., Manev, H., 2000. Inflammatory 5-LOX mRNA and protein are increased in brain of aging rats. Neurobiol. Aging 21, 647–652.

Riddick, C.A., Ring, W.L., Baker, J.R., Hodulik, C.R., Bigby, T.D., 1997. Dexamethasone increases expression of 5-lipoxygenase and its activating protein in human monocytes and THP-1 cells. Eur. J. Biochem. 246, 112–118.

Rogers, J., Kirby, L.C., Hempelman, S.R., Berry, D.L., McGeer, P.L., Kaszniak, A.W., Zalinski, J., Cofield, M., Mansukhani, L., Willson, P., Kogan, F., 1993. Clinical trial of indomethacin in Alzheimer's disease. Neurology 43, 1609–1611.

Sapirstein, A., Bonventre, J.V., 2000. Specific physiological roles of cytosolic phospholipase A(2) as defined by gene knockouts. Biochim. Biophys. Acta 1488, 139–148.

Schwarcz, R., Foster, A.D., French, E.D., Whetsell, W.O. Jr., Kohler, C., 1984. Excitotoxic models for neurodegenerative disorders. Life Sci. 35, 19–32.

Shelton, D.N., Chang, E., Whittier, P.S., Choi, D., Funk, W.D., 1999. Microarray analysis of replicative senescence. Curr. Biol. 9, 939–945.

Shimizu, T., Takusagawa, Y., Izumi, T., Ohishi, N., Seyama, Y., 1987. Enzymic synthesis of leukotriene B4 in guinea pig brain. J. Neurochem. 48, 1541–1546.

Shishido, Y., Furushiro, M., Hashimoto, S., Yokokura, T., 2001. Effect of nordihydroguaiaretic acid on behavioral impairment and neuronal cell death after forebrain ischemia. Pharmacol. Biochem. Behav. 69, 469–474.

Simmet, T., Tippler, B., 1990. Cysteinyl-leukotriene production during limbic seizures triggered by kainic acid. Brain Res. 515, 79–86.

Smith, W.L., DeWitt, D.L., Garavito, R.M., 2000. Cyclooxygenases: structural, cellular, and molecular biology. Annu. Rev. Biochem. 69, 145–182.

Song, S.H., Jong, H.S., Choi, H.H., Inoue, H., Tanabe, T., Kim, N.K., Bang, Y.J., 2001. Transcriptional silencing of cyclooxygenase-2 by hyper-methylation of the 5' CpG islands in human gastric carcinoma cells. Cancer Res. 61, 4628–4635.

Steinhilber, D., Brungs, M., Werz, O., Wiesenberg, I., Danielsson, C., Kahlen, J.P., Nayeri, S., Schrader, M., Carlberg, C., 1995. The nuclear receptor for melatonin represses 5-lipoxygenase gene expression in human B lymphocytes. J. Biol. Chem. 270, 7037–7040.

Steinhilber, D., 1999. 5-Lipoxygenase: a target for antiinflammatory drugs revisited. Curr. Med. Chem. 6, 71–85.

Stewart, L.R., White, A.R., Jobling, M.F., Needham, B.E., Maher, F., Thyer, J., Beyreuther, K., Masters, C.L., Collins, S.J., Cappai, R., 2001. Involvement of the 5-lipoxygenase pathway in the neurotoxicity of the prion peptide PrP106–126. J. Neurosci. Res. 65, 565–572.

Stewart, W.F., Kawas, C., Corrada, M., Metter., E.J., 1997. Risk of Alzheimer's disease and duration of NSAID use. Neurology 48, 626–632.

Szyf, M., 2001. Towards pharmacology of DNA methylation. Trends Pharmacol. Sci. 22, 350–354.

Tucker, K.L., 2001. Methylated cytosine and the brain: a new base for neuroscience. Neuron 30, 649–652.

Uhl, J., Klan, N., Rose, M., Entian, K.D., Werz, O., Steinhilber, D., 2001. The 5-lipoxygenase promoter is regulated by DNA methylation. J. Biol. Chem. 277, 4374–4379.

Uryu, S., Harada, J., Hisamoto, M., Oda, T., 2002. Troglitazone inhibits both post-glutamate neurotoxicity and low-potassium-induced apoptosis in cerebellar granule neurons. Brain Res. 924, 229–236.

Uz, T., Manev, H., 1998. Circadian expression of pineal 5-lipoxygenase mRNA. Neuroreport 9, 783–786.

Uz, T., Longone, P., Manev, H., 1997. Increased hippocampal 5-lipoxygenase mRNA content in melatonin-deficient, pinealectomized rats. J. Neurochem. 69, 2220–2223.

Uz, T., Pesold, C., Longone, P., Manev, H., 1998. Aging-associated up-regulation of neuronal 5-lipoxygenase expression: putative role in neuronal vulnerability. FASEB J. 12, 439–449.

Uz, T., Dwivedi, Y., Savani, P.D., Impagnatiello, F., Pandey, G., Manev, H., 1999. Glucocorticoids stimulate inflammatory 5-lipoxygenase gene expression and protein translocation in the brain. J. Neurochem. 73, 693–699.

Uz, T., Manev, R., Manev, H., 2001a. 5-Lipoxygenase is required for proliferation of immature cerebellar granule neurons *in vitro*. Eur. J. Pharmacol. 418, 15–22.

Uz, T., Dwivedi, Y., Qeli, A., Peters-Golden, M., Pandey, G., Manev, H., 2001b. Glucocorticoid receptors are required for up-regulation of neuronal 5-lipoxygenase (5LOX) expression by dexamethasone. FASEB J. 15, 1792–1794.

Uz, T., Dimitrijevic, N., Tueting, P., Manev, H., 2002. 5-Lipoxygenase (5LOX)-deficient mice express reduced anxiety-like behavior. Restor. Neurol. Neurosci. 20, 15–20.

van Rossum, G.S.A.T., Bijvelt, J.J.M., van den Bosch, H., Verkleij, A.J., Boonstra, J., 2002. Cytosolic phospholipase A_2 and lipoxygenase are involved in cell cycle progression in neuroblastoma cells. Cell. Mol. Life Sci. 59, 181–188.

Wallace, J.L., 1999. Selective COX-2 inhibitors: is the water becoming muddy? Trends Pharmacol. Sci. 20, 4–6.

Wainwright, P.E., 2001.The role of nutritional factors in behavioural development in laboratory mice. Behav. Brain Res. 125, 75–80.

Watanabe, T., Medina, J.F., Haeggstrom, J.Z., Radmark, O., Samuelsson, B., 1993. Molecular cloning of a 12-lipoxygenase cDNA from rat brain. Eur. J. Biochem. 212, 605–612.

Weggen, S., Eriksen, J.L., Das, P., Sagi, S.A., Wang, R., Pietrzik, C.U., Findlay, K.A., Smith, T.E., Murphy, M.P., Bulter, T., Kang, D.E., Marquez-Sterling, N., Golde, T.E., Koo, E.H., 2001. A subset of NSAIDs lower amyloidogenic Abeta42 independently of cyclooxygenase activity. Nature 414, 212–216.

Whitney, L.W., Ludwin, S.K., McFarland, H.F., Biddison, W.E.J., 2001. Microarray analysis of gene expression in multiple sclerosis and EAE identifies 5-lipoxygenase as a component of inflammatory lesions. Neuroimmunol. 121, 40–48.

Willemsen, M.A., Lutt, M.A., Steijlen, P.M., Cruysberg, J.R., van der Graaf, M., Nijhuis van der Sanden, M.W., Pasman, J.W., Mayatepek, E., Rotteveel, J.J., 2001. Clinical and biochemical effects of zileuton in patients with Sjogren-Larson syndrome. Eur. J. Pediatr. 160, 711–717.

Yamagata, K., Andreasson, K.I., Kaufmann, W.I., Barnes, C.A., Worley, P.F., 1993. Expression of a mitogen-inducible cyclooxygenase in brain neurons: regulation by synaptic activity and glucocorticoids. Neuron 11, 371–386.

Yasojima, K., Schwab, C., McGeer, E.G., McGeer, P.L., 1999. Distribution of cyclooxygenase-1 and cyclooxygenase-2 mRNAs and proteins in human brain and peripheral organs. Brain Res. 830, 226–236.

Yermakova, A.V., O'Banion, M.K., 2001. Downregulation of neuronal cyclooxygenase-2 expression in end stage Alzheimer's disease. Neurobiol. Aging 22, 823–836.

Youdim, K.A., Martin, A., Joseph, J.A., 2000. Essential fatty acids and the brain: possible health implications. Int. J. Devl. Neuroscience 18, 383–399.

**Advances in
Cell Aging and
Gerontology**

Cellular cholesterol, membrane signaling, and disease

John P. Incardona

*Program in Ecotoxicology and Environmental Fish Health, Environmental Conservation Division,
National Oceanic and Atmospheric Administration, Northwest Fisheries Science Center,
2725 Montlake Blvd E, Seattle 98112, USA. Tel.: 206-860-3347;
E-mail address: john.incardona@noaa.gov*

Contents

Abbreviations

CDPX2: X-linked dominant chondrodysplasia punctata; CFP: cyan fluorescent protein; CHILD: congenital hemidysplasia with ichthyosiform erythroderma and

Advances in Cell Aging and Gerontology, vol. 12, 137–162

limb defects; CHO: Chinese hamster ovary; 7DHCR: 7-dehydrocholesterol
reductase; DiI: dialkylindocarbocyanine; GFP: green fluorescent protein;
GPI: glycosyl-phosphatidylinositol; Hh: Hedgehog; HMGR: 3-hydroxy-3-methy-
glutaryl coenzyme A reductase; Ihh: Indian hedgehog; LBPA: lysobisphosphatidic
acid; NPC: Niemann-Pick Type C disease; NPC1: Niemann-Pick C1 protein;
Ptc: Patched; SCAP: SREBP cleavage activating protein; Shh: Sonic hedgehog;
SREBP: sterol response element binding protein; SLOS: Smith-Lemli-Opitz
syndrome; Smo: Smoothened; SSD: sterol-sensing domain; TGN: trans-Golgi
network; YFP: yellow fluorescent protein

1. Introduction

The roles for membrane cholesterol in cellular signaling have become increasingly diverse, yet appear to be interconnected by a common theme of the
organization of cellular functions around membrane microdomains. Seemingly
disparate aspects of cell biology such as biosynthetic and endocytic membrane
traffic and signal transduction have converged upon this theme, at the center of
which is the raft hypothesis that lateral segregation of sphingolipid/cholesterol
microdomains in the bilayer serve as platforms for protein sorting and the
assembly of signaling components. Two classes of hereditary disorders appear to
be good candidates for consideration as diseases of rafts and membrane signaling:
lysosomal storage disorders in which lipids accumulate, and the main clinical
features involve neural degeneration; and inborn errors of sterol synthesis, in
which the main clinical features are dysmorphogenesis and in some cases
neurodevelopmental defects. The former is exemplified by Niemann–Pick C
disease, and the latter by the Smith–Lemli–Opitz syndrome. Importantly, the
phenotypes associated with these disorders are in many ways dissimilar, and thus
make distinct contributions to our understanding of the various roles of
membrane cholesterol and rafts in cellular biology. While these disorders may be
unfamiliar to many in the field of aging research, the insight they provide into
fundamental functions of cellular cholesterol will have relevance for classical
disorders of aging in which alterations of cholesterol metabolism are important
aspects of pathogenesis or therapy.

2. Distribution of cholesterol in cellular membranes

The distribution of cholesterol is inhomogeneous and tightly controlled in
cellular membranes (for a detailed review of cellular cholesterol homeostasis,
see Liscum and Munn, 1999). The biophysical properties of cholesterol and its
interaction with other lipids determine both large- and small-scale differences in
cellular lipid distributions. Although estimates vary, most cellular cholesterol
(65–80%) resides in the plasma membrane, but recent studies indicate
that endosomal compartments are relatively cholesterol enriched. Cells derive

cholesterol either by de novo synthesis in the endoplasmic reticulum (ER) or receptor-mediated endocytosis of lipoproteins. Cells in the central nervous system are nearly completely dependent on de novo synthesis, although within the CNS there appears to be cycling of cholesterol via apolipoprotein E-containing lipoproteins. Homeostatic mechanisms exist to maintain the levels of cellular free cholesterol. A change in cholesterol level is sensed at the ER, the site of regulation for both biosynthetic enzymes and transcription factors involved in synthesis of the proteins that function in synthesis and uptake (reviewed by Goldstein and Brown, 1990; Horton et al., 2002). For example, enzymes such as 3-hydroxy-3-methylglutaryl coenzyme A reductase (HMGR) undergo feedback inhibition through sterol-induced proteolytic degradation. Sterols similarly inhibit the proteolytic activation of the sterol response element binding proteins (SREBPs) by controlling their transport via the SREBP cleavage activating protein (SCAP). Sterol-mediated control of both HMGR and SCAP is dependent on a sterol-sensing domain (SSD) present in each protein, a conserved series of five transmembrane segments now identified in a larger family of membrane proteins (Fig. 1), most but not all of which are clearly involved in some aspect of sterol metabolism (reviewed by Kuwabara and Labouesse, 2002). Newly synthesized cholesterol appears to arrive at the cell surface in cholesterol/sphingolipid enriched microdomains, and subsequently is distributed to other portions of the bilayer. Cholesterol derived by the endocytosis of lipoproteins is hydrolyzed in late endosomes/lysosomes and is redistributed to other parts of the cell. In addition to cholesterol derived from internalized lipoprotein, free cholesterol that enters the endocytic pathway through general membrane traffic must be redistributed. Excess cholesterol is transported to the ER for esterification by acyl CoA/cholesterol acyltransferase (ACAT) for storage in lipid droplets.

3. Cholesterol, raft forms and functions

In relation to cellular signaling, by far the most prominent feature of cholesterol is its role in the formation of small microdomains or rafts. While rafts really are defined operationally by biochemical isolation, they represent small laterally segregated domains within a lipid bilayer (for detailed reviews of raft formation, see Brown and London, 2000; Maxfield, 2002). Lipids of different chain length and degree of saturation have different gel-fluid phase transition temperatures (T_m) at which the hydrophobic chains will become tightly packed. Thus binary mixtures of purified lipids in model membranes will show phase separations according to the T_m. The addition of cholesterol to membranes affects these phase separations such that the two types of fluid phase domains can form in more complex mixtures. Specifically, the association of cholesterol with lipids with long saturated chains results in the lateral separation of a "liquid-ordered" phase typically enriched in cholesterol, sphingomyelin, and glycosphingolipids, from a "liquid-disordered" phase enriched in unsaturated phospholipids. Rafts were originally defined as cholesterol/sphingolipid enriched complexes isolated from cellular membranes after lysis in cold non-ionic detergents such as Triton

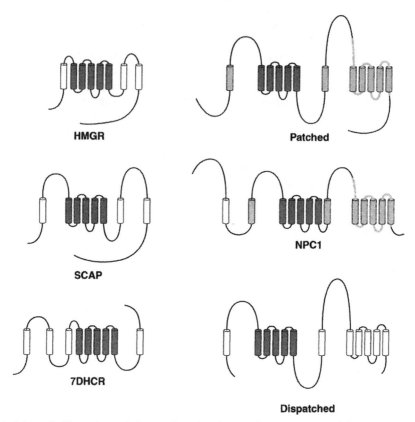

Fig. 1. Schematic illustrations of the sterol-sensing domain family of proteins discussed in the text. Transmembrane domains are indicated by the barrels. The SSD is indicated by dark gray, and the regions of homology between NPC1 and Ptc (including some extramembrane regions in the carboxy terminal portion) are indicated by light gray. Non-homologous segments are indicated in white.

X-100, and it is assumed that rafts in cellular membranes arise by the same processes that promote microdomain formation in model membranes. In biological membranes, raft formation depends largely on the physical interaction between cholesterol and sphingomyelin. Since sphingolipids in cells are localized to the outer leaflet of the bilayer, there is currently little information on the nature of rafts in the corresponding phospholipid-rich inner leaflet. Although intracellular proteins can associate with rafts containing extracellular proteins, the mechanism linking the inner and outer leaflets is unknown. The nomenclature for rafts is variable and often confusing [e.g. Triton rafts, detergent-resistant membranes (DRMs), glycolipid-enriched membranes (GEM), buoyant density membrane fractions, Triton-insoluble floating fractions (TIFF)], because there are now several methods for biochemical isolation, some not involving detergents. In general, rafts can be pelleted by centrifugation of a cold Triton X-100 cellular lysate. Due to their lipid composition, raft membranes will float in a sucrose gradient, and hence are found in the lighter density fractions.

Another prominent feature of rafts are the suite of membrane proteins which are co-isolated with them specifically. The proteins most widely associated with rafts are those with lipid modifications, including glycosyl-phosphatidylinositol (GPI)-anchored and doubly-acylated proteins, such as Src-family kinases. Some transmembrane proteins (which may or may not be palmitoylated) can be raft associated (e.g. influenza hemagglutinin), as are members of the Hedgehog morphogen family, which are covalently modified with cholesterol. The long saturated chains of GPI-anchored and acylated proteins would prefer liquid-ordered domains, providing an obvious means for raft association. The biophysical basis for raft association of other proteins is less clear, but the concept of "lipid shells" was recently advanced by which many proteins might be targeted through specific protein–lipid interactions (Anderson and Jacobson, 2002). As discussed further below, a few proteins known to interact directly with cholesterol also associate with rafts.

Because of the difficulties of studying lipids in intact cells, the behavior of rafts, or even their existence, in live cells has been controversial. Over the last few years, several studies using a variety of techniques have provided convincing evidence that cholesterol-dependent rafts exist in the plasma membrane of live cells. However, the size estimated for individual rafts as well as the total area occupied by raft membranes varies significantly. Studies of GPI-anchored proteins produced estimates in the range of 50–70 nm (Varma and Mayor, 1998; Pralle et al., 2000), whereas studies employing lipid probes give five to ten times larger estimates (Schutz et al., 2000; Dietrich et al., 2002). Evidence for raft association of intracellular proteins in live cells was provided by analysis of membrane targeting doubly-acylated, non-dimerizing forms of cyan (CFP) and yellow fluorescent protein (YFP). Fluorescence resonance energy transfer (FRET) measurements indicated that acylated YFPs and CFPs clustered in a cholesterol-dependent manner, supporting the idea that rafts exist on the inner leaflet as well (Zacharias et al., 2002). However, this study did not address how the outer and inner leaflets communicate.

Caveolae are a morphological structure of many cell types that can be isolated biochemically in raft preparations. The difficulty of linking biochemically-defined rafts to visible features of cells and cellular function are highlighted by the controversy surrounding the function of these small (50–80 nm) "cave-like" cell surface invaginations. Formation of caveolae requires the expression of caveolin-1, a 22 kDa cholesterol-binding protein that forms a coat around the cytoplasmic face of the invagination. Isolated caveolae are enriched with raft lipids and proteins, and may represent stabilized rafts organized by the caveolin structural coat. Many raft proteins involved in signal transduction have also been found to undergo protein–protein interactions with a "scaffolding domain" within caveolin-1 (Okamoto et al., 1998). A functional role in endocytosis and signal transduction for caveolae has been suggested by numerous studies (reviewed by Smart et al., 1999).

Despite a very large number of proteins that are reported to associate with caveolin or caveolae, recent genetic studies indicate that the range of signaling pathways that require functional caveolae is overestimated. Mice homozygous for a targeted deletion of caveolin-1 lack caveolae in all tissues known to express the protein, yet are viable and fertile (Drab et al., 2001; Razani et al., 2001). The predominant defects in these mice were associated with endothelial cells in the lungs and vasculature, and were related to hyperproliferation and nitric oxide signaling, respectively. Remarkably, the animals have no developmental defects, and nitric oxide signaling is the only pathway for which ample biochemical evidence for caveolar function correlates with genetic evidence. Caveolae thus are not required for global raft function in a variety of cell types.

4. Raft heterogeneity and the fate of rafts after internalization

Since most of the information about rafts is derived from studies of cellular lysates, the degree of complexity is most likely underestimated. Information about raft heterogeneity or rafts in intracellular compartments was generally lacking until recently. A study utilizing a variety of non-ionic detergents identified a novel type of cholesterol-based raft that was solubilized by Triton but detected as Lubrol WX-insoluble complexes. The polytopic membrane protein prominin preferentially associates with plasmalemmal protrusions, such as apical microvilli in epithelial cells (Weigmann et al., 1997). Prominin interacted directly with cholesterol and associated with Lubrol-insoluble rafts during transport and incorporation into microvillar membrane microdomains in MDCK cells, which were distinct from classical Triton rafts containing a GPI-anchored protein (Roper et al., 2000). Lubrol rafts may be associated generally with membranes requiring a high degree of curvature, as suggested by similar studies demonstrating an interaction between synaptophysin and cholesterol during synaptic vesicle biogenesis (Thiele et al., 2000). However, the ATP-binding cassette transporter A1 (ABCA1), which functions in cellular cholesterol efflux, was enriched in Lubrol-insoluble rafts isolated from fibroblasts and macrophages, suggesting a more general function (Drobnik et al., 2002). In this study, Lubrol rafts were found to have higher levels of phosphatidylcholine than Triton rafts. The lipid composition of Lubrol rafts from MDCK cells was not reported, although isolated synaptic vesicles are cholesterol-rich and sphingolipid-poor, with phosphatidylcholine levels similar to that reported for Triton rafts (Deutsch and Kelly, 1981).

The distributions of lipids in the endocytic pathway and the fate of internalized rafts have received considerably more attention. Studies of cholesterol distribution in fixed cells with filipin and live cells with a fluorescent cholesterol analog indicated that recycling endosomes, or the endocytic recycling compartment, contain considerable cholesterol pools, while the trans-Golgi network (TGN) was cholesterol enriched in some cell types but not others (Mukherjee et al., 1998; Wustner et al., 2002). An analysis of purified recycling endosomes from MDCK cells also demonstrated an enrichment of raft components (Gagescu et al., 2000). It is likely that these compartments contribute to biochemical raft preparations that are generally thought to represent the plasma membrane. Indeed, caveolin, a prototypical cell surface raft protein, was identified as a TGN component (Kurzchalia et al., 1992), and rafts were first proposed to function in the polarized sorting of lipids in the TGN (van Meer and Simons, 1988). Analyses of

GPI-anchored proteins and several types of lipids with different chain lengths and degrees of saturation, mostly in CHO cells, allowed some predictions about the behavior of raft-preferring lipids after endocytosis, but have also generated some contradictions. Synthetic lipid-mimetic dialkylindocarbocyanine (DiI) molecules with short or unsaturated fatty acid chains were transported on the recycling endosome pathway and excluded from late endosomes (Mukherjee et al., 1999), consistent with prior studies with short-chained sphingomyelin and phosphatidylcholine analogs (Koval and Pagano, 1989). Similarly, GPI-anchored proteins were transported through the endocytic recycling compartment, with kinetics slower than a transmembrane protein, but dependent on cellular cholesterol and sphingolipid levels (Mayor et al., 1998; Chatterjee et al., 2001). In contrast, DiI derivatives with long unsaturated acyl chains, which preferably associate with rafts, were targeted to late endosomes (Mukherjee et al., 1999), where glycosphingolipids are normally degraded (Sandhoff and Klein, 1994). Thus, overall, rafts seem to be targeted along the recycling pathway, yet some raft components must be targeted to the late endocytic pathway.

Many of the discrepancies about intracellular raft traffic may be due to cell type-specific differences. A more recent analysis on a range of cell types demonstrated that GPI-anchored proteins were differentially sorted into the recycling (CHO, primary astrocytes) or late endocytic pathway (BHK, primary skin fibroblasts) (Fivaz et al., 2002). Biochemical analysis indicated that despite enrichment in cholesterol, sphingomyelin, and GPI-anchored proteins, membranes from the CHO recycling compartment did not exhibit the properties of typical rafts. In contrast, generation of small GPI-anchored protein clusters resulted in enhanced raft association and late endocytic targeting in CHO cells.

Since sorting in the endocytic pathway is a major determinant of signal duration and so many signaling pathways have been associated with rafts, clarification of these processes will be important for a mechanistic understanding of control of signaling by rafts. These studies may provide just a glimpse of the increasing complexity underlying the cell biology of lipids. In fact, the identification of the gene product that is defective in Niemann–Pick C disease has accelerated the study of lipid trafficking in the endocytic pathway and spawned many of the studies described above.

5. Niemann–Pick C disease, rafts in the endocytic pathway, and neurodegeneration

5.1. Clinical aspects and basic pathology of NPC in humans and animal models

A key player in the control of intracellular cholesterol distribution was identified through the study of the Niemann–Pick type C (NPC) lysosomal storage disorder. The main feature of NPC is a rapid and widespread degeneration of the CNS that typically begins at preschool age and may be accompanied by hepatosplenomegaly (clinical aspects are reviewed in detail by Fink et al., 1989). Depending on the tissue, patients show variable accumulations of lipids in lysosomes, with cholesterol and sphingomyelin predominating in the liver and spleen, and glycosphingolipids in the CNS. Children with NPC do not show developmental defects, and development is likewise normal in the mouse model (Loftus et al., 1997), which quite consistently reproduces the human condition. Hence, the disease is particularly tragic since it largely strikes children who were apparently healthy infants and toddlers. While the severity of disease and rate of progression varies among early- and late-onset forms, it is invariably fatal, typically within the second decade of life. Neurologic signs include vertical supranuclear gaze paresis and ataxia, and with progression results in seizures and dementia. Neuropathologic findings include widespread degeneration involving cerebellar Purkinje cells, the basal ganglia, thalamus, and brainstem. Neurofibrillary tangles without amyloid deposition were observed in chronic progressive cases (Auer et al., 1995; Love et al., 1995; Suzuki et al., 1995).

NPC skin fibroblasts demonstrated an underlying defect in intracellular cholesterol transport (Pentchev et al., 1985; Pentchev et al., 1987). NPC cells that are cholesterol starved and then loaded with lipoprotein accumulate high levels of cholesterol in late endosomes/lysosomes, and have delayed delivery of free cholesterol for esterification. Notably, this effect can be phenocopied in normal cells by treatment with a number of steroidal and amphiphilic compounds, such as progesterone (at superphysiologic levels) and the synthetic steroidal amine U18666A (Liscum and Faust, 1989; Rodriguez-Lafrasse et al., 1990; Lange and Steck, 1994). The cholesterol accumulation is not limited to that derived from internalized lipoproteins, but also includes cholesterol synthesized *de novo* derived from plasma membrane bilayer internalization (Cruz and Chang, 2000; Lange et al., 2000).

The NPC mouse has provided a system for a more detailed analysis of cellular cholesterol homeostasis in the brain (reviewed by Dietschy and Turley, 2001). The neuropathology is very similar in the mouse, although neurofibrillary tangles were not observed (German et al., 2001b). Nevertheless, the mouse model has allowed specific studies to address whether the pathogenesis results from accumulation of cholesterol, sphingolipids, or both. In newborn pups prior to the onset of neurodegeneration, brain levels of cholesterol are increased, but levels are reduced dramatically as degeneration progresses, apparently due to the loss of cholesterol in myelin (Xie et al., 2000). Generation of double mutant mice demonstrated that CNS cholesterol accumulation and neurodegeneration in NPC was independent of either the LDL receptor or apoE (Xie et al., 2000; German et al., 2001a). Although this suggests redundancy in the receptors and/or lipoproteins involved in cholesterol recycling, the large amount of bilayer cholesterol that is internalized during synaptic vesicle recycling could make a significant contribution. The contribution of glycosphingolipid accumulation to neurodegeneration is more complicated. In mice carrying both the NPC1 mutation and a deletion of β1-4GalNAc transferase, GM2 and other complex gangliosides did not accumulate, but clinical improvement was not observed (Liu et al., 2000). However, considerable improvement was achieved in the mouse and feline models with administration of N-butyldeoxynojirimycin, an inhibitor of glucosylceramide synthase that blocks glycosphingolipid synthesis at an earlier step and prevented the accumulation of neutral glycosphingolipids upstream of β1-4GalNAc transferase (Zervas et al., 2001).

5.2. The NPC1 Protein

NPC patients comprise two complementation groups with identical cellular phenotypes. The vast majority (> 95%) have mutations affecting the NPC1 protein (Carstea et al., 1997), while NPC2 is affected in a much smaller group (Naureckiene et al., 2000). NPC1 is a ~ 140 kDa protein with 13 transmembrane segments and significant structural similarity to bacterial permeases (Carstea et al., 1997; Loftus et al., 1997; Tseng et al., 1999). NPC2 is a small, soluble, ubiquitously expressed lysosomal protein previously identified as HE1 and shown to bind cholesterol 1:1 (Okamura et al., 1999; Naureckiene et al., 2000). NPC1 has been studied much more intensively than NPC2. NPC1 contains the SSD common to other polytopic membrane proteins involved in cholesterol metabolism (Fig. 1), as well as a leucine-rich domain implicated in protein–protein interactions. The structural similarity to permeases was recently extended to' function as well with the finding that NPC1 possessed a transbilayer pump activity coupled to a proton motive force. NPC1 accelerated the removal of hydrophobic compounds such as acriflavine from late endosomes/lysosomes, and when expressed in *E. coli*, resulted in the transport of acriflavine and fatty acid across the membrane, but intriguingly, not cholesterol (Davies et al., 2000).

Initial immunolocalization studies of NPC1 in normal cells found that the protein associated with a subset of late endosomes which were cholesterol-poor (Neufeld et al., 1999). In the presence of U18666A or progesterone, or in the case of NPC1 alleles encoding a full-length but non-functional protein, the protein localized to cholesterol-engorged late endosomes/lysosomes (Higgins et al., 1999; Neufeld et al., 1999). Analysis of functional NPC1-green fluorescent protein (GFP) fusions in live cells showed that NPC1 undergoes very rapid transport involving fine tubular extensions of endosomes (Ko et al., 2001; Zhang et al., 2001), and suggests that the static immunolocalization data belie a simple late endosomal distribution. The rapid transport of NPC1 ceases within a short duration of exposure to U18666A, but it is unclear whether this is due to cholesterol accumulation or vice versa. How the pump activity and rapid transport of NPC1 are linked together and to redistribution of cholesterol is unknown. Other consequences of cholesterol accumulation include more generalized defects in the late endocytic pathway, including reduced endosome mobility and homotypic fusion (Neufeld et al., 1999; Zhang et al., 2001; Lebrand et al., 2002), although it is not yet clear how these findings contribute to the pathophysiology.

5.3. Accumulation of Raft Lipids in the Endocytic Pathway

While the precise nature of the organelles in which cholesterol accumulates has been debated (late endosome vs. lysosome vs. a hybrid), it uniquely involves highly convoluted internal membranes enriched with the nonbilayer-preferring lipid, lysobisphosphatidic acid (LBPA). NPC1 may be required to enter the internal membranes of late endosomes to effect cholesterol redistribution, as mutations in the SSD result in localization of the protein to the limiting membrane of the organelle with concomitant cholesterol accumulation (Watari et al., 1999). Moreover, internalization of a monoclonal antibody against LBPA by live cells results in disorganization of the internal membranes and induction of the NPC phenocopy (Kobayashi et al., 1998; Kobayashi et al., 1999). Disruption of the late endosome LBPA domain also produces defects in protein sorting within the compartment, which has implications for NPC1 as well as autoimmune disorders in which LBPA is an autoantigen (Kobayashi et al., 1998). Since the internal membranes of late endosomes are a platform for a second level of sorting both lipids and proteins in the endocytic pathway, their perturbation is likely to have effects on a range of cellular processes that converge upon the late endosome.

The suite of lipids that accumulate in NPC disease suggests that NPC1 may function more generally in the cellular homeostasis of raft lipids. While the study of rafts has largely focused on processes occurring at the cell surface, the behavior of rafts in the endocytic pathway is not that well understood. As discussed above, raft lipids are largely excluded from the late endocytic pathway and at least in some cell types are returned to the cell surface via recycling endosomes (Mayor et al., 1998; Mukherjee et al., 1998; Gagescu et al., 2000) the main recycling pathway characterized by transport of the transferrin receptor. Analysis of the transport of [^{3}H] cholesterol released from internalized lipoprotein or derived by endocytosis of plasma membrane demonstrated that NPC1 is required to deplete raft cholesterol from membranes leading to late endosomes for return to the plasma membrane (Lange et al., 2000; Lusa et al., 2001). Similarly, raft-associated ganglioside GM1 was found to accumulate in early endosomes of NPC1-deficient CHO cells (Sugimoto et al., 2001). While sphingolipids are normally degraded in lysosomes, the consequences of failed recycling may be to overwhelm the degradative capacity by delivering substrate too rapidly. Intriguingly, cholesterol was also found to accumulate in lysosomes of other storage diseases which accumulate sphingomyelin or glycosphingolipids due to deficiencies of degradative enzymes (Puri et al., 1999). This suggests that the inability to properly handle one raft lipid results in the perturbation of transport of other raft components.

If lipid storage diseases represent general disorders of raft homeostasis, then what are the consequences for cellular signaling? Do defects in signaling contribute to pathophysiology, or is the disease more simply a consequence of disordered lysosomal catabolism? This area is largely unexplored. A recent study showed that NPC cells have normal levels of cholesterol in the plasma membrane (Lange et al., 2002), suggesting that plasma membrane rafts may be normal. One would expect that mutations overtly affecting plasma membrane raft formation or function would have much more profound affects on the developing organism, and likely even be cell lethal. The lethality resulting from replacement of cholesterol with its enantiomeric form in sterol-auxotrophic C. *elegans* is a tantalizing suggestion that this may be the case (Crowder et al., 2001). Some studies are suggestive of altered signaling in NPC cells. Changes in the phosphorylation of several kinases were detected in NPC mouse liver (Garver et al., 1999), cultured embryonic neurons showed a reduced response to brain-derived neurotrophic factor (Henderson et al., 2000), and increased activity of cyclin-dependent kinase 5 and hyperphosphorylation of microtubule-associated proteins was detected in brain tissue (Bu et al., 2002). However, raft-dependent cell surface signaling processes are yet to be examined specifically in NPC cells. Despite the rapid expansion in our understanding of raft behavior in the endocytic pathway that has developed since the discovery of NPC1, there is still no clear understanding of how defective raft trafficking contributes to the pathophysiology of neurodegeneration in NPC disease.

6. Inborn errors of sterol synthesis and congenital defects

6.1. Clinical aspects of defective cholesterol biosynthesis in humans and animal models

In marked contrast to NPC disease, children with inborn errors of sterol synthesis have congenital malformations. Smith–Lemli–Opitz syndrome (SLOS) was the first dysmorphic syndrome linked to a defect in cholesterol synthesis (reviewed by Porter, 2000; Kelley and Herman, 2001). SLOS is a multiple-malformation syndrome in which the most commonly affected structures are craniofacial (characteristic facies, microcephaly, cleft palate), limbs (syndactyly of digits 2 and 3, postaxial polydactyly), and external genitalia (hypospadias, ambiguous genitalia), accompanied by general defects such as poor feeding, growth retardation, and developmental disability. Less common but more severe variants had defects in internal organs such as congenital heart and lung defects, renal agenesis, and intestinal aganglionosis. Brain malformations occurred in 20–40% of patients, with about 5% having mild forms of holoprosencephaly, or an undivided forebrain. The biochemical defect is a deficiency of 7-dehydrocholesterol reductase (7DHCR), the enzyme that catalyzes the final step of cholesterol biosynthesis (Fig. 2). SLOS patients have extremely low levels of serum cholesterol, and accumulate high and lower levels of 7- and 8-dehydrocholesterol, respectively, precursors which are barely detected in normal individuals (Tint et al., 1994; Kelley et al., 1996; Fitzky et al., 1998; Wassif et al., 1998).

Since the identification of SLOS as a disorder of sterol synthesis, several other hereditary syndromes of dysmorphism or skeletal dysplasias were linked to mutations affecting different steps of cholesterol biosynthesis (Fig. 2) (Braverman et al., 1999; Konig et al., 2000; Waterham et al., 2001; for a detailed review, see Kelley and Herman, 2001). These include desmosterolosis (3β-hydroxysteroid-Δ^{24}-reductase deficiency), X-linked dominant chondrodysplasia punctata (CDPX2; 3β-hydroxysteroid-Δ^8,Δ^7-isomerase deficiency), and congenital hemidysplasia with ichthyosiform erythroderma and limb defects (CHILD syndrome, NADH steroid dehydrogenase-like/3β-hydroxysteroid dehydrogenase deficiency). Patients with desmosterolosis have increased serum desmosterol levels and share many

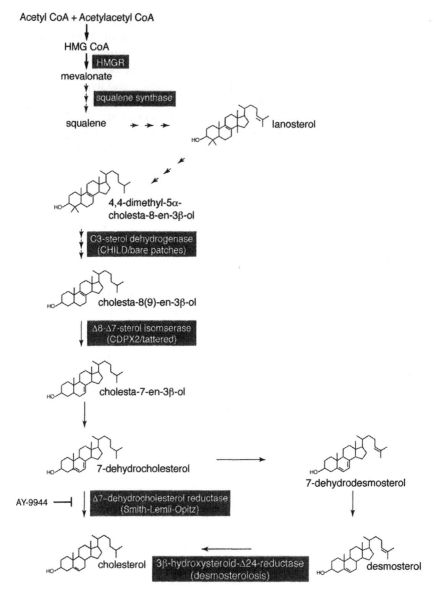

Fig. 2 The cholesterol biosynthetic pathway. Not all steps are shown. Enzymes discussed in the text and corresponding diseases are indicated in the gray boxes.

phenotypic features with SLOS patients, although the former have more severe skeletal dysplasia. The predominant features of CDPX2 and CHILD syndrome are abnormal calcifications in the cartilaginous epiphyses of immature bone and proximal shortening of the limbs, or rhizomelic dwarfism. Both have atrophic defects of skin, with hyperkeratotic lesions (ichthyosiform erythroderma) leading to alopecia. CHILD syndrome has the intriguing feature of unilateral involvement with ipsilateral localization of skin lesions and malformations, while CDPX2 is symmetrical. CHILD syndrome can present with more severe limb reductions and a handful of patients had other malformations including hypoplasia of the brain and spinal cord and renal agenesis. Other defects such as postaxial polydactyly can occur in CDPX2. These phenotypes are closely mirrored by the corresponding mouse mutants, *Tattered* (CDPX2) and *bare patches* (CHILD) (Derry et al., 1999; Liu et al., 1999). In CDPX2, 8-dehydrocholesterol and cholest-8(9)-en-3β-ol accumulate, while the step prior to this is blocked in CHILD syndrome, resulting in accumulation of these precursors containing C-4 methyl groups. CDPX2 and CHILD syndrome are less severe than SLOS and generally do not involve the central nervous system, but this probably reflects the fact that they are X-linked dominant disorders in which tissues of affected females are mosaic for mutant and wild-type chromosomes, and sufficient cholesterol might be provided by normal tissue. Both disorders are male lethal, and the mouse models indicated that male embryos are severely affected, with lethality occurring after mid-gestation for *Tattered* and soon after implantation for *bare patches*. The CDPX2/*Tattered* male mouse showed growth retardation with an abnormally shaped skull, micrognathia, cleft palate, absent intestines, and delayed ossification of the skeleton with severe shortening of all limbs (Derry et al., 1999).

The vast majority of dysmorphic syndromes are linked to mutations affecting components of major signaling pathways involved in pattern formation during development or the transcription factors through which they ultimately act. Of these pathways, which include members of the Wnt, transforming growth factor-β (TGF-β), bone morphogenetic protein (BMP) and Hedgehog (Hh) families, only the latter has components that require cholesterol and are clearly raft-associated. The roles of Hh proteins in patterning of the CNS and skeletal elements has led to a widely published hypothesis that defective Hh signaling is concomitant to defective cholesterol biosynthesis. The following sections examine the evidence that defects in Hh signaling or other raft-dependent pathways contribute to the dysmorphogenesis which occurs with these disorders of sterol biosynthesis.

6.2. Hedgehog signaling, known and putative roles for cholesterol

Members of the Hh family of morphogens are the only proteins known to be covalently modified by cholesterol. Named after the *Drosophila* segment polarity gene, the vertebrate Hh proteins Sonic, Indian, and Desert are involved in patterning during development of virtually every organ system (comprehensively reviewed by Ingham and McMahon, 2001). Sonic hedgehog (Shh) plays a particularly crucial role in patterning the CNS, while Indian hedgehog (Ihh) is essential for bone formation. One clear postnatal function of Shh is in the adult hair cycle (Wang et al., 2000), and somatic mutations in components of the signaling pathway underlie the vast majority of follicle-derived sporadic basal cell carcinomas. Cholesterol plays several roles in the pathway, from ligand biogenesis to signal transduction, but the precise nature of these roles are still poorly understood. This complexity was brought to light by phenotypic similarities between loss-of-function mutations in Shh and environmentally-induced or hereditary defects in cholesterol biosynthesis, such as SLOS.

6.2.1. Cholesterol in Hh biogenesis and transport

Hh proteins are synthesized as a precursor that undergoes an intramolecular autoproteolytic cleavage to generate the active signaling molecule (Porter et al., 1995). Cholesterol catalyzes the cleavage reaction, and becomes covalently attached to the new carboxyl terminus (Porter et al., 1996); this is really the only clearly defined role for cholesterol in the pathway. A second post-translational modification is crucial for the protein's activity, addition of a single palmitate to an amino terminal cysteine (Pepinsky et al., 1998; Chamoun et al., 2001; Lee et al., 2001; Lee and Treisman, 2001). Given these lipid modifications, it is not fully understood how Hh proteins act at considerable distances from source cells. In cells transfected with a full-length Shh cDNA, the majority of the protein is associated with rafts (Pepinsky et al., 1998). In *Drosophila*, a significant fraction of endogenous cholesterol-modified Hh was raft associated, but a large fraction was also found with nonraft membranes (Rietveld et al., 1999). An endogenous soluble form was detected in vertebrate embryonic tissue and demonstrated to be a noncovalent hexameric complex of cholesterol-modified monomers (Zeng et al., 2001). Alternatively, lipid-modified Hh might diffuse as a monomeric form through interactions of its hydrophobic regions with other proteins, in particular proteoglycans (Bellaiche et al., 1998; The et al., 1999). Release of Hh from source cells in *Drosophila* requires the activity of the SSD-containing protein Dispatched (Burke et al., 1999), which could either generate a multimeric soluble form from raft-associated Hh, or simply catalyze the release of monomeric Hh from the bilayer. However, it is currently unknown whether a single or multiple biochemical species mediates all of the signaling activities of Hh proteins, but it is clear that the cholesterol modification is required for long distance signaling in vertebrates (Lewis et al., 2001).

6.2.2. Cholesterol in Hh signal transduction

Cholesterol also plays an elusive role in Hh signal transduction. The Hh receptor, Patched (Ptc), has 12 transmembrane segments and shares extensive similarity to NPC1 including the SSD and a second set of five transmembrane segments (Fig. 1). Ptc controls the activity of a second protein, Smoothened (Smo), which is related to G protein-coupled receptors and actually transduces the Hh signal. Smo signaling activity is constitutive, but inhibited by Ptc in the absence of ligand. Hh binding to Ptc relieves this inhibition, allowing Smo to signal. The mechanism by which Ptc controls Smo activity is still unknown, as the steps are immediately downstream from Smo that link signaling to a transcriptional response mediated by members of the Ci/Gli transcription factor family (reviewed by Ingham and McMahon, 2001).

Although initial cotransfection studies suggested that Ptc and Smo formed a receptor complex (Stone et al., 1996; Carpenter et al., 1998; Murone et al., 1999), recent experiments have failed to demonstrate Ptc/Smo complexes *in vivo* or when not overexpressed transiently (Johnson et al., 2000; Incardona et al., 2002). Instead, Ptc appears to have an intrinsic activity that results in Smo inhibition indirectly, ultimately affecting Smo protein levels and activity through control of its phosphorylation state (Denef et al., 2000; Ingham et al., 2000). However, the inhibition of Smo by Ptc appears to involve some type of intimate interaction, as the two proteins colocalize in the absence of ligand, but are induced to segregate after ligand binding and internalization (Incardona et al., 2002). Smo is internalized along with Ptc/Hh complexes, but the latter are targeted for lysosomal degradation while Smo is recycled. This suggests that Hh inactivates Ptc through lysosomal targeting, and that active Smo is generated in a ligand-dependent manner after passing with Ptc through endosomes. Shh signaling is blocked by disruption of LBPA-rich late endosomal membranes, implicating the late endocytic sorting pathway in the segregation of Smo from Ptc/Shh complexes, and further underscoring the relationship between Ptc and NPC1 (Incardona et al., 2002). The specific subcellular location where Smo signals is presumed to be the plasma membrane, but this has not been determined yet.

Some role for cholesterol became evident with the study of teratogens that cause brain malformations; whether this reflects a function for rafts in Hh signal transduction is still unclear. One consequence of loss of Shh function is a failure of anterior neural tube patter resulting in undivided cerebral hemispheres, or holoprosencephaly, extreme forms of which have coincident cyclopia (Chiang et al., 1996; Roessler et al., 1996). Cyclopamine is a steroidal alkaloid derived from a ubiquitous meadow plant that was responsible for outbreaks of cyclopia in large lambing operations in the Western United States (reviewed by James, 1999). An unrelated teratogen shown to cause holoprosencephaly in rats is the cholesterol synthesis inhibitor AY-9944 (Roux and Aubry, 1966), which blocks the conversion of 7-dehydrocholesterol to cholesterol catalyzed by 7DHCR, the enzyme defective in SLOS. Mild forms of holoprosencephaly occur in a small percentage of SLOS patients (Kelley et al., 1996). Both cyclopamine and AY-9944 cause holoprosencephaly by blocking Shh signaling (Cooper et al., 1998; Incardona et al., 1998). Despite the steroidal nature of cyclopamine, it appears to be a direct antagonist of Shh signal transduction, and its affects are unrelated to cholesterol homeostasis or synthesis (Incardona et al., 1998; Incardona et al., 2000; Taipale et al., 2000). In contrast, the action of AY-9944 is indirect and the inhibition of Shh signaling is most likely due to the accumulation of sterol precursors combined with cholesterol deficit (Incardona et al., 1998; Gofflot et al., 2001). Holoprosencephaly presumably has the same etiology in SLOS, although this has not been demonstrated, and while targeted deletion of 7DHCR in the mouse reproduces many aspects of the human phenotype, holoprosencephaly was not observed (Fitzky et al., 2001; Wassif et al., 2001).

6.3. Cholesterol deficiency, sterol precursors and disruption of developmental signaling

It is unfortunate that the SLOS fibroblast has not received the same attention from cell biologists as has the NPC fibroblast. SLOS and other inborn errors of sterol synthesis may truly represent diseases of rafts, since the associated mutations have much more profound effects on the developing organism than loss of NPC1 function. However, the cellular consequences of replacing cholesterol with precursors such as 7-dehydrocholesterol have been hardly explored. Two studies of the ability of 7-dehydrocholesterol to promote formation of liquid-ordered phase domains with sphingomyelin in model membranes came to opposite conclusions that the precursor promoted domain formation more strongly than cholesterol (Xu et al., 2001) or that 7-dehydrocholesterol/sphingomyelin domains were unstable above 36°C (Wolf and Chachaty, 2000). The effects of sterol precursors on more general membrane properties are likewise largely uncharacterized, although one study observed decrease fluidity and alterations of permeability due to 7-dehydrocholesterol accumulation in intestinal microvillar membranes after 7DHCR inhibition in adult rats (Meddings, 1989). Moreover, it is still unclear how much of the SLOS/AY-9944 phenotype is strictly due to cholesterol deficit, the accumulation of 7-DHC and other precursors, or both.

The pleiotropic effects related to the various disorders of cholesterol synthesis may provide insight into this issue. It is unclear why embryos that accumulate different suites of precursors have phenotypes that do not overlap extensively, although the degree of sterol deficiency likewise is probably deeper in the more severe disorders. A more detailed analysis of male embryos in the X-linked disorders should be informative, since the rather cursory published phenotypic descriptions make attempts at linking malformations to specific signaling pathways very difficult. While holoprosencephaly associated with 7DHCR deficiency has a clear link to Shh signaling, other aspects of the cholesterol deficiency phenotypes do not. Male CDPX2/*Tattered* mice superficially resemble embryos carrying a targeted deletion of Ihh, which are similarly dwarfed due to reduced chondrocyte proliferation and failure of osteoblast development (St-Jacques et al., 1999). However, the CDPX2 male has a cholesterol synthesis defect in every cell, yet does not have the ventral midline patterning defects associated with loss of Shh signaling. Why does the CHILD/*bare patches* male die at an earlier stage of development than CDPX2/*Tattered?*

The pleiotropic nature of cholesterol synthesis defects probably points to the more widespread involvement of cholesterol in cellular and metabolic processes, such as steroid and vitamin D synthesis. What is the etiology of the more generalized defects found in SLOS patients? Areas that have not been explored are the role for cholesterol (and possibly Lubrol rafts) in synaptic vesicle recycling (Thiele et al., 2000), and the synaptogenesis-promoting activity of apoE-cholesterol derived from glial cells (Mauch et al., 2001). Similarly, other developmental signaling pathways that utilize GPI-anchored proteins as ligands or receptor components have not received the same degree of attention as has Hh signaling.

In particular, rafts may be important for ephrin family signaling, which function in CNS and limb development (Wada et al., 1998; Bruckner et al., 1999; Knoll and Drescher, 2002). Similarly, glial cell line-derived neurotrophic factor (GDNF) family members utilize a GPI-anchored coreceptor in combination with the Ret receptor tyrosine kinase, and function both during development of neural crest derivatives (Parisi and Kapur, 2000) and as neuronal survival factors (Saarma, 2000). Defects in proliferation and alterations in interleukin-1 signaling were observed in primary skin fibroblasts from CHILD syndrome ichthyosiform lesions, which also showed abnormal accumulations of lamellar vesicles and multivesicular bodies (Goldyne and Williams, 1989; Emami et al., 1992). However, these observations were made with cells cultured in the presence of lipoprotein-containing serum (prior to identification of the genetic defect), so it is unclear whether they were actually sterol deficient or had increased levels of precursors. The availability of several mouse models should provide the materials with which to address these questions more rigorously.

Since Hh signaling is the only pathway that has been examined in detail in the AY-9944 model, some inferences can be made about the relative roles of cholesterol deficit versus accumulation of precursors. It is often assumed that cholesterol deficit will impair Hh biogenesis. However, the sterol-catalyzed cleavage reaction in cultured cells was insensitive to treatment with cholesterol synthesis inhibitors alone, and was blocked only after drastic, acute reduction of cholesterol levels by combined lipoprotein deprivation, inhibition of synthesis at the level of HMGR, and extraction of cholesterol from the cell surface with a cyclodextrin (Guy, 2000). Moreover, a range of sterols was shown to catalyze the Shh cleavage reaction *in vitro* and undergo covalent attachment (Cooper et al., 1998). The acute sterol deprivation described above typically reduces cellular cholesterol content to 20–30% of untreated cells (Keller and Simons, 1998). Although SLOS patients have markedly depressed levels of plasma cholesterol, lipoprotein-starved cultured SLOS fibroblasts had total sterol levels of 70–80% of normal controls (Honda et al., 1997). Brain tissue from mice with a targeted deletion of 7DHCR showed total sterol levels that were 59% of wild-type littermates (Fitzky et al., 2001). In contrast, rat embryos from AY-9944-treated dams had total tissue sterols at 29% of controls (Gofflot et al., 2001). Cholesterol depression is mild in females with the X-linked disorders such as CDPX2, although sterol levels have not been reported for hemizygous males and are likely to be severely depressed. Skin fibroblasts from a CDPX2 patient with a nonsense mutation had cholesterol levels at 80% of controls, although total sterol levels were at 118% due to the accumulation of cholest-8(9)-en-3β-ol and other precursors (Braverman et al., 1999). Therefore, the AY-9944 model is the only situation in which cells might have the degree of sterol depletion capable of affecting Shh biogenesis. Effects on Shh biogenesis *in vivo* thus cannot be ruled out, since Shh processing has not been examined in AY-9944-treated embryos. Moreover, the activity, release from cells, transport, and multimerization capacity of Shh modified by a cholesterol precursor have not been tested.

The dependence of Hh signal transduction on total membrane sterol levels is unclear as well. Defects due to reduced Shh signaling in AY-9944-treated embryos were rescued by cholesterol supplementation, even in the presence of significant 7-DHC levels (Gofflot et al., 2001). Although this was taken as evidence that Shh signaling is reduced by cholesterol deficit per se rather than the presence of precursors, Shh signaling has not been quantified in embryos that solely have reduced cholesterol levels in the absence of precursor sterols. This issue could be addressed with mice carrying a targeted deletion of squalene synthase. Embryos null for squalene synthase died at post-embryonic day 9.5–10.5 with severe growth retardation and open neural tubes as the predominant malformation (Tozawa et al., 1999). This phenotype may represent the closest to a true sterol deficit, yet open neural tube defects are not associated with reduced Shh signaling. Although the phenotype was not described in great detail, neural plate explants from squalene synthase-deficient embryos might provide a definitive system for testing the sterol requirement in Shh signal transduction. In order to make any strong correlations, the sterol profiles of the most severely affected embryos for each of the sterol synthesis defects need to be examined in parallel.

6.4. Are rafts involved in Hh signal transduction?

No study has definitively demonstrated a role for rafts in organizing or controlling components of the Hh signal transduction cascade. The presence of a SSD in Ptc does not necessarily mean the protein associates with cholesterol or rafts. In fact no member of the SSD protein family has been shown to bind cholesterol directly. SSD proteins such as SCAP and HMGR reside in raft-poor ER or early Golgi membranes. All SSD proteins may not even be sterol sensitive. In *Drosophila*, which obtains dietary sterols and does not have the capacity for biosynthesis, the SSD of SCAP is responsive to phosphatidylethanolamine but not sterols, and the SREBP pathway may function more generally in the control of membrane lipid synthesis (Dobrosotskaya et al., 2002; Seegmiller et al., 2002). NPC1 is variably isolated in raft fractions, depending on the level of sterol loading of cells (Lusa et al., 2001), and in considering its function, it probably associates with raft membranes more dynamically. When transiently over-expressed in COS cells, a kidney-derived fibroblastic line, a small fraction of Ptc localized to buoyant density fractions and coimmunoprecipitated with caveolin-1 (Karpen et al., 2001). The significance of this finding is questionable, however, since caveolins are not expressed in the predifferentiated tissues of the embryo in which Hh signaling is active. Moreover, mice lacking caveolins do not have defects related to Hh signaling (Drab et al., 2001; Razani et al., 2001). When stably expressed at lower levels in KNRK cells, also a kidney-derived fibroblastic line, Ptc was not present at significant levels in Triton-insoluble rafts, although a significant fraction was recovered in Lubrol WX insoluble membranes (Incardona et al., 2002). Similarly, Smo was not strongly associated with Triton rafts in these cells either in the presence or absence of Ptc. However, the activity of Smo could not be analyzed in this system, so it is unclear if changes in Smo activity might affect raft association. Moreover, raft association of endogenous Ptc or Smo in cells with active signaling has not been determined.

The standard methods for assessing raft-dependency of a signaling cascade entail raft disruption in intact cells by cholesterol binding agents, typically modified cyclodextrins or polyene antibiotics such as filipin. Currently, the only assays for Hh signal transduction in vertebrate cells utilize changes in gene expression after a minimum of a day of exposure to ligand. Since embryonic cells generally cannot withstand high concentrations of cholesterol binding agents for this duration, it has been difficult to determine a role for rafts by these methods. The putative role for rafts can and should be tested in *Drosophila* cells, however, where there are more rapid endpoints for Hh signaling, such as phosphorylation of Smo (Denef et al., 2000). However, the sterol auxotrophy of *Drosophila* may have interesting implications. Since *Drosophila* incorporate whatever yeast and plant sterols are in their diets (Bos et al., 1976; Feldlaufer et al., 1995), it is likely that they have evolved a tolerance for variations in membrane sterol composition. However to rule rafts in or out may simply require more detailed biochemical analyses of Ptc and Smo from both insect and vertebrate systems.

7. Sterol-sensing domain proteins and signaling

In *Drosophila*, mutations in the SSD of Ptc render the protein incapable of regulating Smo, but do not affect binding of Hh and transport to endosomes (Martin et al., 2001; Strutt et al., 2001; Johnson et al., 2002). This is the first link of the SSD to regulation of signal transduction, but the mechanism remains a mystery. Ptc SSD mutants also antagonize the activity of the wild-type protein, suggesting that they compete for a factor involved in Smo inhibition. Another possibility is that the SSD is necessary for the colocalization of Ptc and Smo. Alternatively, since many transporters multimerize, and there is genetic evidence for multimerization of Ptc (Johnson et al., 2000), Ptc SSD mutants could antagonize the wild-type through a direct interaction. A difficulty in interpreting these studies is that the nature of an intrinsic activity for Ptc is unknown. The likelihood that Ptc possesses a permease activity similar to NPC1 raises the intriguing possibility that Ptc inhibits Smo by regulating the transbilayer distribution of a lipid or lipophilic substrate required for Smo activity. Although it is unknown how the SSD of NPC1 is linked to its permease activity, such an analysis would be more straightforward for that molecule. The genetic data from both Ptc and NPC1 indicate that the SSD is not involved in the overall targeting of the proteins, since the SSD mutants of each do not have gross changes in their localization to late endocytic compartments (Watari et al., 1999; Martin et al., 2001; Strutt et al., 2001). However, the failure of NPC1 SSD mutants to enter the late endosomal internal membranes and the recent finding that LBPA-rich membranes are important for Shh signaling (Incardona et al., 2002) suggest that an ultrastructural analysis of the mutants' distributions is warranted.

Recent studies of the SSD within SCAP provide insight into how sterols act on SSD proteins. SCAP regulates the activity of SREBPs by transporting them for cleavage when sterol levels are low. SCAP-SREBP complexes are retained in the

ER when sterols are abundant, but are incorporated into ER-Golgi transport vesicles when sterol levels fall. Addition of sterols to ER membranes altered the conformation of SCAP as determined by trypsin sensitivity, and this conformational change did not occur in SCAP SSD mutants (Brown et al., 2002). This raises the possibility that sterols regulate the interaction of SCAP with other proteins (an ER retention protein in this case) through SSD-mediated conformational changes. Additionally, the finding of sterol structural requirements for induction of the conformational change suggests that sterols (or other "sensed" lipids) do in fact interact directly with the SSD.

This raises several implications for Ptc. Although the cholesterol moiety of Shh is not required for binding to Ptc, the fully lipid-modified protein is about 30-fold more potent in activating signaling (and hence inactivating Ptc) (Pepinsky et al., 1998). The enhanced potency could result from a cholesterol-induced conformational change within Ptc that alters its activity or trafficking. Studies indicating that lipid-modified Shh is more efficient at targeting Ptc for lysosomal degradation than unmodified forms are consistent with this idea (J.P.I., unpublished observations). Alternatively, Ptc activity might change upon transport into or out of a sterol-rich microdomain after ligand binding. Whether this would be a Triton raft or Lubrol raft, at the cell surface or in endosomes remains to be determined.

Other receptors have ligand-binding properties that are influenced by membrane sterols through mechanisms unrelated to rafts. The oxytocin receptor showed a stringent sterol requirement for ligand binding and appeared to interact directly with sterols, showing very similar structural specificity as SCAP (Gimpl et al., 1997; Brown et al., 2002). These studies may thus have wider implications for receptor function, particularly in disorders of sterol biosynthesis.

8. Conclusions

The raft theory, the role of cholesterol in the formation of rafts and the function of rafts in signaling and membrane transport, has found wide support in an incredibly diverse array of cellular systems. However, these studies are largely limited to cultured cells *in vitro* and there has been no systematic genetic analysis. Given the complexities of lipid synthesis and the study of lipids *in vivo*, this is not surprising. However, the hereditary disorders described here should provide unique systems for genetic manipulation of rafts and a potential opportunity to disrupt raft-dependent signaling in the whole animal. Moreover, these diseases and their mouse models will provide a more detailed analysis of the wider roles of cholesterol in signaling, and the relationships between signaling and membrane trafficking. In particular, an increased understanding of neuronal cholesterol homeostasis and vesicular transport provided by NPC disease will be important for understanding β-amyloid precursor protein processing in Alzheimer disease, which appears to be linked to both cellular cholesterol homeostasis and raft-rich intracellular compartments (e.g., Koo and Squazzo, 1994; Simons et al., 1998; Walsh et al., 2000; Kojro et al., 2001; Wahrle et al., 2002). Similarly, a greater understanding of the role for sterols in the Hh pathway may help in the development of nonsurgical treatments for basal cell carcinoma, the most widespread tumor in the aging population. While the diseases discussed here will contribute in this way, they also demonstrate how little we really understand about cellular functions of cholesterol. At the same time, these are disorders with significant morbidity and mortality, but essentially no effective treatment. Affected individuals would undoubtedly benefit from a broader scientific audience.

Acknowledgements

J.P.I. is supported by the National Research Council Research Associateships Program.

References

Anderson, R.G., Jacobson, K., 2002. A role for lipid shells in targeting proteins to caveolae, rafts, and other lipid domains. Science 296, 1821–1825.

Auer, I.A. et al. 1995. Paired helical filament tau (PHFtau) in Niemann-Pick type C disease is similar to PHFtau in Alzheimer's disease. Acta Neuropathol. (Berl.). 90, 547–551.

Bellaiche, Y. et al. 1998. Tout-velu is a Drosophila homologue of the putative tumour suppressor EXT-1 and is needed for Hh diffusion. Nature 394, 85–88.

Bos, M. et al. 1976. Development of Drosophila on sterol mutants of the yeast Saccharomyces cerevisiae. Genet. Res. 28, 163–176.

Braverman, N. et al. 1999. Mutations in the gene encoding 3 beta-hydroxysteroid-delta 8, delta 7- isomerase cause X-linked dominant Conradi-Hunermann syndrome. Nat. Genet. 22, 291–294.

Brown, D.A., London, E., 2000. Structure and function of sphingolipid- and cholesterol-rich membrane rafts. J. Biol. Chem. 275, 17221–17224.

Brown, A.J. et al. 2002. Cholesterol addition to ER membranes alters conformation of SCAP, the SREBP escort protein that regulates cholesterol metabolism. Molec. Cell. 10, 237–245.

Bruckner, K. et al. 1999. EphrinB ligands recruit GRIP family PDZ adaptor proteins into raft membrane microdomains. Neuron. 22, 511–524.

Bu, B. et al. 2002. Deregulation of cdk5, hyperphosphorylation, and cytoskeletal pathology in the Niemann-Pick type C murine model. J. Neurosci. 22, 6515–6525.

Burke, R. et al. 1999. Dispatched, a novel sterol-sensing domain protein dedicated to the release of cholesterol-modified hedgehog from signaling cells. Cell 99, 803–815.

Carpenter, D. et al. 1998. Characterization of two patched receptors for the vertebrate hedgehog protein family. Proc. Natl. Acad. Sci. USA 95, 13630–13634.

Carstea, E.D. et al. 1997. Niemann-Pick C1 disease gene: homology to mediators of cholesterol homeostasis. Science 277, 228–231.

Chamoun, Z. et al. 2001. Skinny hedgehog, an acyltransferase required for palmitoylation and activity of the hedgehog signal. Science 293, 2080–2084.

Chatterjee, S. et al. 2001. GPI anchoring leads to sphingolipid-dependent retention of endocytosed proteins in the recycling endosomal compartment. EMBO J. 20, 1583–1592.

Chiang, C. et al. 1996. Cyclopia and defective axial patterning in mice lacking Sonic hedgehog gene function. Nature 383, 407–413.

Cooper, M.K. et al. 1998. Teratogen-mediated inhibition of target tissue response to Shh signaling. Science 280, 1603–1607.

Crowder, C.M. et al. 2001. Enantiospecificity of cholesterol function *in vivo*. J. Biol. Chem. 276, 44369–44372.

Cruz, J.C., Chang, T.Y., 2000. Fate of endogenously synthesized cholesterol in Niemann-Pick type C1 cells. J. Biol. Chem. 275, 41309–41416.

Davies, J.P. et al. 2000. Transmembrane molecular pump activity of Niemann-Pick C1 protein. Science 290, 2295–2298.

Denef, N. et al. 2000. Hedgehog induces opposite changes in turnover and subcellular localization of Patched and Smoothened. Cell 102, 521–532.

Derry, J.M. et al. 1999. Mutations in a delta 8-delta 7 sterol isomerase in the tattered mouse and X-linked dominant chondrodysplasia punctata. Nat. Genet. 22, 286–290.

Deutsch, J.W., Kelly, R.B., 1981. Lipids of synaptic vesicles: relevance to the mechanism of membrane fusion. Biochemistry 20, 378–385.

Dietrich, C. et al. 2002. Relationship of lipid rafts to transient confinement zones detected by single particle tracking. Biophys. J. 82, 274–284.

Dietschy, J.M., Turley, S.D., 2001. Cholesterol metabolism in the brain. Curr. Opin. Lipidol. 12, 105–112.

Dobrosotskaya, I.Y. et al. 2002. Regulation of SREBP processing and membrane lipid production by phospholipids in Drosophila. Science 296, 879–883.

Drab, M. et al. 2001. Loss of caveolae, vascular dysfunction, and pulmonary defects in caveolin-1 gene-disrupted mice. Science 293, 2449–2452.

Drobnik, W. et al. 2002. Apo AI/ABCA1-dependent and HDL3-mediated lipid efflux from compositionally distinct cholesterol-based microdomains. Traffic 3, 268–278.

Emami, S. et al. 1992. Peroxisomal abnormality in fibroblasts from involved skin of CHILD syndrome. Case study and review of peroxisomal disorders in relation to skin disease. Arch. Dermatol. 128, 1213–1222.

Feldlaufer, M.F. et al. 1995. Ecdysteroid production in Drosophila melanogaster reared on defined diets. Insect. Biochem. Mol. Biol. 25, 709–712.

Fink, J.K. et al. 1989. Clinical spectrum of Niemann-Pick disease type C. Neurology 39, 1040–1049.

Fitzky, B.U. et al. 2001. 7-Dehydrocholesterol-dependent proteolysis of HMG-CoA reductase suppresses sterol biosynthesis in a mouse model of Smith-Lemli-Opitz/RSH syndrome. J. Clin. Invest. 108, 905–915.

Fitzky, B.U. et al. 1998. Mutations in the Delta7-sterol reductase gene in patients with the Smith-Lemli-Opitz syndrome. Proc. Natl. Acad. Sci. USA. 95, 8181–8186.

Fivaz, M. et al. 2002. Differential sorting and fate of endocytosed GPI-anchored proteins. EMBO J. 21, 3989–4000.

Gagescu, R. et al. 2000. The recycling endosome of madin-darby canine kidney cells is a mildly acidic compartment rich in raft components. Mol. Biol. Cell. 11, 2775–2791.

Garver, W.S. et al. 1999. The Npc1 mutation causes an altered expression of caveolin-1, annexin II and protein kinases and phosphorylation of caveolin-1 and annexin II in murine livers. Biochim. Biophys. Acta. 1453, 193–206.

German, D.C. et al. 2001a. Degeneration of neurons and glia in the Niemann-Pick C mouse is unrelated to the low-density lipoprotein receptor. Neuroscience 105, 999–1005.

German, D.C. et al. 2001b. Selective neurodegeneration, without neurofibrillary tangles, in a mouse model of Niemann-Pick C disease. J. Comp. Neurol. 433, 415–425.

Gimpl, G. et al. 1997. Cholesterol as modulator of receptor function. Biochemistry 36, 10959–10974.

Gofflot, F. et al. 2001. Expression of Sonic Hedgehog downstream genes is modified in rat embryos exposed in utero to a distal inhibitor of cholesterol biosynthesis. Dev. Dyn. 220, 99–111.

Goldstein, J.L., Brown, M.S., 1990. Regulation of the mevalonate pathway. Nature 343, 425–430.

Goldyne, M.E., Williams, M.L., 1989. CHILD syndrome. Phenotypic dichotomy in eicosanoid metabolism and proliferative rates among cultured dermal fibroblasts. J. Clin. Invest. 84, 357–360.

Guy, R.K., 2000. Inhibition of sonic hedgehog autoprocessing in cultured mammalian cells by sterol deprivation. Proc. Natl. Acad. Sci. USA 97, 7307–7312.

Henderson, L.P. et al. 2000. Embryonic striatal neurons from Niemann-Pick type C mice exhibit defects in cholesterol metabolism and neurotrophin responsiveness. J. Biol. Chem. 275, 20179–20187.

Higgins, M.E. et al. 1999. Niemann-Pick C1 is a late endosome-resident protein that transiently associates with lysosomes and the trans-Golgi network. Mol. Genet. Metab. 68, 1–13.

Honda, A. et al. 1997. Sterol concentrations in cultured Smith-Lemli-Opitz syndrome skin fibroblasts: diagnosis of a biochemically atypical case of the syndrome. Am. J. Med. Genet. 68, 282–287.

Horton, J.D. et al. 2002. SREBPs: activators of the complete program of cholesterol and fatty acid synthesis in the liver. J. Clin. Invest. 109, 1125–1131.

Incardona, J.P. et al. 1998. The teratogenic *Veratrum* alkaloid cyclopamine inhibits Sonic Hedgehog signal transduction. Development 125, 3553–3562.

Incardona, J.P. et al. 2000. Cyclopamine inhibition of Sonic hedgehog signal transduction is not mediated through effects on cholesterol transport. Dev. Biol. 224, 440–452.

Incardona, J.P. et al. 2002. Sonic hedgehog induces the segregation of patched and smoothened in endosomes. Curr. Biol. 12, 983–995.

Ingham, P.W. et al. 2000. Patched represses the Hedgehog signalling pathway by promoting the modification of the Smoothened protein. Curr. Biol. 10, 1315–1318.

Ingham, P.W., McMahon, A.P., 2001. Hedgehog signaling in animal development: paradigms and principles. Genes Dev. 15, 3059–3087.

James, L.F., 1999. Teratological research at the USDA-ARS poisonous plant research laboratory. J. Nat. Toxins. 8, 63–80.

Johnson, R.L. et al. 2000. *In vivo* functions of the patched protein: requirement of the C terminus for target gene inactivation but not Hedgehog sequestration. Mol. Cell. 6, 467–478.

Johnson, R.L. et al. 2002. Distinct consequences of sterol sensor mutations in Drosophila and mouse patched homologs. Dev. Biol. 242, 224–235.

Karpen, H.E. et al. 2001. The Sonic Hedgehog receptor patched associates with caveolin-1 in cholesterol-rich microdomains of the plasma membrane. J. Biol. Chem. 276, 19503–19511.

Keller, P., Simons, K., 1998. Cholesterol is required for surface transport of influenza virus hemagglutinin. J. Cell Biol. 140, 1357–1367.

Kelley, R.I., Herman, G.E., 2001. Inborn errors of sterol biosynthesis. Annu. Rev. Genomics Hum. Genet. 2, 299–341.

Kelley, R.L. et al. 1996. Holoprosencephaly in RSH/Smith-Lemli-Opitz syndrome: does abnormal cholesterol metabolism affect the function of Sonic Hedgehog? Am. J. Med. Genet. 66, 478–484.

Knoll, B., Drescher, U., 2002. Ephrin-As as receptors in topographic projections. Trends Neurosci. 25, 145–149.

Ko, D.C. et al. 2001. Dynamic movements of organelles containing Niemann-Pick C1 protein: npc1 involvement in late endocytic events. Mol. Biol. Cell. 12, 601–614.

Kobayashi, T. et al. 1999. Late endosomal membranes rich in lysobisphosphatidic acid regulate cholesterol transport. Nature Cell Biology. 1, 113–118.

Kobayashi, T. et al. 1998. A lipid associated with the antiphospholipid syndrome regulates endosome structure and function. Nature 392, 193–197.

Kojro, E. et al. 2001. Low cholesterol stimulates the nonamyloidogenic pathway by its effect on the alpha -secretase ADAM 10. Proc. Natl. Acad. Sci. USA 98, 5815–5820.

Konig, A. et al. 2000. Mutations in the NSDHL gene, encoding a 3beta-hydroxysteroid dehydrogenase, cause CHILD syndrome. Am. J. Med. Genet. 90, 339–346.

Koo, E.H., Squazzo, S.L., 1994. Evidence that production and release of amyloid beta-protein involves the endocytic pathway. J. Biol. Chem. 269, 17386–17389.

Koval, M., Pagano, R.E., 1989. Lipid recycling between the plasma membrane and intracellular compartments: transport and metabolism of fluorescent sphingomyelin analogues in cultured fibroblasts. J. Cell Biol. 108, 2169–2181.

Kurzchalia, T.V. et al. 1992. VIP21, a 21-kD membrane protein is an integral component of trans-Golgi-network-derived transport vesicles. J. Cell Biol. 118, 1003–1014.

Kuwabara, P.E., Labouesse, M., 2002. The sterol-sensing domain: multiple families, a unique role? Trends Genet. 18, 193–201.

Lange, Y., Steck, T.L., 1994. Cholesterol homeostasis. Modulation by amphiphiles. J. Biol. Chem. 269, 29371–29374.

Lange, Y. et al. 2000. Cholesterol movement in Niemann-Pick type C cells and in cells treated with amphiphiles. J. Biol. Chem. 275, 17468–17475.

Lange, Y. et al. 2002. Dynamics of lysosomal cholesterol in Niemann-Pick type C and normal human fibroblasts. J. Lipid Res. 43, 198–204.

Lebrand, C. et al. 2002. Late endosome motility depends on lipids via the small GTPase Rab7. EMBO J. 21, 1289–1300.

Lee, J.D. et al. 2001. An acylatable residue of Hedgehog is differentially required in Drosophila and mouse limb development. Dev. Biol. 233, 122–136.

Lee, J.D., Treisman, J.E., 2001. Sightless has homology to transmembrane acyltransferases and is required to generate active Hedgehog protein. Curr. Biol. 11, 1147–1152.

Lewis, P.M. et al. 2001. Cholesterol modification of sonic hedgehog is required for long-range signaling activity and effective modulation of signaling by Ptc1. Cell 105, 599–612.

Liscum, L., Faust, J.R., 1989. The intracellular transport of low density lipoprotein-derived cholesterol is inhibited in Chinese hamster ovary cells cultured with 3-beta-[2-(diethylamino)ethoxy]androst-5-en-17-one. J. Biol. Chem. 264, 11796–11806.

Liscum, L., Munn, N.J., 1999. Intracellular cholesterol transport. Biochim. Biophys. Acta. 1438, 19–37.

Liu, X.Y. et al. 1999. The gene mutated in bare patches and striated mice encodes a novel 3beta-hydroxysteroid dehydrogenase. Nat. Genet. 22, 182–187.

Liu, Y. et al. 2000. Alleviation of neuronal ganglioside storage does not improve the clinical course of the Niemann-Pick C disease mouse. Hum. Mol. Genet. 9, 1087–1092.

Loftus, S.K. et al. 1997. Murine model of Niemann-Pick C disease: mutation in a cholesterol homeostasis gene. Science 277, 232–235.

Love, S. et al. 1995. Neurofibrillary tangles in Niemann-Pick disease type C. Brain 118, 119–129.

Lusa, S. et al. 2001. Depletion of rafts in late endocytic membranes is controlled by NPC1-dependent recycling of cholesterol to the plasma membrane. J. Cell Sci. 114, 1893–1900.

Martin, V. et al. 2001. The sterol-sensing domain of Patched protein seems to control Smoothened activity through Patched vesicular trafficking. Curr. Biol. 11, 601–607.

Mauch, D.H. et al. 2001. CNS synaptogenesis promoted by glia-derived cholesterol. Science 294, 1354–1357.

Maxfield, F.R., 2002. Plasma membrane microdomains. Curr. Opin. Cell. Biol. 14, 483–487.

Mayor, S. et al. 1998. Cholesterol-dependent retention of GPI-anchored proteins in endosomes. EMBO J. 17, 4626–4638.

Meddings, J.B., 1989. Lipid permeability of the intestinal microvillus membrane may be modulated by membrane fluidity in the rat. Biochim. Biophys. Acta. 984, 158–166.

Mukherjee, S. et al. 1998. Cholesterol distribution in living cells: fluorescence imaging using dehydroergosterol as a fluorescent cholesterol analog. Biophys. J. 75, 1915–1925.

Mukherjee, S. et al. 1999. Endocytic sorting of lipid analogues differing solely in the chemistry of their hydrophobic tails. J. Cell. Biol. 144, 1271–1284.

Murone, M. et al. 1999. Sonic hedgehog signaling by the patched-smoothened receptor complex. Curr. Biol. 9, 76–84.

Naureckiene, S. et al. 2000. Identification of HE1 as the second gene of Niemann-Pick C disease. Science 290, 2298–2301.

Neufeld, E.B. et al. 1999. The Niemann-Pick C1 protein resides in a vesicular compartment linked to retrograde transport of multiple lysosomal cargo. J. Biol. Chem. 274, 9627–9635.

Okamoto, T. et al. 1998. Caveolins, a family of scaffolding proteins for organizing "preassembled signaling complexes" at the plasma membrane. J. Biol. Chem. 273, 5419–5422.

Okamura, N. et al. 1999. A porcine homolog of the major secretory protein of human epididymis, HE1, specifically binds cholesterol. Biochim. Biophys. Acta. 1438, 377–387.

Parisi, M.A., Kapur, R.P., 2000. Genetics of Hirschsprung disease. Curr. Opin. Pediatr. 12, 610–617.

Pentchev, P.G. et al. 1985. A defect in cholesterol esterification in Niemann-Pick disease (type C) patients. Proc. Natl. Acad. Sci. USA 82, 8247–8251.

Pentchev, P.G. et al. 1987. Group C Niemann-Pick disease: faulty regulation of low-density lipoprotein uptake and cholesterol storage in cultured fibroblasts. FASEB J. 1, 40–45.

Pepinsky, R.B. et al. 1998. Identification of a palmitic acid-modified form of human Sonic hedgehog. J. Biol. Chem. 273, 14037–14045.

Porter, F.D. 2000. RSH/Smith-Lemli-Opitz syndrome: a multiple congenital anomaly/mental retardation syndrome due to an inborn error of cholesterol biosynthesis. Mol. Genet. Metab. 71, 163–174.

Porter, J.A. et al. 1995. The product of hedgehog autoproteolytic cleavage active in local and long-range signalling. Nature 374, 363–366.

Porter, J.A. et al. 1996. Cholesterol modification of hedgehog signaling proteins in animal development. Science 274, 255–259.

Pralle, A. et al. 2000. Sphingolipid-cholesterol rafts diffuse as small entities in the plasma membrane of mammalian cells. J. Cell. Biol. 148, 997–1008.

Puri, V. et al. 1999. Cholesterol modulates membrane traffic along the endocytic pathway in sphingolipid-storage diseases. Nat. Cell. Biol. 1, 386–388.

Razani, B. et al. 2001. Caveolin-1 null mice are viable but show evidence of hyperproliferative and vascular abnormalities. J. Biol. Chem. 276, 38121–38138.

Rietveld, A. et al. 1999. Association of sterol- and glycosylphosphatidylinositol-linked proteins with *Drosophila* raft lipid microdomains. J. Biol. Chem. 274, 12049–12054.

Rodriguez-Lafrasse, C. et al. 1990. Abnormal cholesterol metabolism in imipramine-treated fibroblast cultures. Similarities with Niemann-Pick type C disease. Biochim. Biophys. Acta. 1043, 123–128.

Roessler, E. et al. 1996. Mutations in the human Sonic Hedgehog gene cause holoprosencephaly. Nat. Genet. 14, 357–360.

Roper, K. et al. 2000. Retention of prominin in microvilli reveals distinct cholesterol-based lipid microdomains in the apical plasma membrane. Nat Cell. Biol. 2, 582–592.

Roux, C., Aubry, M., 1966. Action teratogene chez le rat d'un inhibiteur de la synthese du cholestrol, le AY 9944. C. R. Seances Soc. Biol. Fil. 160, 1353–1357.

Saarma, M., 2000. GDNF – a stranger in the TGF-beta superfamily? Eur. J. Biochem. 267, 6968–6971.

Sandhoff, K., Klein, A., 1994. Intracellular trafficking of glycosphingolipids: role of sphingolipid activator proteins in the topology of endocytosis and lysosomal digestion. FEBS Lett. 346, 103–107.

Schutz, G.J. et al. 2000. Properties of lipid microdomains in a muscle cell membrane visualized by single molecule microscopy. EMBO J. 19, 892–901.

Seegmiller, A.C. et al. 2002. The SREBP pathway in Drosophila: regulation by palmitate, not sterols. Dev. Cell. 2, 229–238.

Simons, M. et al. 1998. Cholesterol depletion inhibits the generation of beta-amyloid in hippocampal neurons. Proc. Natl. Acad. Sci. USA 95, 6460–6464.

Smart, E.J. et al. 1999. Caveolins, liquid-ordered domains, and signal transduction. Mol. Cell. Biol. 19, 7289–7304.

St-Jacques, B. et al. 1999. Indian hedgehog signaling regulates proliferation and differentiation of chondrocytes and is essential for bone formation. Genes Dev. 13, 2072–2086.

Stone, D.M. et al. 1996. The tumour-suppressor gene patched encodes a candidate receptor for Sonic hedgehog. Nature 384, 129–134.

Strutt, H. et al. 2001. Mutations in the sterol-sensing domain of Patched suggest a role for vesicular trafficking in Smoothened regulation. Curr. Biol. 11, 608–613.

Sugimoto, Y. et al. 2001. Accumulation of cholera toxin and GM1 ganglioside in the early endosome of Niemann-Pick C1-deficient cells. Proc. Natl. Acad. Sci. USA 98, 12391–12396.

Suzuki, K. et al. 1995. Neurofibrillary tangles in Niemann-Pick disease type C. Acta Neuropathol. (Berl.). 89, 227–238.

Taipale, J. et al. 2000. Effects of oncogenic mutations in Smoothened and Patched can be reversed by cyclopamine. Nature 406, 1005–1009.

The, I. et al. 1999. Hedgehog movement is regulated through tout velu-dependent synthesis of a heparan sulfate proteoglycan. Mol. Cell. 4, 633–639.

Thiele, C. et al. 2000. Cholesterol binds to synaptophysin and is required for biogenesis of synaptic vesicles. Nat. Cell. Biol. 2, 42–49.

Tint, G.S. et al. 1994. Defective cholesterol biosynthesis associated with the Smith-Lemli-Opitz syndrome. N. Engl. J. Med. 330, 107–113.

Tozawa, R. et al. 1999. Embryonic lethality and defective neural tube closure in mice lacking squalene synthase. J. Biol. Chem. 274, 30843–30848.

Tseng, T.T. et al. 1999. The RND permease superfamily: an ancient, ubiquitous and diverse family that includes human disease and development proteins. J. Mol. Microbiol. Biotechnol. 1, 107–125.

van Meer, G., Simons, K., 1988. Lipid polarity and sorting in epithelial cells. J. Cell. Biochem. 36, 51–58.

Varma, R., Mayor, S., 1998. GPI-anchored proteins are organized in submicron domains at the cell surface. Nature 394, 798–801.

Wada, N. et al. 1998. Glycosylphosphatidylinositol-anchored cell surface proteins regulate position-specific cell affinity in the limb bud. Dev. Biol. 202, 244–252.

Wahrle, S. et al. 2002. Cholesterol-Dependent gamma-Secretase Activity in Buoyant Cholesterol-Rich Membrane Microdomains. Neurobiol. Dis. 9, 11–23.

Walsh, D.M. et al. 2000. The oligomerization of amyloid beta-protein begins intracellularly in cells derived from human brain. Biochemistry 39, 10831–10839.

Wang, L.C. et al. 2000. Conditional disruption of hedgehog signaling pathway defines its critical role in hair development and regeneration. J. Invest. Dermatol. 114, 901–908.

Wassif, C.A. et al. 1998. Mutations in the human sterol delta7-reductase gene at 11q12–13 cause Smith-Lemli-Opitz syndrome. Am. J. Hum. Genet. 63, 55–62.

Wassif, C.A. et al. 2001. Biochemical, phenotypic and neurophysiological characterization of a genetic mouse model of RSH/Smith–Lemli–Opitz syndrome. Hum. Mol. Genet. 10, 555–564.

Watari, H. et al. 1999. Mutations in the leucine zipper motif and sterol-sensing domain inactivate the Niemann-Pick C1 glycoprotein. J. Biol. Chem. 274, 21861–21866.

Waterham, H.R. et al. 2001. Mutations in the 3beta-hydroxysterol delta24-reductase gene cause desmosterolosis, an autosomal recessive disorder of cholesterol biosynthesis. Am. J. Hum. Genet. 69, 685–694.

Weigmann, A. et al. 1997. Prominin, a novel microvilli-specific polytopic membrane protein of the apical surface of epithelial cells, is targeted to plasmalemmal protrusions of non-epithelial cells. Proc. Natl. Acad. Sci. USA 94, 12425–12430.

Wolf, C., Chachaty, C., 2000. Compared effects of cholesterol and 7-dehydrocholesterol on sphingomyelin-glycerophospholipid bilayers studied by ESR. Biophys. Chem. 84, 269–279.

Wustner, D. et al. 2002. Rapid nonvesicular transport of sterol between the plasma membrane domains of polarized hepatic cells. J. Biol. Chem. 277, 30325–30336.

Xie, C. et al. 2000. Cholesterol is sequestered in the brains of mice with Niemann-Pick type C disease but turnover is increased. J. Neuropathol. Exp. Neurol. 59, 1106–1117.

Xu, X. et al. 2001. Effect of the structure of natural sterols and sphingolipids on the formation of ordered sphingolipid/sterol domains (rafts). Comparison of cholesterol to plant, fungal, and disease-associated sterols and comparison of sphingomyelin, cerebrosides, and ceramide. J. Biol. Chem. 276, 33540–33546.

Zacharias, D.A. et al. 2002. Partitioning of lipid-modified monomeric GFPs into membrane microdomains of live cells. Science 296, 913–916.

Zeng, X. et al. 2001. A freely diffusible form of Sonic hedgehog mediates long-range signalling. Nature 411, 716–720.

Zervas, M. et al. 2001. Critical role for glycosphingolipids in Niemann-Pick disease type C. Curr. Biol. 11, 1283–1287.

Zhang, M. et al. 2001. Cessation of rapid late endosomal tubulovesicular trafficking in Niemann-Pick type C1 disease. Proc. Natl. Acad. Sci. USA 98, 4466–4471.

**Advances in
Cell Aging and
Gerontology**

Cholesterol, β-amyloid, and Alzheimer's disease

Miguel A. Pappolla[1], Suzana Petanceska[2], Lawrence Refolo[3] and Nicolas G. Bazan[4]

[1]*Department of Pathology, University of South Alabama Medical Center, Mobile, Alabama, USA.*
[2]*Nathan Kline Institute, Orangeburg, New York, USA.*
[3]*Institute for the Study of Aging, New York, New York, USA.*
[4]*LSU Neuroscience Center, Louisiana State University Health Sciences Center,
New Orleans, Louisiana, USA.*
*Correspondence address: Miguel A. Pappolla, LSU Neuroscience Center, 2020 Gravier Street,
Suite D, New Orleans, LA 70112. Tel.: +1-504-599-0831; fax: +1-504-568-5801.*

Contents

1. Introduction

Many studies have shown that the brains of persons with Alzheimer's disease (AD) are subject to widespread oxidative stress. One important set of data arose from *in vitro* models and suggests that the amyloid peptide causes extensive degeneration and death of neurons by mechanisms that involve free radicals. Another body of work revealed that oxidative stress is increased in regions with amyloid deposition. In addition, emerging experimental data implicate cholesterol in the early steps of amyloid formation. This review discusses the potential roles of oxygen free radicals and cholesterol in the pathogenesis of AD. Results from

Advances in Cell Aging and Gerontology, vol. 12, 163–175

our laboratory suggest that the cholesterol content in plasma and brain of Alzheimer's-transgenic mice is strongly correlated with the rate of development of amyloid pathology. The data suggest two potential mechanisms whereby hyper-cholesterolemia accelerates brain β-amyloid accumulation. One of these is altered APP processing and the other is increased apoE expression. Participation of unscheduled oxidations (oxidative stress) in these processes is possible, as suggested by significant parallels with atherosclerosis.

2. Neuropathology of AD

AD is a progressive neurodegenerative disorder characterized by behavioral pathology and global deterioration of intellectual functions (Selkoe, 2000). A major neuropathologic feature of AD is the neuritic or senile plaque, which, in addition to other components, is comprised primarily of insoluble β-amyloid fibrils of a 40- to 42-amino acid peptide termed Aβ (Selkoe, 2000). Aβ is proteolytically derived, by the action of β- and γ-secretase activities, from the large precursor protein APP (Selkoe, 2000). Alternately, APP is cleaved by α-secretase activities, producing a soluble APP fragment (sAPPα), which precludes Aβ formation (Moghadasian, 1999). Recent studies have shown that increased levels of Aβ peptides are among the earliest detectable abnormalities in the pathophysiology of AD and may mediate downstream events that lead to neuronal degeneration and cognitive decline (Selkoe, 2000). Intracytoplasmic neuronal inclusions called neurofibrillary tangles are another conspicuous neuro-pathologic feature of the disorder (Esiri et al., 1997). Filamentous bundles of hyperphosphorylated tau proteins are, among other cytoskeletal proteins, the principal components of these lesions (Grundke-Iqbal et al., 1986; Kosik, 1993). Another hallmark of AD is the extensive neuronal and synaptic loss in brain; these deficits appear to be the most responsible for patients' symptoms (Hyman et al., 1995; Masliah, 1995). Senile plaques and neurofibrillary tangles are also present (in substantially fewer numbers) in most intellectually normal individuals fortunate to reach an advanced age (Pappolla et al., 1999). In addition, these lesions are consistently observed in the majority of older patients with Down syndrome (Esiri et al., 1997). Most cases of AD are sporadic, but approximately 5% have a familial pattern of inheritance. The neuropathology is identical in both forms (Esiri et al., 1997).

3. Amyloid as the cause of neuronal degeneration in AD

Several lines of evidence strongly implicate Aβ in the widespread neuronal degeneration that occurs in brains with AD, including genetic studies identifying several point mutations within the APP gene. These mutations segregate with a subgroup of patients afflicted with a familial form of AD, strongly suggesting a relationship between the APP gene, Aβ, and AD (Chartier-Harlin et al., 1991; Goate et al., 1991). A second group of studies demonstrated that Aβ is neurotoxic for cultured neurons (Yanker et al., 1990; Breen et al., 1991). A third line of support for the Aβ hypothesis arises from observations that amyloid deposition generally precedes the development of neurofibrillary changes (Pappolla and Robakis, 1995; Tagliavini et al., 1988). Despite some initial controversy emerging from *in vitro* studies, the weight of the evidence currently suggests that Aβ is a critical player in the pathogenesis of AD (Pappolla and Ogden, 2000).

4. The cholesterol-amyloid connection

Several clinical, epidemiologic, and laboratory studies suggest that cholesterol plays a role in the pathogenesis of AD (Bodovitz and Klein, 1996; Mizuno et al., 1998; Simons et al., 1998; Frears et al., 1999; Grant, 1999; Roher et al., 1999; Wood et al., 1999; Bullido and Valdivieso, 2000; Jick et al., 2000; Refolo et al., 2000; Sparks et al., 2000; Wolozin et al., 2000). These include *in vitro* studies indicating that cellular cholesterol content modulates Aβ production and the enzymatic processing of APP (Bodovitz and Klein, 1996; Mizuno et al., 1998; Simons et al., 1998; Frears et al., 1999); animal studies demonstrating that cholesterol modulates Aβ accumulation in the brain (Refolo et al., 2000; Sparks et al., 2000); and several observational, clinical studies suggesting that the prevalence and incidence of probable AD are lower in patients taking cholesterol-lowering drugs (Jick et al., 2000; Wolozin et al., 2000). Taken together these studies suggest that, at least in a subgroup of AD patients, there may be an alteration of cholesterol homeostasis that modulates the disease neuropathology.

In a transgenic mouse model of AD (double APP/presenilin 1 mutant), we have shown that diet-induced hypercholesterolemia resulted in a dramatic acceleration of the neuropathologic and biochemical changes in the animals (Refolo et al., 2000). The hypercholesterolemic mice showed a marked increase in amyloid deposition and significantly greater levels of Aβ40 and Aβ42 peptides in the brain. Levels of total Aβ correlated with the levels of both plasma and total brain cholesterol. In addition, biochemical analyses of brain homogenates revealed that, compared with the control, the hypercholesterolemic mice had significantly decreased levels of secreted soluble APP N-terminus fragments (sAPP, cleaved at the α-secretase site) and increased levels of APP βCTF (cleaved at the β-secretase site) (Refolo et al., 2000). Conversely, we performed experiments blocking cholesterol synthesis in transgenic mice that yielded virtually opposite results: a dramatic decrease in the amyloid load in treated mice accompanied by enhanced α-secretase cleavage of APP (Refolo et al., 2001).

Studies by other groups of investigators also support the hypothesis of a connection between high cholesterol and amyloid pathology *in vivo*. In a study using an APP gene-targeted mouse (expressing the Swedish familial AD mutation), increased dietary cholesterol led to changes in levels of non-amyloidogenic sAPPα in brain that were negatively correlated with serum cholesterol levels (Howland et al., 1998). Reduction in secreted sAPPα leads to amyloidogenic APP processing and ultimately increased Aβ production. These results demonstrate that APP processing and the level of Aβ peptides can be modulated *in vivo* by hypercholesterolemia and provide evidence that cholesterol plays a mechanistic role in the formation of amyloid. Further support for the link between cholesterol and amyloid formation comes from *in vitro* studies that agree in concept with the *in vivo* research. Cholesterol added to the culture medium of APP 751 stably transfected HEK 293 cells lowers sAPP production (Bodovitz and Klein, 1996). In addition, cells depleted of cholesterol by treatment with lovastatin showed reduced secretion of Aβ, a phenomenon that is fully reversible by the re-addition of cholesterol (Simons et al., 1998). Another study showed that the reduction of intracellular cholesterol blocks the production of Aβ by reducing the production of amyloidogenic APP C-terminal fragments by β-secretase. Conversely, supplementing these cells with cholesterol resulted in increased β-secretase cleavage products along with increased secretion of Aβ1-40 and Aβ1-42 (Frears et al., 1999). Taken together, the data suggest that cholesterol causes an increase in amyloid generation by inducing amyloidogenic processing of the amyloid precursor.

However, additional experimental data from human studies suggest that cholesterol content alone (in the brain or plasma) may not be sufficient to account for the extent of the amyloid pathology. In a recent retrospective review of autopsy cases, conducted to determine whether a relationship exists between amyloid deposition and cholesterolemia (Pappolla et al., unpublished observations), we found a highly significant association between cholesterolemia and presence of amyloid; however, no significant correlation (in contrast with the data from mice) was present between cholesterolemia and amyloid load. Therefore, it appears that in addition to higher-than-normal cholesterolemia, amyloid pathology in human brain is dependent on additional factors.

At present, one can only speculate as to how cholesterol levels affect brain Aβ accumulation. Data from our experiments in transgenic mice and from other laboratories suggest two possibilities: one is that alterations in plasma and/or CNS cholesterol levels modulate APP processing and Aβ accumulation. The other possibility would involve Aβ clearance, which as suggested by ongoing studies in our laboratories, may be mediated by higher-than-normal levels of apoE expression.

5. The apoE–amyloid connection

An important connection between cholesterol and AD is the epidemiologic and genetic data implicating apoE, a cholesterol-transport protein, in various aspects of the disease pathogenesis (Tomiyama et al., 1999; Fagan and Holtzman, 2000). Despite the fact that apoE remains the main susceptibility gene in AD, its role in the disease is poorly understood. Several pathogenic mechanisms have been postulated, including functional differences between isoforms (Finch and Sapolsky, 1999), lipoprotein metabolism (Beffert et al., 1998), and levels of apoE expression that may also vary as a response to various forms of injury (Laskowitz et al., 1998). The last of these includes oxidative stress, which can also be triggered or exacerbated by a spectrum of injury modalities including hypercholesterolemia and hyperhomocystinemia, as well as other mechanisms ranging from head injury to reactivation of latent infections.

ApoE is a 299-amino acid protein that binds cholesterol and is a major constituent of several cholesterol-carrying lipoprotein particles (Tomiyama et al., 1999; Fagan and Holtzman, 2000). ApoE binds Aβ peptides and is believed to promote fibrillization of soluble Aβ, affecting amyloid clearance from the brain (Fagan and Holtzman, 2000; Zlokovic et al., 2000). There are three common alleles of apoE in humans: E2, E3, and E4. Genetic and epidemiologic studies indicate that the E4 allele is a major risk factor for late-onset, sporadic AD. Individuals with two copies of the apoE4 allele have elevated plasma cholesterol levels and apoE levels and also increased risk for developing cardiovascular disease (Jarvik et al., 1994; Jarvik et al, 1995; Notkola et al., 1998). The importance of apoE in Aβ deposition has been strongly suggested in apoE-knockout mice, where there is markedly decreased Aβ deposition and little or no fibrillar Aβ (Bales et al., 1999). The three isoforms of ApoE differ in two amino acid positions such that E2 contains two cysteines in those positions E3 contains a cysteine in one of the positions; while E4 does not have cysteines in either position. Interestingly, Pedersen et al. (2000) have provided evidence that E2 and E3 isoforms are much more effective in protecting neurons against oxidative injury because they are able to covalently bind 4-hydroxynonenal, a toxic lipid peroxidation product that covalently modifies proteins on cysteine residues. Previous studies have suggested that lipid peroxidation and 4-hydroxynonenal play important roles in the degeneration of neurons in AD (Mark et al., 1997a, 1997b; Mattson et al., 1997).

To support the hypothesis that apoE levels and cholesterol are related to, and perhaps essential for, amyloid accumulation, we performed preliminary experiments testing whether cholesterol regulates brain apoE levels in AD-transgenic mice (Refolo et al., work in progress). Western and Northern blot methods showed very large increases in mRNA and protein levels of brain apoE in the mice that were placed on different cholesterol treatments. Results showed a strong positive correlation between cholesterol levels, brain apoE expression, and the amyloid pathology. The mean levels of brain apoE in the hypercholesterolemic group were approximately 65% greater than that of the control group. Conversely, the mean apoE levels in mice treated with cholesterol-lowering drugs were approximately 31% less than that of the control group. Although this was the first demonstration of such a regulatory pathway in the brain, it is not the first time that a connection between cholesterol and apoE expression has been demonstrated. In previous work conducted in rats, diet-induced hypercholesterolemia led to increased apoE expression in the liver. While the mechanism for the regulation of brain apoE by cholesterol remains unknown, the finding may explain, in part, the link between cholesterol metabolism, apoE genotype, and an increased risk of AD. In the context of these studies, it appears that cholesterol influences both APP processing and brain apoE levels. Recently published data from another group of investigators appear to support our hypothesis and show that the levels of apoE mRNA and protein are significantly higher in the brains and plasma of patients with AD (Boyt et al., 1999). Most germane to the above discussion are studies in humans demonstrating that certain apoE-promoter polymorphisms are associated with increased apoE expression and increased risk of AD, independent of the risk conveyed by the E4 allele of apoE. Interestingly, some of these promoter polymorphisms appear to co-segregate with the apoE4 isoform. ApoE transcription is very complex and driven by an array of tissue-specific cis-acting elements that are distributed as distal and proximal elements along a 20-kb region spanning the gene. In the liver, apoE expression is regulated by diet (Lin-Lee et al., 1981; Kim et al., 1989), thyroid and growth hormones, insulin, and estrogens (Tam et al., 1986; Oscarsson et al., 1991; Ogbonna et al., 1993).

6. Cholesterol and oxidative stress in AD

Work from many laboratories has demonstrated that brains of patients with AD are subject to a pervasive load of oxidative stress. Among the most important unresolved issues regarding the role of oxidative stress and AD is whether this form of injury is the cause or consequence of amyloidogenesis. Current experimental evidence suggests that both propositions may be operational, i.e., that an initial source of oxidative stress, such as high cholesterol or other forms of injury, may initiate amyloid formation, which in itself is another potent source of oxidative stress in AD. Thus, one could speculate that an initial free radical-induced injury would inaugurate a vicious cycle in which amyloidogenic processing of APP is further enhanced, generating more Aβ, which in turn exacerbates oxidative stress (Pappolla et al., 1995; Yan et al., 1995; Zhang et al., 1997).

Although mounting data from *in vitro* studies appear to support such a mechanism, more information is necessary, since direct evidence of oxidation-mediated amyloidogenesis is still lacking from *in vivo* paradigms. Therefore, the possibility of whether such a self-sustaining cycle is involved in the pathogenesis of AD is intriguing and in need of more supporting research. Figure 1 illustrates the manner in which cholesterol and oxidative stress may be involved in the pathogenesis of AD.

Thus, it is possible that cholesterol, acting through parallel mechanisms to those of atherosclerosis, causes free radical-induced membrane damage to neurons and increased apoE expression. In the case of AD, decreased secreted APP and increased production of amyloidogenic fragments may result from cellular cholesterol causing oxidation of membrane components and a decrease in accessibility of secretases to APP substrate. In addition, increased apoE expression may result from Aβ-mediated oxidative damage and microglial cell activation, leading to acceleration of Aβ aggregation and its decreased removal. As mentioned, both possibilities are supported by experimental data from AD-transgenic mice. In addition, it was previously shown in microglia (which are the main cell type producing apoE in response to damage of the nervous system) that apoE transcription is strongly induced by tumor necrosis factor (Duan et al., 1995), sterols (Duan et al., 1997), oxidized LDL (Cader et al., 1997), and neurotoxic agents (Boschert et al., 1999). In this regard, elevated cholesterol content is associated with increases in markers of oxidative stress (Prasad and Kalra, 1993) that can lead to increases in Aβ generation (Pappolla et al., 1995; Yan et al., 1995;

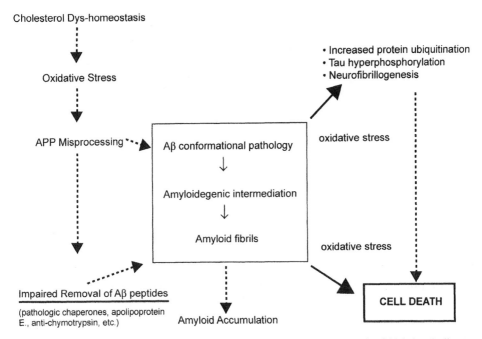

Fig. 1. How cholesterol and oxidative stress may be involved in the pathogenesis of Alzheimer's disease.

Zhang et al., 1997). Superoxide dismutase-1 (SOD-1) is upregulated in vessels and perivascular tissues in the presence of increasing cholesterol levels. In rabbits, a highcholesterol diet increased the lipid peroxidation by-product, malondialdehyde, and increased the oxygen free radical-producing activity of polymorphonuclear leukocytes (Prasad, 1999). These indicators are reduced when vitamin E, a free-radical scavenger, is administered in conjunction with high-cholesterol diets (Prasad and Kalra, 1993). There is one potentially neurotoxic substance arising from altered cholesterol metabolism, which may involve free-radical damage: 24S-hydroxycholesterol, the main cholesterol-elimination product of the brain. 24S-hydroxycholesterol is increased in the serum of patients with Alzheimer's disease, as well as those with vascular dementia, and it is neurotoxic (Lutjohann et al., 2000). Interestingly, the neurotoxicity of 24S-hydroxycholesterol is also prevented by vitamin E (Kolsch et al., 2001). Clinical benefit on slowing time to important endpoints has been seen with vitamin E in AD (Sano et al., 1997).

Identification of the subcellular and molecular targets of oxidative damage is an important area for future study, and is being pursued by our laboratories. APP, Aβ, and the putative γ-secretase have all been found in cholesterol- and glycosphingolipid-enriched lipid rafts, raising the possibility that raft microdomains or their associated proteins (such as caveolins), may be targets of oxidative injury. Rafts are subcellular structures of eukaryotic cells that have been implicated in many cellular processes, such as polarized sorting of apical membrane proteins in epithelial cells and signal transduction (Kurzchalia and Parton, 1999).

7. Cholesterol and AD: the clinical evidence

Several clinical studies have also identified significant associations among total plasma cholesterol and AD risk. Notkola et al. examined the possible role of total plasma cholesterol level and AD in a population-based sample of 444 men aged 70–89 years. The investigators found that a high cholesterolemia measured prior to diagnosis was a significant predictor of the prevalence of AD, after controlling for age and apoE isoforms. In men who subsequently developed AD, the cholesterol level in plasma decreased before the clinical manifestations of AD. These findings led the investigators to conclude that high total cholesterolemia is an independent risk factor for AD and that the effect of the apoE epsilon 4 allele on risk of AD may, at least in part, be mediated through high plasma cholesterol (Notkola et al., 1998). Similarly, Jarvik and colleagues examined the relationship between apoE genotype, cholesterolemia, and AD risk in a community-based study of 206 AD cases and 276 controls. Using a logistic regression model, the authors found that the relationship between apoE genotype and AD was dependent, at least in part, on total plasma cholesterol (Jarvik et al., 1995). In another population-based study of elderly African-Americans, Evans et al. found that high cholesterol level was associated with increased prevalence of AD (Evans et al., 2000). Additional support for this hypothesis is provided in a study by Kivipelto et al. (Kivipelto et al., 2001). These investigators evaluated the impact of elevated midlife plasma cholesterol level and blood pressure on the subsequent development of mild cognitive impairment, a potential precursor of AD. The authors found that elevated midlife cholesterol level was a significant and independent risk factor for the development of late-life cognitive decline. Lastly, observational studies have demonstrated an association between statin treatment and reduced prevalence and incidence of probable AD (Jick et al., 2000; Wolozin et al., 2000), suggesting that reduction of cholesterol levels before the disease becomes manifested may decrease the risk of developing AD.

However, a number of studies regarding cholesterol and AD (Sparks, 1997; Mason et al., 1992) have generated contradictory information. Examining the frontal cortex of AD patients with apoE4 genotypes, one study reported that the cholesterol content was significantly increased compared to nondemented controls (Sparks et al., 1997). Data from another study, however, produced the opposite results (Mason et al., 1992). In addition, there are variable data and no consensus among studies regarding whether the levels of plasma and CNS cholesterol are elevated or decreased in AD patients (Roher et al., 1999). Some studies report a lack of correlation between high cholesterol and incidence of AD (Romas et al., 1999).

There are several possible explanations for these discrepancies. First, studies examining samples of older individuals can be inadvertently biased by the fact that those with the highest levels of plasma cholesterol generally die at younger ages from cardiovascular events and are lost from the sample, introducing a "survivorship effect" into the population (Newschaffer et al., 1992). Second, experimental data suggest that the extent of amyloid deposition may be influenced by events that regulate removal of amyloid peptides from the brain (Zlokovic et al., 2000); therefore, better clearance mechanisms in certain individuals may preclude clinically significant accumulations of amyloid, despite increased cholesterol-mediated amyloidogenesis. Third, population-based studies of patients that subsequently developed AD showed a significant decrease in plasma cholesterol levels preceding development of cognitive symptoms (Notkola et al., 1998), potentially obscuring the past effects of higher cholesterol levels earlier in life. Finally, negative studies correlating the clinical diagnosis of AD with cholesterolemia did not have neuropathologic confirmation of the diagnosis and may have included some cases of non-AD dementia, weakening the association between cholesterol and AD. Because of these reasons we conducted a retrospective review of autopsies and found that a relatively mild elevation of cholesterol was an early risk factor for the development of amyloid deposition. Although this observation established a link between cholesterolemia and amyloid pathology, there was a relatively poor correlation between levels of cholesterol and the extent of the pathology. Thus, a substantial number of individuals with "higher-than-normal" cholesterol levels never developed enough amyloid for this to become clinically significant, which suggested that additional factors (such as poor clearance) may be required in the human brain (a "two-hit" mechanism).

8. Conclusions

Most recent developments in the study of AD pathogenesis resulted from the compositional analysis of senile plaques and neurofibrillary tangles that began over two decades ago. These research efforts have provided intriguing information about the role of several factors, including cholesterol and oxidative stress, in the pathogenesis of this condition. Although many facets of the biology and genetics of this disease remain to be discovered, strides have been made toward developing animal models and elucidating the biological properties of several proteins such as Aβ, APP, presenilins, and apolipoproteins, and of factors that can modulate their interactions. This information has resulted in the development of strategies aimed at either reducing production or aggregation of Aβ or at blocking the neuro-toxicity and inflammation that this peptide appears to cause in brain regions responsible for intellectual functioning. Based on the findings reviewed here, it would seem that interventions with lipid-lowering drugs, antioxidants, and/or anti-amyloidogenic strategies may be therapeutically beneficial in late-onset, sporadic AD.

References

Bales, K.R., Verina, T., Cummins, D.J., Du, Y., Dodel, R.C., Saura, J., Fishman, C.E., DeLong, C.A., Piccardo, P., Petegnief, V., Ghetti, B., Paul, S.M., 1999. Apolipoprotein E is essential for amyloid deposition in the APP^{V717F} transgenic mouse model of Alzheimer's disease. Proc. Natl. Acad. Sci. USA 96, 15233-15238.

Beffert, U., Danik, M., Krzywkowski, P., Ramassamy, C., Berrada, F., Poirier, J., 1998. The neurobiology of apolipoproteins and their receptors in the CNS and Alzheimer's disease. Brain Res. Rev. 27, 119–142.

Bodovitz, S., Klein, W.L., 1996. Cholesterol modulates alpha-secretase cleavage of amyloid precursor protein. J. Biol. Chem. 271, 4436–4440.

Boschert, U., Merlo-Pich, E., Higgins, G., Roses, A.D., Catsicas, S., 1999. Apolipoprotein E expression by neurons surviving excitotoxic stress. Neurobiol. Dis. 6, 508–514.

Boyt, A.A., Taddei, K., Hallmayer, J., Mamo, J., Helmerhorst, E., Gandy, S.E., Martins, R.N., 1999. The relationship between lipid metabolism and plasma concentration of amyloid precursor protein and apolipoprotein E. Alz. Rep. 6, 339-346.

Breen, K.C., Bruce, M., Anderson, B.H., 1991. Beta amyloid precursor protein mediates neuronal cell-cell and cell-surface adhesion. J. Neurosci. Res. 28, 90-100.

Bullido, M.J., Valdivieso, F., 2000. Apolipoprotein E gene polymorphisms in Alzheimer's disease. Microsc. Res. Tech. 50, 261-267.

Cader, A.A., Steinberg, F.M., Mazzone, T., Chait, A., 1997. Mechanisms of enhanced macrophage apoE secretion by oxidized LDL. J. Lipid Res. 38, 981–991.

Chartier-Harlin, M.C., Crawford, F., Houlden, H., Warren, A., Hughes, D., Fidani, L., Goate, A., Rossor, M., Roques, P., Hardy, J., Mullan, M., 1991. Early-onset Alzheimer's disease caused by mutations at codon 717 of the beta-amyloid precursor protein gene. Nature 353, 844–846.

Duan, H., Li, Z., Mazzone, T., 1995. Tumor necrosis factor-alpha modulates monocyte/macrophage apoprotein E gene expression. J. Clin. Invest. 96, 915–922.

Duan, H., Lin, C.Y., Mazzone, T., 1997. Degradation of macrophage ApoE in a nonlysosomal compartment. Regulation by sterols. J. Biol. Chem. 272, 31156–31162.

Esiri, M.M., Hyman, B., Beyreuther, K., Masters, C.L., 1997. Aging and dementia. In: D.I. Graham and P.L. Lantio (Eds.), Greenfield's Neuropathology, Arnold Publishers, London, pp. 151–233.

Evans, R.M., Emsley, C.L., Gao, S., Sahota, A., Hall, K.S., Farlow, M.R., Hendrie, H., 2000. Serum cholesterol, APOE genotype, and the risk of Alzheimer's disease: a population-based study of African Americans. Neurology 54, 240–242.

Fagan, A.N., Holtzman, D.M., 2000. Astrocyte lipoproteins, effects of apoE on neuronal function, and role of apoE in amyloid β deposition *in vivo*. Microsc. Res. Tech. 50, 297–304.

Finch, C.E., Sapolsky, R.M., 1999. The evolution of Alzheimer disease, the reproductive schedule, and apoE isoforms. Neurobiol. Aging 20, 407–428.

Frears, E.R., Stephens, D.J., Walters, C.E., Davies, H., Austen, B.M., 1999. The role of cholesterol in the biosynthesis of beta-amyloid. Neuroreport 10, 1699–1705.

Goate, A., Chartier-Harlin, M.C., Mullan, M., Brown, J., Crawford, F., Fidani, L., Giuffra, L., Haynes, A., Irving, N., James, L., Mant, R., Newton, P., Rooke, K., Roques, P., Talbot, C., Pericak-Vance, M., Roses, A., Williamson, R., Rossor, M., Owen, M., Hardy, J., 1991. Segregation of a missense mutation in the amyloid precursor protein gene with familial Alzheimer's disease. Nature 349, 704–706.

Grant, W.B., 1999. Dietary links to Alzheimer's disease: 1999 update. J. Alz. Dis. 1, 197–201.

Grundke-Iqbal, I., Iqbal, K., Tung, Y.C., Quinlan, M., Wiesniewski, H.M., Binder, L.I., 1986. Abnormal phosphorylation of the microtubule-associated protein tau in Alzheimer's cytoskeletal pathology. Proc. Natl. Acad. Sci. USA 83, 4913–4917.

Howland, D.S., Trusko, S.P., Savage, M.J., Reaume, A.G., Lang, D.M., Hirsch, J.D., Maeda, N., Siman, R., Greenberg, B.D., Scott, R.W., Flood, D.G., 1998. Modulation of secreted beta-amyloid precursor protein and amyloid beta-peptide in brain by cholesterol. J. Biol. Chem. 273, 16576–16582.

Hyman, B.T., West, H.L., Gomez-Isla, T., Mui, S., 1995. Quantitative neuropathology in Alzheimer's disease: neuronal loss in high-order association cortex parallels dementia. In: K. Iqbal, J.A. Mortimer, B. Winblad, H.M. Wiesniewski (Eds.), Research Advances in Alzheimer's disease and Related Disorders. John Wiley and Sons, New York, pp. 453–460.

Jarvik, G.P., Austin, M.A., Fabsitz, R.R., Auwerx, J., Reed, T., Christian, J.C., Deeb, S., 1994. Genetic influences on age related change in total cholesterol, low density lipoprotein cholesterol, triglyceride levels: longitudinal apolipoprotein E genotype effects. Genet. Epidemiol. 11, 375–384.

Jarvik, G.P., Wijsman, E.M., Kukull, W.A., Schellenberg, G.D., Yu, C., Larson, E.B., 1995. Interactions of apolipoprotein E genotype, total cholesterol level, age, and sex in prediction of Alzheimer's disease: a case-control study. Neurology 45, 1092–1096.

Jick, H., Zornberg, G.L., Jick, S.S., Drachman, D., 2000. Statins and the risk of dementia. Lancet 356, 1627–1631.

Kim, M.H., Nakayama, R., Manos, P., Tomlinson, J.E., Choi, E., Ng, J.D., Holten, D., 1989. Regulation of apolipoprotein E synthesis and mRNA by diet and hormones. J. Lipid Res. 30, 663–671.

Kivipelto, M., Helkala, E.L., Hanninen, T., Laakso, M.P., Hallikainen, M., Alhainen, K., Soininen, H., Tuomilehto, J., Nissinen, A., 2001. Midlife vascular risk factors and late-life mild cognitive impairment: a population-based study. Neurology 56, 1683–1689.

Kolsch, H., Ludwig, M., Lutjohann, D., Rao, M.L., 2001. Neurotoxicity of 24-hydroxycholesterol, an important cholesterol elimination product of the brain, may be prevented by vitamin E and estradiol-17beta. J. Neural. Transm. 108, 475–488.

Kosik, K.S., 1993. The molecular and cellular biology of Tau. Brain Pathol. 3, 39–43.

Kurzchalia, T.V., Parton, R.G., 1999. Membrane microdomains and caveolae. Curr. Opin. Cell Biol. 11, 424–431.

Laskowitz, D.T., Horsburgh, K., Roses, A.D., 1998. Apolipoprotein E and the CNS response to injury. J. Cereb. Blood Flow Metab. 18, 465–471.

Lin-Lee, Y.C., Tanaka, Y., Lin, C.T., Chan, L., 1981. Effects of an atherogenic diet on apolipoprotein E biosynthesis in the rat. Biochemistry 20, 6474–6480.

Lutjohann, D., Papassotiropoulos, A., Bjorkhem, I., Locatelli, S., Bagli, M., Oehring, R.D., Schlegel, U., Jessen, F., Rao, M.L., von Bergmann, K., Heun, R., 2000. Plasma 24S-hydroxycholesterol (cerebrosterol) is increased in Alzheimer and vascular demented patients. J. Lipid Res. 41, 195–198.

Mark, R.J., Lovell, M.A., Markesbery, W.R., Uchida, K., Mattson, M.P., 1997. A role for 4-hydroxynonenal, an aldehydic product of lipid peroxidation, in disruption of ion homeostasis and neuronal death induced by amyloid beta-peptide. J. Neurochem. 68, 255–264.

Mark, R.J., Pang, Z., Geddes, J.W., Uchida, K., Mattson, M.P., 1997b. Amyloid beta-peptide impairs glucose transport in hippocampal and cortical neurons: involvement of membrane lipid peroxidation. J. Neurosci. 17, 1046–1054.

Masliah, E., 1995. Mechanisms of synaptic dysfunction in Alzheimer's disease. Histol. Histopathol. 10, 509–519.

Mason, R.P., Shoemaker, W.J., Shajenko, L., Chambers, T.E., Herbette, L.G., 1992. Evidence for changes in the Alzheimer's disease brain cortical membrane structure mediated by cholesterol. Neurobiol. Aging 13, 413–419.

Mattson, M.P., Fu, W., Waeg, G., Uchida, K., 1997. 4-Hydroxynonenal, a product of lipid peroxidation, inhibits dephosphorylation of the microtubule-associated protein tau. Neuroreport 8, 2275–2281.

Mizuno, T., Haass, C., Michikawa, M., Yanagisawa, K., 1998. Cholesterol-dependent generation of a unique amyloid beta-protein from apically missorted amyloid precursor protein in MDCK cells. Biochim. Biophys. Acta 1373, 119–130.

Moghadasian, M., 1999. Clinical pharmacology of 3-hydroxy-methylglutaryl coenzyme A reductase inhibitors. Life Sci. 65, 1329–1337.

Newschaffer, C.J., Bush, T.L., Hale, W.E., 1992. Aging and total cholesterol levels: cohort, period and survivorship effects. Am. J. Epidemiol. 136, 23–34.

Notkola, I.L., Sulkava, R., Pekkanen, J., Erkinjuntti, T., Ehnholm, C., Kivinen, P., Tuomilehto, J., Nissinen, A., 1998. Serum total cholesterol, apolipoprotein E epsilon 4 allele, and Alzheimer's disease. Neuroepidemiology 17, 14–20.

Ogbonna, G., Theriault, A., Adeli, K., 1993. Hormonal regulation of human apolipoprotein E gene expression in HepG2 cells. Int. J. Biochem. 25, 635–640.

Oscarsson, J., Carlsson, L.M., Bick, T., Lidell, A., Olofsson, S.O., Eden, S., 1991. Evidence for the role of the secretory pattern of growth hormone in the regulation of serum concentrations of cholesterol and apolipoprotein E in rats. J. Endocrinol. 128, 433–438.

Pappolla, M.A., Ogden, M., 2000. Oxidative stress and the amyloid conundrum. What is the connection? J. Alz. Dis. 2, 1–7.

Pappolla, M.A., Robakis, N.K., 1995. Neuropathology and molecular biology of Alzheimer's disease. In: M. Stein and M. Baum (Eds.), Perspectives in Behavioral Medicine Alzheimer's Disease. Academic Press, San Diego, pp. 3–20.

Pappolla, M.A., Chyan, Y.J., Poeggeler, B., Bozner, P., Ghiso, J., LeDoux, S.P., Wilson, G.L., 1999. Aβ mediated oxidative damage of mitochondrial DNA: prevention by melatonin. J. Pineal Res. 27, 226–229.

Pappolla, M.A., Sambamurti, K., Efthimiopoulos, S., Refolo, L., Omar, R.A., Robakis, N.K., 1995. Heat-shock induces abnormalities in the cellular distribution of amyloid precursor protein (APP) and APP fusion proteins. Neurosci. Lett. 192, 105–108.

Pedersen, W.A., Chan, S.L., Mattson, M.P., 2000. A mechanism for the neuroprotective effect of apolipoprotein E: isoform-specific modification by the lipid peroxidation product 4-hydroxynonenal. J. Neurochem. 74, 1426–1433.

Prasad, K., 1999. Reduction of serum cholesterol and hypercholesterolemic atherosclerosis in rabbits by secoisolariciresinol diglucoside isolated from flaxseed. Circulation 99, 1355–1362.

Prasad, K., Kalra, J., 1993. Oxygen free radicals and hypercholesterolemic atherosclerosis: effect of vitamin E. Am. Heart J. 125, 958–973.

Refolo, L.M., Malester, B., LaFrancois, J., Bryant-Thomas, T., Wang, R., Tint, G.S., Sambamurti, K., Duff, K., Pappolla, M.A., 2000. Hypercholesterolemia accelerates the Alzheimer's amyloid pathology in a transgenic mouse model. Neurobiol. Dis. 7, 321–331.

Refolo, L.M., Pappolla, M.A., LaFrancois, J., Malester, B., Schmidt, S.D., Bryant-Thomas, T., Tint, G.S., Wang, R., Mercken, M., Petanceska, S.S., Duff, K.E., 2001. A cholesterol-lowering drug reduces beta-amyloid pathology in a transgenic mouse model of Alzheimer's disease. Neurobiol. Dis. 8, 890–899.

Roher, A.E., Kuo, Y.M., Kokjohn, K.M., Emmerling, M.R., Gracon, S., 1999. Amyloid and lipids in the pathology of Alzheimer disease. Amyloid 6, 136–145.

Romas, S.N., Tang, M.X., Berglund, L., Mayeux, R., 1999. APOE genotype, plasma lipids, lipoproteins, and AD in community elderly. Neurology 53, 517–521.

Sano, M., Ernesto, C., Thomas, R.G., Klauber, M.R., Schafer, K., Grundman, M., Woodbury, P., Growdon, J., Cotman, C.W., Pfeiffer, E., Schneider, L.S., Thal, L.J., 1997. A controlled trial of selegiline, alpha-tocopherol, or both as treatment for Alzheimer's disease. N. Engl. J. Med. 336, 1216–1222.

Selkoe, D.J., 2000. The genetics and molecular pathology of Alzheimer's disease: roles of amyloid and the presenilins. Neurol. Clin. 18, 903–922.

Simons, M., Keller, P., De Strooper, B., Beyreuther, K., Dotti, C.G., Simons, K., 1998. Cholesterol depletion inhibits the generation of beta-amyloid in hippocampal neurons. Proc. Natl. Acad. Sci. USA 95, 6460–6464.

Sparks, D.L., 1997. Coronary artery disease, hypertension, ApoE and cholesterol: a link to Alzheimer's disease? Ann. NY Acad. Sci. 826, 128–146.

Sparks, D.L., Martin, T.A., Gross, D.R., Hunsaker, J.C., 3rd. 2000. Link between heart disease, cholesterol and Alzheimer's disease: a review. Microsc. Res. Tech. 50, 287–290.

Tagliavini, F., Giaccone, G., Frangione, B., Bugiani, O., 1988. Preamyloid deposits in the cerebral cortex of patients with Alzheimer's disease and nondemented individuals. Neurosci. Lett. 93, 191–196.

Tam, S.P., Archer, T.K., Deeley, R.G., 1986. Biphasic effects of estrogen on apolipoprotein synthesis in human hepatoma cells: mechanism of antagonism by testosterone. Proc. Natl. Acad. Sci. USA 83, 3111–3115.

Tomiyama, T., Corder, E.H., Mori, H., 1999. Molecular pathogenesis of apolipoprotein E-mediated amyloidosis in late onset Alzheimer's disease. Cell. Mol. Life Sci. 56, 268–279.

Wolozin, B., Kellman, W., Ruosseau, P., Celesia, G.G., Siegel, G., 2000. Decreased prevalence of Alzheimer disease associated with 3-hydroxy-3-methylglutaryl coenzyme A reductase inhibitors. Arch. Neurol. 57, 1439–1443.

Wood, W.G., Schroeder, F., Avdulov, N.A., Chochina, S.V., Igbavboa, U., 1999. Recent advances in brain cholesterol dynamics: transport, domains and Alzheimer's disease. Lipids 34, 225–234.

Yan, S.D., Yan, S.F., Chen, X., Fu, J., Chen, M., Kuppusamy, P., Smith, M.A., Perry, G., Godman, G.C., Nawroth, P., 1995. Non-enzymatically glycated tau in Alzheimer's disease induces neuronal oxidant stress resulting in cytokine gene expression and release of amyloid β peptide. Nature Med. 1, 693–699.

Yanker, B.A., Duffy, L.K., Kirschner, D.A., 1990. Neurotrophic and neurotoxic effects of amyloid β protein: Reversal by tachykinin neuropeptides. Science 270, 279–282.

Zhang, L., Zhao, B., Yew, D.T., Kusiak, J.W., Roth, G.S., 1997. Processing of Alzheimer's amyloid precursor protein during H_2O_2-induced apoptosis in human neuronal cells. Biochem. Biophys. Res. Commun. 235, 845–848.

Zlokovic, B.V., Yamada, S., Holtzman, D., Ghiso, J., Frangione, B., 2000. Clearance of amyloid beta-peptide from brain: transport or metabolism? Nat. Med. 6, 718.

Advances in
Cell Aging and
Gerontology

Phospholipase A$_2$ in the pathogenesis of cardiovascular disease

Eva Hurt-Camejo[1,3], Peter Sartipy[2], Helena Peilot[3]
Birgitta Rosengren[1], Olov Wiklund [3]
Germán Camejo[1,3]

[1]*AstraZeneca, R&D, Mölndal.*
[2]*The Scripps Research Institute, La Jolla, CA, USA.*
[3]*Wallenberg Laboratory for Cardiovascular Research, Göteborg University, Sweden.*
Correspondence address: Eva Hurt-Camejo, AstraZeneca, R&D Mölndal 431 83, Sweden.
Tel: +46-31-776-2967; Fax: +46-31-776-3736.
E-mail address: eva.hurt-camejo@astrazeneca.com

Contents

1. Introduction

Phospholipases A$_2$ (PLA$_2$) are enzymes that hydrolyze the *sn*-2 ester bond in the glyceroacyl phospholipids present in lipoproteins and cell membranes, forming nonesterified fatty acids (NEFA) and lysophospholipids (Dennis and Six, 2000). These products may either act as intracellular secondary messengers or be further metabolized into mediators of a broad range of cellular processes (for review see: Brash, 2001; Fitzpatrick and Seberman, 2001; Funk, 2001; Hla et al., 2001). This

Advances in Cell Aging and Gerontology, vol. 12, 177–204

chapter is motivated by evidences accumulated in recent years suggesting that PLA$_2$ activity present in arterial intima-media tissue and in plasma may contribute to atherosclerosis, the main pathological process behind cardiovascular diseases (CVD) (Hurt-Camejo and Camejo, 1997; Sartipy and Hurt-Camejo, 1999; Hurt-Camejo et al., 2000; Kovanen et al., 2000). *In vivo* and *in vitro* results indicate that PLA$_2$ may hydrolyze the phospholipids of the apolipoprotein (apo) B-100-containing lipoproteins retained in the arterial intima (Camejo et al., 1985; Daugherty et al., 1988; Tailleux et al., 1993; Keaney et al., 1995). The products of this hydrolysis, NEFA and lysophospholids, can trigger a variety of proinflammatory actions, which may contribute to atherosclerotic plaque development (Sparrow et al., 1988; Gimbrone, 1995; Berliner and Heinecke, 1996; Olsson et al., 1999; Perrella et al., 2001). Furthermore, PLA$_2$-modified apoB-100 lipoproteins, which are more susceptible to further enzymatic and nonenzymatic modifications, induce accumulation of intracellular lipids in macrophages and bind strongly to extracellular matrix proteoglycans (Menschikowski et al., 1995; Hurt-Camejo et al., 1997; Romano et al., 1998; Sartipy et al., 2000). Furthermore, PLA$_2$ can induce aggregation and fusion of the matrix-bound lipoproteins and hence further increase their binding strength to the matrix proteoglycans (Hakala et al., 2001). Thus, in atherosclerosis-prone regions of arteries, PLA$_2$ activity may contribute to both intra- and extracellular accumulation of apoB-100 lipoproteins, a hallmark of atherosclerotic lesions. In addition to the potential modification of lipoproteins, PLA$_2$ is also reported to induce cycloxygenase-2 expression, cytokine release by macrophages and phagocytosis of injured cells thereby enhancing inflammation and tissue damage (Hack et al., 1997; Cai et al., 1999; Bidgood et al., 2000; Kaneko et al., 2000; Hernández et al., 2002; Triggiani et al., 2002). On the other hand, circulating PLA$_2$ activity can modify low-density lipoprotein (LDL) and high-density lipoproteins (HDL) in plasma and may contribute to the generation of an atherogenic lipoprotein profile associated with cardiovascular disease (Leitinger et al., 1999; Sartipy et al., 1999; De et al., 2001). A high incidence of cardiovascular disease is reported in patients with rheumatoid arthritis who have prolonged periods of high extracellular secretory PLA2 type IIA activity (sPLA$_2$-IIA) in the plasma (Wållberg-Jonsson et al., 1999). Furthermore, recent clinical studies indicate that an elevated plasma level of PLA$_2$ is a strong independent risk factor for coronary heart disease(Kugiyama et al., 1999; Packard et al., 2000; Porela et al., 2000). Whether this is associated with the proatherogenic mechanisms of PLA$_2$ activity in plasma or with their actions in the arterial wall remains to be elucidated. The hypothesis of the potential involvement of PLA$_2$ activity in the pathogenesis of atherosclerosis is reinforced by *in vivo* data showing that transgenic mice expressing human sPLA$_2$-IIA have increased susceptibility to atherosclerosis(Ivandic et al., 1999; Leitinger et al., 1999). In this chapter we present evidence supporting the proatherogenic properties of PLA$_2$ activity contributing to cardiovascular diseases with a main focus on sPLA$_2$-IIA enzyme. We also discuss the expression of different PLA$_2$ enzymes in the heart and their function in myocardial lipid metabolism associated with ischemic heart disease. Finally, in future directions we discuss the need for further research to elucidate the physiological and pathological functions of different PLA2 enzymes recently discovered and how this knowledge may represent a novel approach for the development of therapies in cardiovascular disease treatment.

2. The family of PLA$_2$ enzymes

The phospholipase A$_2$ comprises a rapidly growing family of intracellular and secreted enzymes, which hydrolyze the acyl-group at the *sn-2* position of glycerophospholipids to release fatty acids and lysophospholipids (Six, 2000). Mammalian and Dennis secreted phospholipases A$_2$ (sPLA$_2$) encompass 10 groups of isoenzymes: IB, IIA, IIC, IID, IIE, IIF, III, V, X and XII with a partial overlapping tissue distribution (Ishizaki et al., 1999; Lambeau and Lazdunski, 1999; Valentin et al., 1999; Ho et al., 2001). The homology between these sPLA$_2$ enzymes varies, however they have in common a low molecular weight, 13–20 kDa, except the type III which is 55.3 kDa (Valentin et al., 2000). They also share the presence of 6–8 disulfide bridges, a Ca^{2+}-dependent catalytic mechanism, and a conserved three-dimensional structure. Group III and group XII sPLA$_2$s each have unique structural features, and are homologous to group I, II, V and X sPLA$_2$s only in the Ca^{2+}-binding site (Gelb et al., 2000, Valentin et al., 2000). Although the biological function of each of these sPLA$_2$ enzymes is not clearly defined, mammalian sPLA$_2$ appear to be implicated in a variety of physiological and pathological processes. These include lipid digestion, release of potent lipid mediators in response to cytokines (Murakami et al., 1998), cell proliferation (Hernández et al., 1998; Pruzanski et al., 2001), control of virus and bacterial infection (Fenard et al., 1999; Koduri et al., 2002), removal of apoptotic or injured cells (Hack et al., 1997; Cummings et al., 2000), phospholipid repair, lipoprotein catabolism (de et al., 1997; Sartipy et al., 1999; Menschikowski et al., 2000; Min et al., 2001), blood coagulation (Mounier et al., 2000) and inflammation (Han et al., 1999) (for review see: Uhl et al., 1997). Interestingly, sPLA$_2$ activity is also reported to facilitate cholesterol absorption (Mackay et al., 1997). More recently, some reports indicate that sPLA$_2$ may activate cells through direct interaction with cell surface receptors independently of the enzymatic activity of sPLA$_2$(Rufini et al., 1999; Triggiani et al., 2002).

3. SPLA$_2$-IIA in the arterial tissue

The human secretory non-pancreatic PLA$_2$ belongs to the group II-A (sPLA$_2$.IIA) and is characterized by a low molecular weight (14.4 kDa), a pH optimum of 7–9, and a requirement of mM calcium concentrations for activity. The enzyme was first isolated and purified from rheumatoid synovial fluid and is often referred to in the literature as synovial fluid sPLA$_2$. The human sPLA$_2$-IIA was cloned in 1989 by Kramer and co-workers. Interestingly, the gene for sPLA$_2$-IIA resides on chromosome 1 and lies within a sPLA$_2$ gene cluster of about 300 kbp that also contains the genes for group IIC, IID,IIE, IIF and V sPLA$_2$ (Kramer et al., 1989; Valentin et al., 2000), suggesting that these genes arise from a common ancestor by gene duplication (Murakami et al., 2002). The crystal structure of the human recombinant sPLA$_2$ was established (Wery et al., 1991; Berg et al., 2001). SPLA$_2$-IIA contains 7 disulfide bridges resulting in a rigid structure, and there are 23 cationic amino acid residues, Arginine (Arg) and Lysine (Lys), which contribute to the high positive charge of the protein (pI = 10.5). Specific cationic residues, mainly Arg-7, Lys-10 and Arg-16, are involved in the interfacial binding of the enzyme to its substrate. However, the functions of other positively charged patches remain unclear. Several evidences suggest two different functions. First, to facilitate the interaction with negatively charged proteoglycans in cell membranes and extracellular matrices, an interaction that may contribute to the accretion and modulation of local sPLA$_2$-IIA activity (Sartipy et al., 2000; Murakami et al., 2001). Second, to be involved in binding with factor Xa and prothrombinase inhibition (Mounier et al., 2000).

In nonatherosclerotic human arteries, the sPLA$_2$-IIA is mainly associated with the smooth muscle cells of the media, whereas in atherosclerotic plaques the enzyme is also found in macrophage-rich regions, in the acellular lipid core of atheromas, and in the extracellular matrix of the diseased intima in association with collagen fibers (Menschikowski et al., 1995; Hurt-Camejo et al., 1997; Romano et al., 1998). Immunohistochemical studies performed by several research groups show that sPLA$_2$-IIA is found at all stages of atherosclerotic lesion development (Menschikowski et al., 1995; Elinder et al., 1997; Hurt-Camejo et al., 1997; Romano et al., 1998; Schiering et al., 1999; Menschikowski et al., 2000). Cytoplasmic PLA$_2$ (Group IV) is also present in atherosclerotic lesions containing macrophages. However, the activity of sPLA$_2$-IIA is more prominent than that of the cytoplasmic enzyme in the same human plaque (Elinder et al., 1997). Expression of sPLA$_2$ *in vitro* is reported to be regulated by differentiation of monocytes into macrophages and by further exposure to mildly oxidized low-density lipoprotein (Anthonsen et al., 2000).

In summary, immunohistochemical studies show that sPLA$_2$-IIA is present in normal arteries, and that, in early and late atherosclerotic lesions, its extracellular distribution and level of cell expression is increased, suggesting that the enzyme may be implicated in atherogenesis.

Most studies on sPLA$_2$-IIA expression in vascular cells have been performed with rat smooth muscle cells (Pfeilschifter et al., 1997; Rufini et al., 1999; Couturier et al., 2000). Although some of these results can be extrapolated to human arterial smooth muscle cells, recent studies suggest that regulation of sPLA$_2$-IIA gene expression is cell- and species-specific (Anderson et al., 1997; Andreani et al., 2000; Peilot et al., 2000). The expression of both sPLA$_2$-IIA mRNA and protein by human arterial smooth muscle cells from the aorta and the coronary and uterine arteries require conditions that promote cell differentiation in *vitro* (Peilot et al., 2000). These *in vitro* results agree with immunohistochemical data showing that, in arteries, the main source of sPLA$_2$-IIA are smooth muscle cells (Menschikowski et al., 1995; Elinder et al., 1997; Hurt-Camejo et al., 1997; Romano et al., 1998; Schiering et al., 1999; Menschikowski et al., 2000; Peilot et al., 2000; Sartipy et al., 2000). Electron microscopy shows that sPLA$_2$-IIA is stored intracellularly inside vesicles close to the smooth muscle cell membrane (Romano et al., 1998). Cytokines, *in vitro*, differentially modulate cell secretion and mRNA levels of sPLA$_2$-IIA. Thus, interferon-gamma (IFN-γ) increases the expression of mRNA and sPLA$_2$-IIA protein secretion 2- to 6-fold, and, after addition of IFN-γ, this effect lasts up to 48 hours. On the other hand, tumor necrosis factor-alpha (TNF-α) stimulates sPLA$_2$-IIA secretion for only 4 hours, without detectable changes in mRNA levels(Peilot et al., 2000). Interestingly, a similar effect of TNF-α is seen in sPLA$_2$-IIA transgenic mice(Laine et al., 1999). Interleukin-10 (IL-10), an anti-inflammatory cytokine, downregulates IFN-γ, but not TNF-α, induction of sPLA$_2$-IIA secretion. In contrast to what is reported with rat smooth muscle cells, we found that interleukin-1-β (IL1-β) alone is not a strong inducer of sPLA$_2$-IIA in human arterial smooth muscle cells. Similar results are also reported in human vein smooth muscle cells (Beasley, 1999). Colocalization of sPLA$_2$-IIA with the messenger RNA transcript for IFN-γ, IL1-β, and TNF-α in human atherosclerotic lesions supports a possible *in vivo* involvement of these cytokines in the regulation of sPLA$_2$-IIA gene expression and protein secretion in atherosclerotic plaques (Menschikowski et al., 2000).

Cytokines acting on tissues may also indirectly regulate the circulating levels of sPLA$_2$-IIA in plasma. sPLA$_2$-IIA is an acute-phase reactant and, in diseases that involve systemic inflammation, such as sepsis, rheumatoid arthritis (Nevalainen, 1993), osteoarthritis (Pruzanski et al., 1991), cardiovascular disease (Kugiyama et al., 1999; Kugiyama et al., 2000; Porela et al., 2000), its plasma levels are increased. Recently, plasma levels of sPLA$_2$-IIA were reported to increase with increasing adiposity in Pima Indians, an obesity- and diabetes-prone population. Thus, obesity may be associated with a low-grade inflammation (Weyer et al., 2002). Hepatocytes synthesize and secrete sPLA$_2$-IIA and other acute-phase proteins in response to cytokines such as IL-6, TNF-α, and IL-1β (Vadas et al., 1997), but not IFN-γ (Peilot et al., 2000). In addition, peritoneal injections of IL-6, TNF-α, and IL-1β increase plasma levels of sPLA$_2$-IIA in transgenic mice expressing the gene of human sPLA$_2$-IIA (Laine et al., 1999). Together, these data suggest that hepatocytes and vascular smooth muscle cells may contribute to the bulk of circulating sPLA$_2$-IIA in plasma modulated by systemic inflammatory conditions.

4. Potential atherogenic actions of secretory phospholipase A$_2$ group II-A

Atherosclerosis is the response of arterial tissue to local accumulation and modification of apoB-100-containing lipoproteins in the intima of the arterial wall (Williams and Tabas, 1995; Camejo et al., 1998). Specific interactions of apoB-100-containing lipoproteins with chondroitin-sulfate proteoglycans such as versican or decorin appear to be important mechanisms contributing to their retention and modification (Camejo et al., 1993). This early phenomenon is documented in *in vivo* studies showing accumulation of aggregated/fused lipoprotein particles in the subendothelial extracellular matrix of atherosclerotic-prone regions in aorta preceding endothelial activation and macrophages appearance (Tamminen et al., 1999). Furthermore, early lesions in human aortas from fetuses of hypercholesterolemic women frequently contain apoB-lipoproteins and oxidized-apoB-lipoproteins in the absence of monocyte-macrophages (Napoli et al., 1997). Together, these observations suggest that modification of the lipoproteins retained in the extracellular matrix is an essential step triggering an inflammatory response associated with lesion formation. In the arterial intima, there are proteolytic and lipolytic enzymes and prooxidant conditions capable of modifying the apoB-containing lipoproteins retained in the extracellular matrix (Bhakdi et al., 1995; Schissel et al., 1996; Hurt-Camejo et al., 1997; Öörni et al., 2000; Sugiyama et al., 2001; Pentikäinen et al., 2002). Indirect evidence indicates that once apoB-100 lipoproteins are in the intima, hydrolysis of phosphatidylcholine by PLA_2-like activity takes place (Camejo et al., 1985; Daugherty et al., 1988, Tailleux et al., 1993). In addition, the concentration of lysophospholipids in rabbit atherosclerotic aorta is reported to be higher than in control tissue (Keaney et al., 1995). Furthermore, active $sPLA_2$-IIA isolated from human arterial tissue is able to hydrolyze the phospholipids in low-density lipoprotein (LDL), an apoB-lipoprotein (Hurt-Camejo and Camejo, 1997). In the arterial wall, $sPLA_2$-IIA may exert proatherogenic and proinflammatory effects contributing to the pathogenesis of atherosclerosis. PLA_2-activity may be proatherogenic by three mechanisms. First, it may induce release of relatively high concentrations of lipid mediators which may trigger and sustain a local inflammatory reaction affecting the function and properties of vascular cells at sites of apoB-100-lipoprotein accumulation (for review see: Hurt-Camejo and Camejo, 1997; Camejo et al., 1999). Second, $sPLA_2$-IIA may modify apoB-100-lipoproteins to a more atherogenic form by increasing their binding affinity toward proteoglycans and also by making the PLA_2-treated lipoproteins more susceptible to further oxidative and enzymatic modifications (Sparrow et al., 1988; Neuzil et al., 1998; Sartipy et al., 1999), which may contribute to macrophage-foam cell formation. Third, extensive lipolysis by PLA2 may enhance lipoprotein accumulation by inducing aggregation and fusion of the proteoglycan-bound apoB-100-lipoproteins (Öörni et al., 2000; Pentikäinen et al., 2000). Treatment of LDL with $sPLA_2$-IIA in the presence of physiologic albumin concentration leads to the formation of small dense LDL particles with increased affinity for glycosaminoglycans and proteoglycans (Sartipy et al., 1998). As discussed above, this may take place in the plasma or in the arterial wall, where extracellular active $sPLA_2$ can hydrolyze the phospholipids on the lipoprotein particles. Because PLA_2-treated LDL particles have an increased affinity for extracellular proteoglycans, their residence time in the arterial wall is likely to increase. This provides the possibility for further modifications of these particles. In fact, PLA_2-treated LDL particles are more susceptible to lipid peroxidation (Neuzil et al., 1998), generation of bioactive phospholipids (Leitinger et al., 1999), and hydrolysis by secretory sphingomyelinase (Schissel et al., 1998). In addition, treatment of LDL with PLA_2 facilitates sphingomyelinase-induced aggregation and fusion of LDL particles (Öörni et al., 1998). PLA_2 can also directly induce aggregation and fusion of LDL particles: treatment of proteoglycan-bound LDL with $sPLA_2$-IIA leads to aggregation and subsequent fusion of the modified LDL particles (Hakala et al., 2001). Interestingly, PLA_2 induces fusion of LDL particles only if the particles are bound to glycosaminoglycans either before, during, or after lipolysis (Hakala et al., 1999). The interaction between LDL and glycosaminoglycans can apparently overcome the rigidifying effect of PLA_2-induced hydrolysis on LDL particles. Thus, it can be hypothesized that if the small dense LDL particles generated by the action of $sPLA_2$-IIA either in plasma or in the arterial intima bind to arterial proteoglycans, this interaction will then trigger aggregation and fusion of the bound LDL particles. Since each aggregate or fused particle contains at least two copies of apoB-100, it is not surprising that the aggregated/fused particles bind to proteoglycans even more tightly than do the small dense PLA_2-treated LDL (Öörni et al., 2000). Thus, aggregation and fusion of LDL particles in the arterial intima are likely to lead to accumulation of LDL-derived lipid particles within the extracellular matrix. Such progressive deposition of lipid within the extracellular matrix of the arterial intima is a central feature of atherogenesis, and thus makes the arterial $sPLA_2$-IIA a strong candidate acting as one of the key enzymes in the extracellular space during the development of atherosclerotic lesions (Guyton, 2001). These proatherogenic mechanisms may be potentiated by the colocalization of $sPLA_2$-IIA and apoB-100-lipoproteins through their interaction with extracellular matrix proteoglycans. It should be stressed that the total hydrolysis of the phosphoglycerides from one LDL particle may generate more than 500 molecules of lysophospholipids and nonesterified fatty acids. Thus, at sites of retention of apoB-100-lipoproteins in the arterial intima, nonesterified fatty acids and lysophospholipids may reach high local concentrations. These bioactive products can induce different proatherogenic cellular processes and cell membrane perturbation. Human arterial smooth muscle cells exposed to albumin-bound nonesterified free fatty acids upregulate the synthesis of matrix proteoglycans (Olsson et al., 1999). This results in a matrix that has a higher affinity for LDL, suggesting the possibility of a noxious cycle in which $sPLA_2$-IIA products from apoB-100-lipoproteins entrapped in the intima trigger matrix changes, which, in turn, lead to further accumulation of lipoproteins. Peroxisome proliferator-activated receptor gamma (PPAR-γ) agonists oppose this action of nonesterified fatty acids, suggesting that the improved metabolism of nonesterified fatty acids down-regulates the increase in matrix synthesis *in vivo*. Furthermore, the presence of products of hydrolysis of lipoproteins, LDL and HDL, markedly enhances mitogenic activity of vascular smooth muscle cells as compared with unhydrolyzed lipoproteins (Pruzanski et al., 2001). Since lipoproteins, $sPLA_2$ and smooth muscle cells are known to colocalize in the vascular wall (Ishikawa et al., 2001), this should be considered as another potential proatherogenic effect of $sPLA_2$.

In addition to the proatherogenic properties described above, $sPLA_2$ in the arterial intima-media may also exert proinflammatory effects by inducing cyclooxygenase-2 expression, cytokine release and enhancing the overall inflammatory capacity of leukocytes (Cai et al., 1999; Marshall et al., 1999; Bidgood et al., 2000). The central function of $sPLA_2$-IIA in inflammation is supported by a large body of *in vitro* and *in vivo* data. However, there was a need for an *in vivo* model to study sPLA$_2$-IIA in atherosclerosis. For this purpose a transgenic mouse overexpressing human sPLA$_2$-IIA was backcrossed with mouse strain C57BL/6 which is a natural knockout since it has a nonfunctional sPLA$_2$-IIA allele and therefore provide a clean background for the overexpression of human sPLA$_2$-IIA (Grass et al., 1996). On the other hand, C57BL/6 mice express sPLA$_2$ type V, an isoenzyme closely related to sPLA$_2$-IIA. This enzyme is induced by proinflammatory stimuli, such as lipopolysaccharides, and is expressed in various tissues (Sawada et al., 1999). The genetically engineered transgenic mice express abundant sPLA$_2$-IIA in liver, lung, kidney and skin, and the serum levels were several times higher than in the nontransgenic littermates. Interestingly, there were no signs of systemic inflammation in these mice based on the levels of serum amyloid P and A (Grass et al., 1996). Notably, the sPLA$_2$-IIA transgenic mice develop aortic lesions compared with nontransgenic littermates when exposed to a high-fat diet. Immunohistochemical staining indicates that sPLA$_2$ was present in atherosclerotic lesions in aorta. Interestingly, some sPLA$_2$-IIA transgenic mice develop atherosclerosis even in the absence of an atherogenic diet (Ivandic et al., 1999; Leitinger et al., 1999). These results support the proatherogenic contribution of sPLA$_2$-IIA. This transgenic model was also important for studying the potential role of sPLA$_2$ in lipoprotein metabolism and atherosclerosis as discussed further in a separate section.

5. Molecular basis and effects of the sPLA$_2$-IIA interaction with proteoglycans

Extracellular sPLA$_2$-IIA is associated with collagen fibers and proteoglycans in atherosclerotic lesions from human coronary arteries (Romano et al., 1998). The enzyme binds to collagen via a specific interaction with decorin (Sartipy et al., 2000), a proteoglycan containing the chondroitin and dermatan sulfate chains of the serine- and leucine-rich protein (SLRP) family that is associated with collagen fibers (Iozzo, 1998). Studies from our laboratory show that in physiological salt and pH conditions, sPLA$_2$-IIA is able to interact with both the glycosaminoglycan moiety and the core protein of decorin. In addition, sPLA$_2$-IIA binds to the major proteoglycans of the arterial intima, versican and biglycan. However, with these molecules, the interaction takes place only through the glycosaminoglycan moiety. These associations with versican, biglycan, and decorin increase the hydrolytic activity of sPLA$_2$-IIA toward the phospholipids on apoB-lipoproteins or in micelles (Sartipy et al., 1996; Sartipy et al., 1998; Sartipy et al., 2000). Together with the remarkable stability of the enzyme, these results suggest that sPLA$_2$-IIA could remain sequestered in an active form for prolonged periods in the extracellular environment of the arterial intima. This may be important for the normal functioning of the enzyme but, when matrix-bound sPLA$_2$-IIA is released in excess, it may potentiate its inflammatory actions. For example, heparin injection induces release of extracellular-bound PLA$_2$ into circulation during cardiopulmonary bypass. This high PLA$_2$ activity may be responsible for the increased production of eicosanoids and, thus, may be implicated in various pathophysiological events associated with cardiac surgery (Kern et al., 2000).

The possibility that the pericellular extracellular matrix acts as a boundary to protect adherent cells from the action of basic sPLA$_2$ enzymes such as type IIA, V and X, has been proposed (Koduri et al., 1998). Immunofluorescent microscopy of cytokine-stimulated cells revealed that sPLA$_2$-IIA is associated with glypican, a heparan sulfate-containing cell surface proteoglycan, and is then internalised by caveolae. This process may allow sPLA$_2$-IIA to come in contact with a substrate interface to release free fatty acids for coupling to oxygenating enzymes for eicosanoid production (Murakami et al., 1999; Kim et al., 2001). However, the biological relevance of this ability of sPLA$_2$s to bind to cell surface proteoglycans has been challenged (Bezzine et al., 2000; Murakami et al., 2002). Thus, there is much to be learned about how interaction with cell surface proteoglycans is involved in the cellular functions of sPLA$_2$s and about how intact mammalian cell membranes and lipoproteins protect themselves against sPLA$_2$ hydrolysis or become modified to allow interfacial catalysis by different sPLA$_2$ enzymes.

6. SPLA$_2$ modification of LDL and HDL forms atherogenic small dense particles

The presence of a lipoprotein profile with abundance of small, dense low-density lipoproteins (LDL), low levels of high-density lipoprotein (HDL), and elevated levels of triglyceride-rich very low-density lipoproteins (VLDL) is associated with an increased risk for coronary heart disease (CHD) and the insulin resistance syndrome (Gardner et al., 1996; Frieldlander et al., 2000). The atherogenicity of small, dense LDL is believed to be one of the main reasons for this association (Packard et al., 1997; Griffin, 1999). This particle contains less phospholipids (PL) and unesterified cholesterol (UC) than large LDL (Hurt-Camejo et al., 1990; McNamara et al., 1996). The molecular or metabolic reasons for the association between increased risk of CHD and the presence of high concentrations of small, dense LDL remain to be established. It is not obvious why a reduced number of PL and UC molecules at the surface monolayer could be more atherogenic than the situation with larger buoyant LDL. One possibility is the dissimilar affinity of different LDL subclasses for extracellular and pericellular arterial proteoglycans (Anber et al., 1996; Olsson et al., 1997). Results with rabbit aortic segments indicate that small, dense LDL is better retained in the extracellular matrix intima than buoyant LDL (Björnheden et al., 1996). This increased residence time or retention of LDL is an early marker of atherosclerotic lesion progress (Schwenke and Carew, 1989). We recently reported that a specific reduction of approximately 50% of the PL content of normal buoyant LDL sPLA$_2$ (type III) produces smaller and denser particles (Sartipy et al., 1999). These smaller LDL particles displayed a higher tendency to form nonsoluble complexes with proteoglycans and glycosaminoglycans than the parent LDL. Binding parameters of LDL and glycosaminoglycans and proteoglycans produced by human arterial smooth muscle cells were measured at near physiological conditions. The sPLA$_2$-modified LDL has about two fold higher affinity for the sulfated polysaccharides than the control, larger LDL. In addition, by nuclear magnetic resonance spectroscopy analysis we found that incubation of human plasma in the presence of sPLA$_2$ generated smaller LDL and HDL particles compared with the control plasma incubated without sPLA$_2$. These *in vitro* results suggest a new process by which sPLA$_2$ reduction of surface PL, characteristic of small, dense LDL subfractions, besides contributing to its small size and density, may indeed enhance its tendency to be retained by proteoglycans and therefore its atherogenecity.

This sPLA$_2$-mediated change in LDL properties can be related to previous studies by our laboratory where we found that human arterial proteoglycans can discriminate at least four LDL subfractions, binding more efficiently those small, dense LDL that were poor in PL and UC in its surface monolayer(Hurt-Camejo et al., 1990). LDL subfractions with differential binding capacity to proteoglycans were also obtained by density-gradient centrifugation. The gradual increase in affinity to proteoglycan was associated with a gradual decrease in the PL content which was reflected in an almost two fold decrease in the total area of the lipoprotein occupied by the monolayer of polar lipids (4.7 to 2.5 nm^2 × 10^{-2}). As a consequence, the area that is accessible to the apoB in the LDL with highest affinity for proteoglycans is 64.4%, but only 36.7% in the LDL with lowest proteoglycan affinity. This suggests that, in the subfraction of LDL with higher reactivity for the proteoglycans, the apoB is more extended in the surface (small LDL), whereas it is more compact in the LDL subfraction with lower affinity (large LDL). Interestingly, McNamara and coworkers using similar calculations evaluated the differences in lipid composition of eight subclasses of LDL isolated by gradient ultracentrifugation. These authors found also that a decrease in LDL size was associated with a gradual decrease in PL and UC of the surface mono-layer. They also observed that the surface area requiring coverage by protein increases from 2.2 to 3.46 Å^2 × 10^4, indicating that on the smaller particles there is more area available for protein coverage and therefore a less compact arrangement of the apoprotein (McNamara et al., 1996). A similar situation is obtained by sPLA$_2$-mediated changes in LDL. The reduction of phosphatidylcholine (PC) in the surface monolayer of LDL by sPLA$_2$ *in vitro* reduces the size and increases the density of the LDL particle to values close to those observed for the main peak of small, dense LDL, in LDL-III, in plasma. With the established lipid composition of the sPLA$_2$-treated LDL it is possible to estimate the surface area covered by the apoB-100 based on the structural assumptions, which McNamara and we used. As expected, a reduction in 50% of the PL content after sPLA$_2$ modifica-tion of LDL resulted in a 25% decreased surface area (Sartipy et al., 1999). Such new free area should be covered by a more extended or less compact apoB-100 in the sPLA$_2$-treated LDL, assuming also that there is no movement from the lipid core components to the surface monolayer. This last possibility is not thermodynamically favored (Hurt-Camejo et al., 1990; McNamara et al., 1996).

The decrease in the surface area of LDL particles by reduction of PL is associated with an increase in the affinity of the LDL particle for proteoglycans and glycosaminoglycans. This interaction depends strongly on ionic interactions between arg- and lys-rich segments of the apoB-100 with the sulfated groups of

the glycosaminoglycan chains(Camejo et al., 1998). However, the lipid comple-
ment modulates the affinity of apoB-100 toward proteoglycans(Camejo et al.,
1998). Using computer molecular modeling Segrest et al. divided the structure of
apoB-100 into five domains, NH_2-α_1-β_2-α_3-β_4-α_5-COOH, with alternating amphi-
patic α-helixes and β-strands (Segrest et al., 2001). Based on *in vitro* studies, the
β-strands are tightly associated with the neutral lipid core of LDL and serves as a
rigid backbone of the molecule. The α-domains, on the other hand, are flexible
lipid binding regions that allow for changes in particle size due to variations in
lipid composition, especially variations in PL content, during LDL metabolism.
Particularly the α_3-domain was very sensitive to changes in native LDL particle
diameter, notably the same region that previously was found to be sensitive to
PLA_2-modificaion (Kleinman et al., 1988). One may speculate that the phospholipid depletion of LDL by $sPLA_2$ may induce changes in the conformation in the
α_3-domain that leads to more exposure of PG-binding regions in apoB-100. Our
hypothesis is that when the LDL particles become smaller two potential
proteoglycan-binding segments in the apoB, 3147–3157 and 3359–3367, coalesce
(Olsson et al., 1993; Camejo et al., 1998). This could account for the finding that
PLA_2-modified LDL binds with higher affinity to proteoglycans/glycosaminogly-
cans. Which $sPLA_2$ enzyme in plasma may induce this type of modification on
LDL is still unknown. We have shown that human recombinant $sPLA_2$-IIA is
able to hydrolyze LDL, however its catalytic activity on LDL is low if compared
with the hydrolysis induced by the type III $sPLA_2$ enzyme from bee venom. This
difference in catalytic activity between $sPLA_2$ isoenzymes appears to be due to the
presence of trypthopan residues in $sPLA_2$ type III, V and X that are absent in
$sPLA_2$-IIA, enhancing their caoacity to interact with phosphatidylcholine-rich
membranes (Gelb et al., 1999). Recently a new human $sPLA_2$ with homology to
the group III bee venom enzyme was cloned (Valentin et al., 2000). Northern blot
analysis showed the presence of a 4.4 kilobase hGIII transcript in kidney, heart,
liver and skeletal muscle. The biological function of this $sPLA_2$ type III is
unknown and whether it is involved in lipid metabolism remains to be
investigated. Another possible enzyme with phospholipase A_2-like activity is
hepatic lipase. In a recent study Zambon and coworkers studied the relationship
among LDL density, hepatic lipase and CHD progression (Zambon et al., 1999).
The authors found that therapy-associated decrease in plasma hepatic lipase
activity correlated positively with the presence of large buoyant LDL particles,
which favorably influences CHD progression. The authors do not have any
explanation for the molecular mechanism(s) behind their data. However, their
results support the hypothesis of the possible participation of specific or unspecific
phospholipase A_2 activity in mediating physicochemical changes in LDL particles.
Further work is necessary to understand the possible participation of $sPLA_2$
activity in the metabolism and catabolism of LDL.

The action of phospholipases on the surface monolayer of LDL depends on the
surface concentration of the substrates and on the surface pressure of the
monolayer (Ibdah et al., 1989; Cai et al., 1999). The surface monolayer of LDL
appears to have a high lateral pressure that does not allow the association of

other apolipoporoteins except apoB-100 (McNamara et al., 1996). This may explain why HDL is a better substrate for sPLA$_2$-IIA than LDL (McNamara et al., 1996). Therefore, reduction of PL by sPLA$_2$ or hepatic lipase in plasma or by sPLA$_2$ in the intima may facilitate the action of other enzymes like sphingomyelinase because of the increase in the surface concentration of sphingomyelin and the reduction in surface pressure. This hypothesis received support from experiments showing increased hydrolysis of LDL by sphingomyelinase after treatment with human recombinant sPLA$_2$ (Schissel et al., 1998) (Fig. 1). This indicates that further reduction in the intima of LDL surface-polar components by lipases could make its entrapment irreversible by aggregation and increased affinity for extracellular proteoglycans (Öörni et al., 1998; Pentikäinen et al., 2002).

Lp(a) is also susceptible to PLA$_2$-modification *in vitro*. The biological effects of Lp(a) modification by PLA$_2$ include an increase in lysine binding (Hoover-Plow et al., 1998) without changing the capacity of apo(a) to bind to apoB-100 (Fless

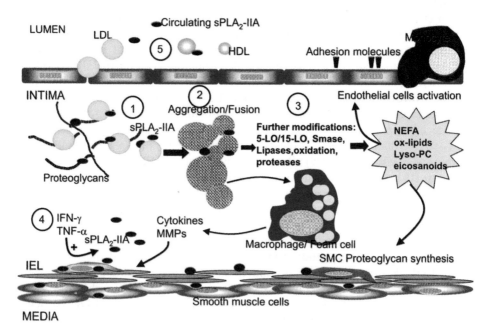

Fig. 1. Diagram illustrating the different pro-atherogenic mechanisms how sPLA$_2$ activity may contribute to atherosclerosis. (1) Co-localization of apoB-lipoproteins with extracellular sPLA$_2$ through interaction with proteoglycans may facilitate the hydrolysis of lipoproteins. This process may contribute to the aggregation of apoB-lipoproteins, (2) and further enhance modification of retained apoB-lipoproteins by non-enzymatic and enzymatic processes. (3) Together this may contribute to a local release of pro-inflammatory lipid mediators and oxidative stress at places of apoB-lipoprotein deposition in the arterial intima. Inflammatory cytokines increase secretion of sPLA$_2$ by vascular smooth muscle cells sustaining a local inflammatory response. (4) Circulating sPLA$_2$ in plasma may be pro-atherogenic by modifying lipoproteins, especially high-density lipoproteins. (5) This process may be enhanced during a systemic inflammation or acute phase response.

et al., 1999). This lipoprotein facilitates the binding of LDL to extracellular matrix proteoglycans, furthermore, the hydrolysis of Lp(a) by sPLA$_2$-IIA increases its binding to proteoglycans (Hoover-Plow et al., 1998; Lundstam et al., 1999). Taken together, these data suggest that sPLA$_2$ activity either in the arterial wall or in plasma may contribute to the atherogenicity of apoB-containing lipoproteins.

Secretory PLA$_2$-IIA can also hydrolyse phospholipids in high-density lipoproteins. HDL levels in transgenic mice overexpressing human sPLA$_2$-IIA are reported to be markedly decreased. HDL in the transgenics are smaller in size and contain significantly less phospholipids compared to HDL in nontransgenic littermates (de et al., 1997; Menschikowski et al., 2000). Analysis of biologically active phospholipid-derived mediators showed increased levels in transgenic animals. In addition, the activity of HDL-associated paraxonase is decreased three-fold in the transgenic mice. This loss of the protective antiinflammatory activity of HDL may be one reason for accumulation of oxidized phospholipids that potentially lead to increase atherosclerosis observed in this sPLA$_2$-IIA transgenic mice (Ivandic et al., 1999; Leitinger et al., 1999).

HDL cholesterol levels are decreased in humans during infections and inflammation (Khovidhunkit et al., 2000). On the other hand, low plasma levels of HDL is associated with high risk for coronary artery disease due to atherosclerosis. However, the mechanism for the decrease in HDL levels during infections and inflammation is not well established. One possible mechanism could be the hydrolysis of HDL phospholipids by sPLA$_2$-IIA, which plasma level is increased in infections and inflammation. *In vivo* and *in vitro* studies showed that sPLA$_2$-IIA in plasma is able to induce marked changes in HDL properties (de et al., 1997; Pruzanski et al., 1998; Menschikowski et al., 2000). These changes will most likely affect the antiatherosclerotic functions of HDL, namely its role in protecting LDL against oxidation, its antiinflammatory property and reverse cholesterol transport from cells (Rothblat et al., 1999; Navab et al., 2001), Thus increasing the risk for cardiovascular disease due to atherosclerosis. These proatherogenic changes of lipoproteins taking place during infections and inflammation may be the mechanism behind the epidemiologic observations linking chronic infections and inflammatory conditions and atherosclerosis (Khovidhunkit et al., 2000; Van et al., 2001). Several proteins, besides lipolytic enzymes, present in HDL particles and in plasma, may contribute to HDL remodeling and turnover (Rye et al., 1999). The function of different proteins is affected by the phospholipid content of the lipoproteins. HDL-modification by other sPLA$_2$ enzymes reported to be induced by inflammation (Murakami et al., 2002) is unknown and deserves future research.

7. PLA$_2$ and cardiovascular diseases

As discussed above, modification of lipoproteins by lipolytic enzymes including sPLA$_2$s is a plausible mechanism for generation of atherogenic lipoprotein particles *in vivo* (Hevonoja et al., 2000; Hurt-Camejo et al., 2000). However, we still have no direct evidence that this may be of physiological relevance in humans.

Rheumatoid arthritis (RA) patients have high cardiovascular disease mortality (Wållberg-Jonsson et al., 1999). Characteristic for RA is a systemic inflammatory response accompanied by elevated levels of C-reactive protein, serum amyloid protein A, fibrinogen, and sPLA$_2$-IIA in plasma that correlate with the disease severity. Atherosclerosis and RA have common pathological features (Pasceri and Yeh, 1999). Foam cells present in early arterial lesions due to upregulated cellular uptake of modified lipoproteins, and a modified form of LDL, similar to oxidized LDL, are found in RA synovial fluid but not in control synovial fluid (Dai et al., 1997). One additional similarity between RA patients and patients with cardiovascular disease is the presence of extracellular sPLA$_2$-IIA and C-reactive protein (Kugiyama et al., 2000). These two acute-phase proteins, may serve to promote phagocytosis of injured cells and tissue debris, thereby enhancing inflammation and tissue damage (Hack et al., 1997). We recently found that RA patients have high levels of small dense LDL with high affinity to chondroitin-6-sulfate glycosaminoglycan. All other lipoprotein parameters of the RA patients were normal (Hurt-Camejo et al., 2001). We hypothesized that this atherogenic lipoprotein marker combined with chronic elevated inflammatory markers may contribute to the high susceptibility for cardiovascular disease of RA patients.

Strong evidence for the importance of the association of sPLA$_2$-IIA with human CVD was presented in a recent case–control study from Japan (Kugiyama et al., 2000). The authors demonstrated that in patients with CVD, defined by angiographically documented coronary stenosis, the plasma level of sPLA$_2$-IIA was an independent risk factor. Furthermore, a higher level of sPLA$_2$ was also shown to be a significant independent predictor of developing coronary events and mortality due to atherosclerosis in two- and four-year follow up period studies (Kugiyama et al., 1999; Porela et al., 2000). Recently two clinical studies with hypercholesterolemic patients reported positive correlations between sPLA$_2$-IIA and soluble adhesion molecules, CRP, and antibody titers to oxidized-LDL in plasma (Hulthe et al., 2000; Wiklund et al., 2001) (Table 1). *In vitro* hydrolysis of LDL-phospholipids makes the lipoprotein particles more susceptible to oxidative modification (Parthasarathy et al., 1985; Sparrow et al., 1988; Neuzil et al., 1998). Thus, one may hypothesize that a similar process in plasma may generate oxidation-susceptible LDL and a possible circulating antigen.

These clinical studies cited suggest that in addition to its proatherogenic properties, sPLA$_2$-IIA levels in plasma may represent a new inflammatory marker for cardiovascular diseases, similar to CRP, serum amyloid A protein (SAA), soluble adhesion molecules, IL-6 and circulating oxidized LDL (Holvoet et al., 2001).

8. Other phospholipase A$_2$ enzymes of interest in cardiovascular disease

Other secretory PLA$_2$ enzymes may have relevance for the physiology and pathology of the arterial wall (Table 2). Plasma platelet-activating factor (PAF) acetylhydrolase (PAF-AH or Group VII) is a serine lipase that hydrolyzes the *sn-2* ester bond of PAF and oxidized-phospholipids, thus attenuating their bioactivity

Table 1
Partial correlation coefficients (adjusted for sex) of sPLA$_2$-IIA, inflammatory markers and auto-antibodies against modified low density lipoproteins levels in plasma from patients with hypercholesterolaemia and control group. Modified from Hulthe, 2000 and Wiklund et al., 2001

	CRP	sICAM-1	sVCAM-1	E-selectin	sPLA$_2$-IIA
Patients (n = 102)	1	0.20	0.14	0.04	0.35***
CRP		1	0.30**	0.39***	0.44***
sICAM-1			1	0.04	0.09
sVCAM-1				1	0.07
IgG Ox-LDL					0.27*
IgG MDA-LDL					0.28*
sPLA$_2$-IIA					
Controls (n = 102)					
CRP	1	0.17	0.17	0.09	0.20*
sICAM-1		1	0.19	0.40***	0.16
sVCAM-1			1	0.11	0.18
IgG Ox-LDL				1	0.02
IgG MDA-LDL					0.03
sPLA$_2$-IIA					0.07
					1

CRP: C-reactive protein; sICAM: soluble intercellular adhesion molecule; sVCAM: soluble vascular cell adhesion molecule; IgG ox-LDL: immunoglobulin type G against oxidized-modified low density lipoprotein; IgG MDA-LDL: immunoglobulin type G against malondialdehyde-modified low density lipoprotein.
*$P < 0.05$, **$P < 0.01$; ***$P < 0.001$.

Table 2
Different groups of PLA2 enzymes in mouse and humans with potential relevance for atherosclerosis

Group PLA$_2$	Human/mouse	kD	Ca++	Catalytic Residue	Chromosome Human/mouse	Binding to proteoglycans
IIA,B,C,D,E	+/+	13–15	mM	Histidine	1p34–36/4	Yes
V	+/+	14	mM	Histidine	1p34–36/4	Yes
X	+/+	14	mM	Histidine	16p12–13/16.34	No
XII	+/+	20	mM	Histidine	4q25/3	
IV, A,B,C	+/+	85	μM	Serine	1q25;15;19/1;2;7	No
VII	+/+	45	None	Serine	6p21.2–p12/17	No
EL	+/+	40	None	Serine	18q21.1/18	Yes

(Stemler et al., 1991; Lee et al., 1999). Experimental and clinical studies suggest that PAF acetylhydrolase has antiinflammatory properties (Tjoelker and Stafforini, 2000). In contrast, this enzyme was suggested to be a pro-inflammatory agent because of its capacity to hydrolyze oxidized phospholipids releasing inflammatory lipid mediators such as lysophospholipids and oxidized fatty acids from oxidized-modified lipoproteins retained in the arterial intima (Tew et al., 1996). PAF-AH circulating in plasma is associated with lipoproteins, mainly LDL, and, because of its phospholipase A_2 property, it is also known as lipoprotein-associated PLA_2. In addition, this enzyme is expressed by macrophages in human and rabbit atherosclerotic lesions (Häkkinen et al., 1999; Brochériou et al., 2000). Recently, lipoprotein-associated PLA_2 was reported to be an independent risk factor for coronary heart diseases in hypercholesterolemic patients (Packard et al., 2000). Furthermore, specific inhibition of the enzyme diminishes macrophage death induced by oxidized lipids and slows down atherogenesis in LDL receptor-defective Watanabe rabbits (Carpenter et al., 2001; Leach et al., 2001). This suggests that lipoprotein-associated PLA_2 could become a new target for therapeutic antiatherosclerotic intervention. However, this hypothesis has been challenged recently with reports showing that expression of PAF-AH activity in arterial tissue and associated to lipoproteins has antiatherosclerotic properties (Lee et al., 1999; Theilmeier et al., 2000). There are no studies that could help us to conclude which of these two phospholipases, $sPLA_2$-IIA or the lipoprotein-associated enzyme has the greater potential for atherogenesis. However, compared with the lipoprotein-associated PLA_2, $sPLA_2$-IIA hydrolyzes intact as well as oxidizes phospholipids, thus generating NEFA, oxidized-NEFA, and lysophospholipids from a broader spectrum of substrates.

Low-molecular weight, calcium-dependent secretory PLA_2 genes, including $sPLA_2$-IIA, are linked and map to homologous chromosome regions in mouse and human (Tischfield et al., 1996). $SPLA_2$s are tightly linked to chromosome 1p34–p36 in human, and to the distal part of chromosome 4 in mouse, a region exhibiting synteny with human 1p34–p36. Interestingly, a recent study on quantitative trait locus analysis in the LDLr knockout model of atherosclerosis, mapped a susceptibility loci for atherosclerosis in chromosome 4 (*Athsq1*). This loci contains the genes for five $sPLA_2$ isoenzymes. Overexpression of human $sPLA_2$-IIA in the mouse resulted in increased atherosclerosis, and $sPLA_2$-IIA plasma levels in human predict coronary events in patients with well-defined atherosclerosis (Ivandic et al., 1999; Kugiyama et al., 1999). These genetic and clinical data support the potential contribution of $sPLA_2$-IIA activity in atherosclerosis. Therefore, $sPLA_2$-IIA may also be a potential target for therapeutic antiatherogenic agents. *In vivo* studies in models of atherosclerosis with specific inhibitors of each of these enzyme will be necessary to evaluate their relative relevance for cardiovascular disease *in vivo*.

Although not yet directly implicated in arterial disease, the Groups V and X secretory PLA_2 enzymes are expressed in macrophages and mast cells (Group V) and in spleen and thymus (Group X) respectively. (Rantapää-Dahlqvist et al., 1991; Hanasaki et al., 1999; Cho, 2000). A novel secretory PLA_2, Group XII, is found in stimulated type 2 helper T cells (Ho et al., 2001). Therefore, these enzymes may be prominent at sites where macrophages, lymphocytes, and mast cells accumulate in atherosclerotic lesions. Groups V and X are more efficient than sPLA$_2$-IIA in hydrolyzing phosphatidylcholine vesicles and the outer plasma membrane of intact mammalian cells (Bezzine et al., 2000). Thus, their capacity to release fatty acids and lysophospholipids, which in turn induce cellular eicosanoid production, may also be higher than that of sPLA$_2$-IIA (Hanasaki et al., 1999; Bezzine et al., 2000). The presence of tryptophan and less basic residues in the interface-binding region of these enzymes appears to contribute to their efficient hydrolysis of cell-membrane phospholipids (Hanasaki et al., 1999; Bezzine et al., 2000). However, the presence of sPLA$_2$-V, sPLA$_2$-X, sPLA$_2$-XII and other sPLA isoenzymes in normal and atherosclerotic tissue, their possible involvement and their potential overlapping function with sPLA$_2$-IIA in sustaining a chronic inflammation in atherogenesis remains to be studied.

In addition to the secretory phospholipases mentioned above, an endothelial lipase (EL) was discovered recently (Hirata et al., 1999; Jaye et al., 1999). EL belongs to the family of triglyceride lipases. However, in contrast to other lipases, EL has substantial phospholipase activity and lipoprotein-phospholipids are its major substrate. EL is synthesized by macrophages, and its expression by coronary artery endothelial cells is upregulated by inflammatory cytokines (Jaye et al., 1999; Hirata et al., 2000). Thus, this enzyme provides the arterial intima with another regulated lipolytic mechanism for local release of fatty acids and lysophospholipids from lipoproteins that may influence atherogenesis (Rader and Jaye, 2000).

9. PLA$_2$ expression and function in the heart

The cardiomyocyte obtains about 70% of its energy from fatty acid β-oxidation and under normal circumstances the fatty acid level in the tissue is low. During ischemia and reperfusion, (I/R), the fatty acid homeostasis in the heart is altered as a result of increased activity of phospholipases and other phospholipid degrading enzymes, including ceramidase and sphingomyelinase. Reduced mito-chondrial β-oxidation (Ford, 2002) and decreased synthesis of phospholipids also occur, contributing also to the altered fatty acid metabolism (Otani et al., 1989). This leads to changes in membrane integrity resulting in reduced activity of important membrane bound proteins including ion channels and transporters. In addition, accumulation of fatty acids and lysophospholipids activate intracellular signaling cascades. It is generally believed that degradation of membrane phospholipids is correlated with irreversible damage of heart tissue and the importance of PLA$_2$ in this process has gained increasing interest (Weglicki et al., 1973; van et al., 1992; De et al., 2001). At least six types of PLA$_2$ are expressed in the heart, cPLA$_2$, iPLA$_2$, sPLA$_2$-II, sPLA$_2$-V, sPLA$_2$-III and sPLA$_2$-XII (Clark et al., 1991; Chen et al., 1994; Tangt et al., 1997; Gelb et al., 2000; Suzuki et al., 2000; Valentin et al., 2000). So far the exact functions of these different types of PLA$_2$ during I/R are not clear. Calcium independent PLA$_2$ accounts for the main phospholipase activity in cardiac tissue, and its activity increases in tissue preparations from acutely ischemic rabbits (Hazen et al., 1991). In addition, pretreatment with an $iPLA_2$-inhibitor, bromoenol lactone (BEL), before I/R reduce the infarct size remarkably (Williams and Gottlieb, 2002). However, the contribution of the $sPLA_2$-IIA in I/R induced damage of the cardiac tissue is not fully understood. Perfusion of antibodies against $sPLA_2$-IIA in isolated rat hearts before I/R could completely prevent phospholipid degradation, improving the recovery of the heart function (Prasad et al., 1991). These results are difficult to correlate with data from a study comparing three different mice strains, swiss mice, C57BL/6 $sPLA_2$-IIA $(-/-)$ and C57BL/Ks $sPLA_2$-IIA $(+/+)$ in which no differences in I/R induced PL-degradation, cellular damage or cardiac function could be ·observed between the different mice strains. The latter results indicate that $sPLA_2$-IIA does not play an important function in these processes. However we should be aware that there are other isoforms of $sPLA_2$; for instance $sPLA_2$-V that could possibly compensate for the loss of $sPLA_2$-IIA in c57BL/6 regarding phospholipid degradation (de Windt et al., 2001). Finally, although there are several studies showing a protective effect of drugs like quinacrine and mepacrine on the cardiac tissue after I/R, it is controversial if the PLA_2 inhibitory effect of these drugs is responsible for the observed benefit (Chiariello et al., 1987; van et al., 1990; Bugge et al., 1997). Clearly there is need for studies using specific inhibitors to be able to resolve the issue of the importance of different types of PLA_2 in the cardiac tissue hypoxic damage and oxygen reperfusion.

10. Future directions

Cardiovascular diseases represent the main cause of mortality and morbidity in industrialized countries. Despite treatments and awareness, by the year 2020 cardiovascular disease will be the main cause of death worldwide (López and Murray, 1998). Although statins, HMG-CoA reductase inhibitors, represent a big success among drugs classes for prevention and treatment of atherosclerotic CVD, they only reduce 30% of CVD events over a five-year period (Libby et al., 2000). Thus, a significant medical need remains unmet for prevention of cardiovascular disease. Today, the search for new therapeutic targets for treatment of atherosclerotic lesions that act more directly on the response of vascular cells to lipoprotein deposition and modification, of which $sPLA_2$ secretion is a component, is of relevance. For many years, researchers and the pharmaceutical industry have studied the possibility of using inhibitors of $sPLA_2$ to treat proinflammatory conditions (Yedgar et al., 2000). However, new knowledge about the existence of many types of $sPLA_2$ highlights the need for specific inhibitors. The recent elucidation of the human and other genomes may add new PLA_2 genes to this already growing family (Six et al., 2000). Therefore, further research is needed in order to increase our knowledge of the specific and potentially overlapping roles of individual $sPLA_2$ enzymes as mediators of physiological and pathological processes. Hopefully, such understanding will enable the development of specific agents aimed at decreasing the potential contribution of individual sPLA$_2$ enzymes to cardiovascular diseases.

Acknowledgement

This work was supported by the Medical Research Council (Project No 12129 to E.H.-C.), the Swedish Heart and Lung Foundation (Project No 41224 to E.H.- C.), and AstraZeneca R & D, Mölndal, Sweden.

References

Anber, V., Griffin, B., McConnell, M., Packard, C., Sheperd, J., 1996. Influence of plasma lipid and LDL-subfraction profile on the interaction between low density lipoprotein with human arterial wall proteoglycans. Atherosclerosis 124, 261–271.

Anderson, K.M., Roshak, A., Winkler, J.D., McCord, M., Marshall, L.A., 1997. Cytosolic 85-kDa phospholipase A2-mediated release of arachidonic acid is critical for proliferation of vascular smooth muscle cells. J. Biol. Chem. 272, 30504–30511.

Andreani, M., Olivier, J.L., Berenbaum, F., Raymondjean, M., Béréziat, G., 2000. Trancriptional regulation of inflammatory secreted phospholipases A2. Biochim. Biophys. Acta 1488, 149–158.

Anthonsen, M., Stengel, D., Hourton, D., Ninio, E., Johansen, B., 2000. Mildly oxidized LDL induces expression of group IIa secretory phospholipase A2 in human monocyte-derived macrophages. Arterioscler. Thromb. Vasc. Biol. 20, 1276–1282.

Beasley, D., 1999. COX-2 and cytosolic PLA2 mediate IL-1beta-induced cAMP production in human vascular smooth muscle cells. Am. J. Physiol. 276, H1269–H1377.

Berg, O., Gelb, M., Tsai, M., Jain, M., 2001. Interfacial enzymology: the secreted phospholipase a(2)-paradigm. Cardiovas. Pathol. 101, 2613–2654.

Berliner, J., Heinecke, J., 1996. The role of oxidized lipoproteins in atherogenesis. Free Rad. Biol. Med. 20, 707–727.

Bezzine, S., Koduri, R.S., Valentin, E., Murakami, M., Kudo, I., Ghomashchi, F., Sadilek, M., Lambeau, G., Gelb, M.H., 2000. Exogenously added human group X secreted phospholipase A2 but not the group IB, IIA, and V enzymes efficiently release arachidonic acid from adherent mammalian cells. J. Biol. Chem. 275, 3179–3191.

Bhakdi, S., Dorweiler, B., Kirchmann, R., Torzewski, J., Weise, E., Tranum-Jensen, J., Walev, I., Wieland, E., 1995. On the pathogenesis of atherosclerosis: enzymatic transformation of human low density lipoprotein to an atherogenic moiety. J. Exp. Med. 182, 1959–1971.

Bidgood, M.J., Jamal, O.S., Cunningham, A.M., Brooks, P.M., Scott, K.F., 2000. Type IIA secretory phospholipase A2 up-regulates cyclooxygenase-2 and amplifies cytokine-mediated prostaglandin production in human rheumatoid synoviocytes. J. Immunol. 165, 2790–2797.

Björnheden, T., Babyi, A., Bondjers, G., Wiklund, O., 1996. Accumulation of lipoprotein fractions and subfractions in the arterial wall, determined in an in vitro perfusion system. Atherosclerosis 123, 43–56.

Brash, A.R., 2001. Arachidonic acid as a bioactive molecule. J. Clin. Invest. 107, 1339–1345.

Brochériou, I., Stengel, D., Mattsson-Hultén, L., Stankova, J., Rola-Pleszcynski, M., Koskas, F., Wiklund, O., Charpentier, Y.L., Ninio, E., 2000. Expression of platelete-activating factor receptor in human carotid atherosclerotic plaques. Relevance to progression of atherosclerosis. Circulation 102, 2569–2575.

Bugge, E., Gamst, T., Hegstad, A., Andreasen, T., Ytrehus, K., 1997. Mepacrine protects the isolated rat heart during hypoxia and reoxygenation-but not by inhibition of phospholipase A2. Basic Res. Cardiol. 92, 17–24.

Cai, T.-Q., Thieblemont, N., Wong, B., Thieringer, R., Kennedy, B.P., Wright, S.D., 1999. Enhancement of leukocyte response to lipopolysaccharide by secretory group IIA phospholipase A2. J. Leukoc. Biol. 65, 750–756.

Camejo, G., Hurt, E., Romano, M., 1985. Properties of Lipoprotein Complexes Isolated by Affinity Chromatography from Human Aorta. Biomedica. Biochimica. Acta. 44, 389–401.

Camejo, G., Fager, G., Rosengren, B.E.H.-C., Bondjers, G., 1993. Binding of low density lipoproteins by proteoglycans synthesized by proliferating and quiescent human arterial smooth muscle cells. J. Biol. Chem. 268, 14131–14137.

Camejo, G., Hurt-Camejo, E., Wiklund, O., Bondjers, G., 1998. Association of apoB lipoproteins with arterial proteoglycans: pathological significance and molecular basis. Atherosclerosis 139, 205–222.

Camejo, G., Hurt-Camejo, E., Olsson, U., Bondjers, G., 1999. Lipid mediators that modulate the extracellular matrix structure and the function in vascular cells. Curr. Ather. Rep. 1, 142–149.

Carpenter, K., Dennis, I., Challis, I., Osborn, D., McPhee, C., Leake, D., Arends, M., Mitchinson, M., 2001. Inhibition of lipoprotein-associated phospholipase A2 diminishes the death-inducing effects of oxidized LDL on human monocyte-macrophages. FEBS Lett. 505, 357–363.

Chen, J., Engle, S., Seilhamer, J., Tischfield, J., 1994. Cloning and recombinant expression of a novel low molecular weight Ca(2+)-dependent phospholipase A2. J. Biol. Chem. 269, 2365–2368.

Chiariello, M., Ambrosio, G., Cappelli-Bigazzi, M., Nevola, E., Perrone-Filardi, P., Marone, G., 1987. Inhibition of ischemia-induced phospholipase activation by quinacrine protects jeopardized myocardium in rats with coronary artery occlusion. J. Pharmacol. Exp. Ther. 241, 560–568.

Cho, W., 2000. Structure, function, and regulation of group V phospholipase A2. Biochim. Biophys. Acta. 1488, 48–58.

Clark, J., Lin, L., Kriz, R., Ramesha, C., Sultzman, L., Lin, A., 1991. A novel arachidonic acid-selective cytosolic PLA2 contains a Ca(2+)-dependent translocation domain with homology to PKC and GAP. Cell 65, 1043–1051.

Couturier, C., Antonio, V., Brouillet, A., Bereziat, G., Raymondjean, M., Andreani, M., 2000. Protein kinase A-dependent stimulation of rat type II secreted phospholipase A2 gene transcription involves C/EBP-beta and -delta in vascular smooth muscle cell. Arterioscler. Thromb. Vasc. Biol. 20, 2559–2565.

Cummings, B.S., Mchowat, J., Schellmann, R.G., 2000. Phospholipase A2 in cell injury and death. J. Pharmacol. Exp. Therapeutics 294, 793–799.

Dai, L., Zhang, Z., Winyard, P.G., Gaffney, K., Jones, H., Blake, D.R., Morris, C.J., 1997. A modified form of low density lipoprotein with increased electronegative charge is present in rheumatoid synovial fluid. Free Rad. Biol. Med. 22, 705–710.

Daugherty, A., Zweifel, B.S., Sobel, B.E., Schonfeld, G., 1988. Isolation of Low Density Lipoprotein from Atherosclerotic Vascular Tissue of Watanabe Heritable Hyperlipidemic Rabbits. Arteriosclerosis 8, 768–777.

de, B.F.C., de, B.M.C., Westhuyzen, D.R.v.d., Castellani, L.W., Lusis, A.J., Swanson, M.E., Grass, D.S., 1997. Secretory non-pancreatic phospholipase A2: influence on lipoprotein metabolism. J. Lipid Res. 38, 2232–2239.

De Windt, L.J. Willems, J., Roemen, T., Coumans, W.A., Reneman, R.S.S.R., van, G.J.V.d., 2001. Ischemic-reperfused isolated working mouse hearts: membrane damage and type IIA phospholipase A2. Am. J. Physiol. Heart Circ. Physiol. 280, H2572–2580.

Dennis, E.A., Six, D.A., 2000. The expanding superfamily of phospholipase A2 enzymes: classification and characterization. Biochim. Biophys. Acta. 1488, 1–19.

Elinder, L.S., Dumitrescu, A., Larsson, P., Hedin, U., Fostergård, J., Claesson, H.-E., 1997. Expression of phosholipase A2 isoforms in human normal and atherosclerotic arterial wall. Arterioscler. Thromb. Vasc. Biol. 17, 2257–2263.

Fenard, D., Lambeau, G., Valentin, E., Lefebvre, J.-C., Lazdunski, M., Doglio, A., 1999. Secreted phospholipases A2, a new class of HIV inhibitors that block virus entry into host cells. J. Clin. Invest. 104, 611–618.

Fitzpatrick, F.A., Soberman, R., 2001. Regulated formation of eicosanoids. J. Clin. Invest. 107, 1347–1351.

Fless, G.M., Kirk, E.W., Klezovitch, O., Santiago, J.Y., Edelstein, C., Hoover-Plow, J., Scanu, A.M., 1999. Effect of phospholipase A2 digestion on the conformation and lysine/fibrinogen binding properties of human lipoprotein(a). J. Lipid Res. 40, 583–592.

Ford, D.A., 2002. Alterations in myocardial lipid metabolism during myocardial ischemia and reperfusion. Prog. Lipid Res. 41, 6–26.

Frieldlander, Y., Kidron, M., Caslake, M., Lamb, T., McConnell, M., Bar-On, H., 2000. Low density lipoprotein particle size and risk factors of insulin resistance syndrome. Atherosclerosis 148, 141–149.

Funk, C., 2001. Prostaglandins and leukotriens: advances in ecosanoid biology. Science 294, 1871–1875.

Gardner, C.D., Fortman, S.P., Krauss, R.M., 1996. Association of small low-density lipoprotein particles with the incidence of coronary artery disease in men and women. JAMA 276, 875–881.

Gelb, M.H., Cho, W., Wilton, D.C., 1999. Interfacial binding of secreted phospholipases A2: more than electrostatics and a major role for tryptophan. Curr. Opin. Struct. Biol. 9, 428–432.

Gelb, M.H., Valentin, E., Ghomashchi, F., Lazdunski, M., Lambeau, G., 2000. Cloning and recombinant expression of a structurally novel human secreted phospholipase A2. J. Biol. Chem. 275, 39823–39826.

Gimbrone, M.A.J., 1995. Vascular endothelium: an integrator of pathophysiological stimuli in atherogenesis. Ann. NY Acad. Sci. 748, 122–132.

Grass, D.S., Felkner, R.H., Chiang, M.-Y., Wallace, R.E., Nevalainen, T.J., Bennett, C.F., 1996. Expression of human group II PLA2 in transgenic mice results in epidermal hyperplasia in the absence of inflammatory infiltrate. J. Clin. Invest. 97, 2233–2241.

Griffin, B.A., 1999. Lipoprotein atherogenicity: an overview of current mechanisms. Proc. Nutr. Soc. 58, 163–169.

Guyton, J.R., 2001. Phospholipid hydrolytic enzymes in a cesspool of arterial intimal lipoproteins. Arterioscler. Thromb. Vasc. Biol. 21, 884–886.

Hack, C.E., Wolbink, G.-J., Schalkwijk, C., Speijer, H., Hermens, W.T., van, H.d.B., 1997. A role for secretory phospholipase A2 and C-reactive protein in the removal of injured cells. Immunol. Today 18, 111–115.

Hakala, J.K., Öörni, K., Ala-Korpela, M., Kovanen, P.T., 1999. Lipolytic modification of LDL by phospholipase A2 induces partice aggregation in the absence and fusion in the presence of heparin. Arterioscler. Thromb. Vasc. Biol. 19, 1276–1283.

Hakala, J.K., Öörni, K., Pentikäinen, M.O., Hurt-Camejo, E., Kovanen, P.T., 2001. Lipolysis of LDL by human secretory phospholipase A2 induces particle fusion and enhances the retention of LDL to human aortic proteoglycans. Arterioscler. Thromb. Vasc. Biol. 21, 1053–1058.

Han, S., Kim, K., Koduri, R., Bittova, L., Munoz, N., Leff, A., Wilton, D., Gelb, M., Cho, W., 1999. Roles of Trp31 in high membrane binding and proinflammatory activity of human group V phospholipase A2. J. Biol. Chem. 274, 11881–11888.

Hanasaki, K., Ono, T., Saiga, A., Morioka, Y., Ikeda, M., Kawamoto, K., Higashino, K.-i., Nakano, K., Yamada, K., Ishizaki, J., 1999. Purified group X secretory phospholipase A2 induced prominent release of arachidonic acid from human myeloid leukemia cells. J. Biol. Chem. 274, 34203–34211.

Hazen, S., Ford, D., Gross, R., 1991. Activation of a membrane-associated phospholipase A2 during rabbit myocardial ischemia which is highly selective for plasmalogen substrate. J. Biol. Chem. 266, 5629–5633.

Hernández, M., Burillo, S.L., Crespo, M.S., Nieto, M.L., 1998. Secretory phospholipase A2 activates the cascade of mitogen-activated protein kinases and cytosolic phospholipase A2 in the human astrocytoma cell line 1321N1. J. Biol. Chem. 273, 606–612.

Hernández, M., Fuentes, L., Fernández, F.J., Crespo, M.S., Nieto, M.L., 2002. Secretory phospholipase A2 elicits proinflammatory changes and upregulates the surface expression of Fas ligand in monocytic cells. Potential relevance for atherogenesis. Cir. Res. 90, 38–45.

Hevonoja, T., Pentikäinen, M., Hyvönen, M., Kovanen, P., Ala-Korpela, M., 2000. Structure of low density lipoprotein (LDL) particles: basis for understanding molecular changes in modified LDL. Biochim. Biophys. Acta. 1488, 189–210.

Hirata, K.-i., Dichek, H.L., Cioffi, J.A., Choi, S.Y., Leeper, N.J., Quintana, L., Kronmal, G.S., Cooper, A.D., Quertermouse, T., 1999. Cloning of a unique lipase from endothelial cells extends the lipase gene family. J. Biol. Chem. 274, 14170–14175.

Hirata, K.-i., Ishida, T., Matsushita, H., 2000. Regulated expression of endothelial cell-derived lipase. Biochem. Biophys. Res. Commun. 272, 90–93.

Hla, T., Lee, M.-J., Ancellin, N., Paik, J., Kluk, M., 2001. Lysophospholipids – receptor relevations. Science 294, 1875–1878.

Ho, I.C., Arm, J.P., Bingham, C.O., Choi, A., Austen, K.F., Glimcher, L.H., 2001. A novel group of phospholipase A2s preferentially expressed in type 2 helper T cells. J. Biol. Chem. 276, 18321–18326.

Holvoet, P., Mertens, A., Verhamme, P., Bogaerts, K., Beyens, G., Verhaeghe, R., Collen, D., Muls, E., Werf, F.V.d., 2001. Circulating oxidized LDL is a useful marker for identifying patients with coronary artery disease. Arterioscler. Thromb. Vasc. Biol. 21, 844–848.

Hoover-Plow, J., Khaitan, A., Fless, G.M., 1998. Phospholipase A2 modification enhances lipoprotein(a) binding to the subendothelial matrix. Thromb. Haemost. 79, 640–648.

Hulthe, J., Wiklund, O., Bondjers, G., Hurt-Camejo, E., 2000. Antibodies against oxidized LDL in relation to cell-adhesion molecules, snpPLA2 and carotid atherosclerosis in patients with hypercholesterolemia. Arterioscler. Thromb. Vasc. Biol. 21, 269–274.

Hurt-Camejo, E., Camejo, G., Rosengren, B., López, F., Wiklund, O., Bondjers, G., 1990. Differential Uptake of Proteoglycan-Selected Subfractions of Low Density Lipoprotein by Human Macrophages. J. Lipid Res. 31, 1387–1398.

Hurt-Camejo, E., Anderssen, S., Standal, R., Rosengren, B., Sartipy, P., Stadberg, E., Johansen, B., 1997. Localization of Non Pancreatic Secretory Phospholipase A2 in Normal and Atherosclerotic Arteries: Activity of the Isolated Enzyme on Low Density Lipoprotein. Arterioscler. Thromb. Vasc. Biol. 17, 300–309.

Hurt-Camejo, E., Camejo, G., 1997. Potential involvement of type II phospholipase A2 in atherosclerosis. Athersoclerosis 132, 1–8.

Hurt-Camejo, E., Camejo, G., Sartipy, P., 2000. Phospholipase A2 and small, dense low-density lipoprotein. Curr. Opin. Lipid 11, 465–471.

Hurt-Camejo, E., Paredes, S., Masana, L., Camejo, G., Sartipy, P., Rosengren, B., Pedreno, J., Vallve, J.C., Benito, P., Wiklund, O., 2001. Rheumatoid arthritis patients have elevated levels of small low density lipoprotein with high affinity for arterial matrix components: possible contribution of phoshoplipase A2 to this atherogenic profile. Arthritis Rheumatism 44, 2761–2767.

Häkkinen, T., Luoma, J.S., Hiltunen, M.O., Macphee, C.H., Millinser, K.J., Patel, L., Rice, S.Q., Tew, D.G., Karkola, K., Ylä-Herttuala, S., 1999. Lipoprotein-associated phospholipase A2, platelet-activating factor acetylhydrolase, is expressed by macrophages in human and rabbit atherosclerotic lesions. Arterioscler. Thromb. Vasc. Biol. 19, 2909–2917.

Ibdah, J.A., Lund-Katz, S., Phillips, M.C., 1989. Molecular packing of high-density and low-density lipoprotein surface lipids and apolipoprotein A-I binding. Biochemistry 28, 1126–1133.

Iozzo, R.V., 1998. Matrix proteoglycans: from molecular design to cellular function. Annu. Rev. Biochem. 67, 609–652.

Ishikawa, Y., Ishii, T., Akasaka, Y., Masuda, T., Strong, J.P., Zieske, A.W., Takei, H., Malcom, G.T., Taniyama, M., Choi-Miura, N.-H., Tomita, M., 2001. Immunolocalization of apolipoproteins in aortic atherosclerosis in American youths and young adults: findings from the PDAY study. Atherosclerosis 158, 215–225.

Ishizaki, J., Suzuki, N., Higashino, K.-I., Yokota, Y., Ono, T., Kawamoto, K., Fujii, N., Arita, H., Hanasaki, K., 1999. Cloning and characterization of novel mouse and human secretory phospholipase A2. J. Biol. Chem. 274, 24973–24979.

Ivandic, B., Castellini, L.W., Wang, X.-P., Qiao, J.-H., Mehrabian, M., Navab, M., Fogelman, A.M., Grass, D.S., Swanson, M.E., Beer, M.C.d., Beer, F.d., Lusis, A.J., 1999. Role of Group II Secretoty Phospholipase A2 in Atherosclerosis. 1. Increased atherogenesis and altered lipoproteins in transgenic mice expressing group IIa phospholipase A2. Arterioscler. Thromb. Vasc. Biol. 19, 1284–1290.

Jaye, M., Lynch, K.J., Krawiec, J., Marchadier, D., Maugeais, C., Doan, K., South, V., Amin, D., Perrone, M., Rader, D.J., 1999. A novel endothelial-derived lipase that modulates HDL metabolism. Nature Genetics 21, 424–428.

Kaneko, K., Sakai, M., Matsumura, T., Biwa, T., Furukawa, N., Shirotani, T., Kiritoshi, S., Anami, Y., Matsuda, K., Sasahara, T., Shichiri, M., 2000. Group-II phospholipase A2 enhances oxidized low density lipopotein-induced macrophage growth through enhancement of GM-CSF release. Atherosclerosis 153, 37–46.

Keaney, J.F., Xu, A., Cunningham, D., Jackson, T., Frei, B., Vita, J.A., 1995. Dietary probucol preserves endothelial function in cholesterol-fed rabbits by limiting vascular oxidative stress and superoxide generation. J. Clin. Invest. 95, 2520–2529.

Kern, H., Johnen, W., Braun, J., Frey, B., Rüstow, B., Kox, W., Schlame, M., 2000. Heparin induces release of phospholipase A2 into the spanchnic circulation. Anesth. Analg. 91, 528–532.

Khovidhunkit, W., Memon, R., Feingold, K., Grunfeld, C., 2000. Infection and inflammation-induced proatherogenic changes of lipoproteins. Am. J. Cardiol. 181, S462–S472.

Kim, K., Rafter, J., Bittova, L., Han, S., Snikto, Y., Munoz, N., Leff, A., Cho, W., 2001. Mechanism of human group V phospholipase A2 (PLA2) induced leukotriens biosynthesis in human neutrophils. A potential role of heparn sulfate binding in PLA2 internalization and degradation. J. Biol. Chem. 276, 11126–11134.

Kleinman, Y., Krul, E.S., Burnes, M., Aronson, W., Pfleger, B., Schonfeld, G., 1988. Lipolysis of LDL with phospholipase A2 alters the expression of selected apoB-100 epitopes and the interaction of LDL with cells. J. Lipid Res. 29, 729–743.

Koduri, R., Gronroos, J., Laine, V., Calvez, C.L., Lambeau, G., Nevalainen, T., Gelb, M., 2002. Bactericidal properties of human and murine groups I, II, V, X, and XII secreted phospholipase A(2). J. Biol. Chem. 277, 5849–5857.

Koduri, R.S., Baker, S.F., Snitko, Y., Han, S.K., Cho, W., Wilton, D.C., Gelb, M.H., 1998. Action of human group IIa secreted phospholipase A2 on cell membranes. Vesicle but not heparinoid binding determines rate of fatty acid release by exogenously added enzyme. J. Biol. Chem. 273, 32142–32153.

Kovanen, P.T., Pentikäinen, M.O., 2000. Secretory group II phospholipase A2. A newly recognized acute-phase reactant with a role in atherogenesis. Cir. Res. 86, 610–612.

Kramer, R.M., Hession, C., Johansen, B., Hayes, G., McGray, P., Chow, E.P., Tizard, R., Pepinsky, R.B., 1989. Structure and Properties of a Human Non-Pancreatic Phospholipase A$_2$. J. Biol. Chem. 264, 5768–5775.

Kugiyama, K., Ota, Y., Takazoe, K., Moriyama, Y., Kawano, H., Miyao, Y., Sakamoto, T., Soejima, H., Ogawa, H., Doi, H., Sugiyama, S., Yasue, H., 1999. Circulating levels of secretory type II phospholipase A2 predict coronary events in patients with coronary artery disease. Circulation 100, 1280–1284.

Kugiyama, K., Ota, Y., Kawano, H., Soejima, H., Ogawa, H., Sugiyama, S., Doi, H., Yasue, H., 2000. Increase in plasma levels of secretory type II phospholipase A2 in patients with coronary spastic angina. Cardiovas. Res. 47, 159–165.

Laine, V.J.O., Grass, D.S., Nevalainen, T.J., 1999. Protection by group II phospholipase A2 against *Staphylococcus aureus*. J. Immunol. 162, 7402–7408.

Lambeau, G., Lazdunski, M., 1999. Receptors for growing family of secreted phospholipases A2. TiPS 20, 162–170.

Leach, C.A., Hickey, D.M.B., Ife, R.J., Macphee, C.H., Smith, S.A., Tew, D.G., 2001. Lipoprotein-associated PLA2 inhibition – a novel, non-lipid lowering strategy for atherosclerosis therapy. Il Farmaco. 56, 45–50.

Lee, C., Sigari, F., Segrado, T., Horkko, S., Hama, S., Subbaiah, P., Miwa, M., Navab, M., Witztum, J., Reaven, P., 1999. All apoB-containing lipoproteins induce monocyte chemotaxis and adhesion when minimally modified. Modulation of lipoprotein bioactivity by platelet-activating factors acetylhydrolase. Arterioscler. Thromb. Vasc. Biol. 19, 1437–1446.

Leitinger, N., Watson, A.D., Hama, S.Y., Ivandic, B., Qiao, J.-H., Huber, J., Faull, K.F., Grass, D.S., Navab, M., Fogelman, A.M., Beer, F.C.d., Lusis, A.J., Berliner, J.A., 1999. Role of group II secretory phospholipase A2 in atherosclerosis. 2. Potential involvement of biologically active oxidized phospholipids. Arterioscler. Thromb. Vasc. Biol. 19, 1291–1298.

Libby, P., Aikawa, M., Schönbeck, U., 2000. Cholesterol and atherosclerosis. Biochim. Biophys. Acta. 1529, 299–309.

López, A., Murray, C.C., 1998. The global burden of disease, 1990–2020. Nature Medicine 4, 1241–1243.

Lundstam, U., Hurt-Camejo, E., Olsson, G., Sartipy, P., Camejo, G., Wiklund, O., 1999. Proteoglycans contribution to association of Lp(a) and LDL with smooth muscle cell extracellular matrix. Arterioscler. Thromb. Vasc. Biol. 19, 1162–1167.

Mackay, K., Starr, J.R., Lawn, R.M., Ellsworth, J.L., 1997. Phosphatidylcholine hydrolysis is required for pancreatic cholesterol esterase- and phospholipase A2-facilitated cholesterol uptake into intestinal Caco-2 cells. J. Biol. Chem. 272, 13380–13389.

Marshall, L., Bolognese, B., Roshak, A., 1999. Respective roles of the 14 kDa and 85 kDa phospholipase A2 enzymes in human monocyte eicosanoid formation. Adv. Exp. Med. Biol. 469, 215–219.

McNamara, J.R., Small, D.M., Li, Z., Schaefer, E.J., 1996. Differences in LDL subspecies involve alterations in lipid composition and conformational changes in apolipoprotein B. J. Lipid Res. 37, 1924–1935.

Menschikowski, M., Kasper, M., Lattke, P., Schiering, A., Schiefer, S., Stockinger, H., Jaross, W., 1995. Secretory group II phospholipase A_2 in human atherosclerotic plaques. Atherosclerosis 118, 173–181.

Menschikowski, M., Eckey, R., Pietsch, J., Auffenanger, J., Kumpf, R., Nelz, P., Jaross, W., 2000. Expression of human secretory group IIA phospholipase A2 is associated with reduced concentrations of plasma cholesterol in transgenic mice. Inflammation 24, 227–237.

Min, J., Wilder, C., Aoki, J., Arai, H., Inoue, K., Paul, L., Gelb, M., 2001. Platelet-activating factor acetylhydrolases: broad substrate specificity and lipoprotein binding does not modulate the catalytic properties of the plasma enzyme. Biochemistry 40, 4539–4549.

Mounier, C., Luchetta, P., Lecut, C., Koduri, R., Faure, G., Lambeau, G., Valentin, E., Singer, A., Ghomashchi, F., Beguin, S., Gelb, M., Bon, C., 2000. Basic residues of human group IIA phospholipase A2 are important for binding to factor Xa and prothrombinase inhibition comparison with other mammalian secreted phospholipases A2. Eur. J. Biochem. 267, 4960–4969.

Murakami, M., Shimbara, S., Kambe, T., Kuwatat, H., Winstead, M.V., Tischfield, J.A., Kudo, I., 1998. The functions of five distinct mammalian phospholipase A2 in regulating arachidonic acid release. J. Biol. Chem. 273, 14411–14423.

Murakami, M., Kambe, T., Shimbra, S., Yamamoto, S., Kuwata, H., Kudo, I., 1999. Functional association of type IIA secretory phospholipase A2 with the glycosylphosphatidylinositol-anchored heparan sulfate proteoglycan in the cyclooxygenase-2-mediated delayed prostanoid-biosynthetic pathway. J. Biol. Chem. 274, 29927–29936.

Murakami, M., Koduri, R., Enomoto, A., Shimbara, S., Seki, M., Yoshihara, K., Singer, A., Valentin, E., Ghomashchi, F., Lambeau, G., Gelb, M., Kudo, I., 2001. Distinct arachidonate-releasing functions of mammalian secreted phospholipase A2s in human embryonic kidney 293 and rat mastocytoma RBL-2H3 cells through heparan sulfate shuttling and external plasma membrane mechanisms. J. Biol. Chem. 276, 10083–10096.

Murakami, M., Yoshihara, K., Shimbara, S., Lambeau, G., Gelb, M.H., Singer, A.G., Sawada, M., Inagaki, N., Nagai, H., Ishihara, M., Ishikawa, Y., Ishii, T., Kudo, I., 2002. Cellular arachidonate-releasing function and inflammation-associated expression of group IIF secretory phospholipase A2. J. Biol. Chem. 277, 19145–19155.

Napoli, C., DArmiento, F.P., Mancini, F.P., Postiglione, A., Witzum, J.L., Palumbo, G., Palinski, W., 1997. Fatty streak formation occurs in human fetal aortas and is greatly enhanced by maternal hypercholesterolemia. Intimal accumulation of low density lipoprotein and its oxidation precede monocyte recruitment into early atherosclerotic lesions. J. Clin. Invest. 100, 2680–2690.

Navab, M., Berliner, J., Subbanagounder, G., Hama, S., Lusis, A., Castellani, L., Reddy, S., Shih, D., Shi, W., Watson, A., Lenten, B.v., Vora, D., Fogelman, A., 2001. HDL and the inflammatory response induced by LDL-derived oxidized phospholipids. Arterioscler. Thromb. Vasc. Biol. 21, 481–488.

Neuzil, J., Upston, J.M., Witting, P.K., Scott, K.F., Stocker, R., 1998. Secretory phospholipase A2 and lipoprotein lipase enhance 15-lipoxygenase-induced enzymic and nonenzymic lipid paroxidation in low density lipoproteins. Biochemistry 37, 9203–9210.

Nevalainen, T.J., 1993. Serum phospholipase A2 in inflammatory diseases. Clin. Chem. 39, 2453–2459.

Olsson, U., Camejo, G., Bondjers, G., 1993. Binding of a synthetic apolipoprotein B-100 peptide analogues to chondroitin-6-sulfate: effects of the lipid environment. Biochemistry 32, 1858–1865.

Olsson, U., Camejo, G., Hurt-Camejo, E., Elfsberg, K., Wiklund, O., Bondjers, G., 1997. Possible functional interactions of apolipoprotein B-100 segments that associate with cell proteoglycans and the apoB/apoE receptor. Arterioscler. Thromb. Vasc. Biol. 17, 149–155.

Olsson, U., Bondjers, G., Camejo, G., 1999. Fatty acids modulate the composition of extracellular matrix in cultured human arterial smooth muscle cells by altering the expression of genes for proteoglycan core proteins. Diabetes 48, 616–622.

Öörni, K., Hakala, J.K., Annila, A., Ala-Korpela, M., Kovanen, P.T., 1998. Sphingomyelinase induces aggregation and fusion, but phospholipase A2 only aggregation, of low density lipoprotein particles. J. Biol. Chem. 273, 29127–29134.

Öörni, K., Pentikäinen, M.O., Ala-Korpela, M., Kovanen, P.T., 2000. Aggregation, fusion, and vesicle formation of modified low density lipoprotein particles: molecular mechanisms and effect on matrix interactions. J. Lipid Res. 41, 1703–1713.

Otani, H., Prasad, M.R., Jones, R., Das, D.K., 1989. Mechanism of membrane phospholipid degradation in ischemic-reperfused rat hearts. Am. J. Physiol. 257, H252–1258.

Packard, C., O'Reilly, D.S.J., Caslake, M.J., McMahon, A.D., Ford, I., Cooney, J., Macphee, C.H., Suckling, K.E., Krishna, M., Wilkinson, F.E., Rumley, A., Lowe, G.D.O., 2000. Lipoprotein-associated phospholipase A2 as an independent predictor of coronary heart disease. N. Engl. J. Med. 343, 1148–1154.

Packard, C.J., Shepherd, J., 1997. Lipoprotein heterogeneity and apolipoprotein B metabolism. Arterioscler. Thromb. Vasc. Biol. 17, 3542–3556.

Parthasarathy, S., Steinbrecher, U., Barnett, J., Wittum, J., Steinberg, D., 1985. Essential role of phospholipase A2 activity in endothelial cell-induced modification of low density lipoprotein. Proc. Natl. Acad. Sci. USA 82, 3000–3004.

Pasceri, V., Yeh, E.T.H., 1999. A tale of two diseases. Atherosclerosis and rheumathoid arthritis. Circulation 100, 2124–2126.

Peilot, H., Rosengren, B., Bondjers, G., Hurt-Camejo, E., 2000. IFN-gamma induces secretory group IIA phospholipase A2 in human arterial smooth muscle cells. Involvement of cell differentiation, STAT-3 activation and modulation by other cytokines. J. Biol. Chem. 275, 22895–22904.

Pentikäinen, M.O., Öörni, K., Ala-Korpela, M., Kovanen, P.T., 2000. Modified LDL-trigger of atherosclerosis and inflammation in the arterial intima. J. Int. Med. 247, 359–370.

Pentikäinen, O.M., Oksjoki, R., Öörni, K., Kovanen, P.T., 2002. Lipoprotein lipase in the arterial wall: linking LDL to the arterial extracellular matrix and much more. Arterioscler. Thromb. Vasc. Biol. 22, 211–217.

Perrella, M.A., Pellacani, A., Layne, M.D., Patel, A., Zhao, D., Schreiber, B.M., Storch, J., Feinberg, M.W., Hsieh, C.-M., Haber, E., Lee, M.-E., 2001. Absence of adipocyte fatty acid binding protein prevents the development of accelerated atherosclerosis in hypercholesterolemic mice. FASEB J. 10.1096/fj.01–0017fje,

Pfeilschifter, J., Walker, G., Kunz, D., Pignat, W., Bosch, H.v.d., 1997. Control of phospholipase A2 gene expression. In: Uhl, W., Nevalainen, T.J., Büchler, M.W. (Eds.), Phospholipase A2 basic and clinical aspects in inflammatory diseases. Karger, Basel, pp. 31–37.

Porela, P., Pulkki, K., Voipio-Pulkki, L.-M., Pettersson, K., Leppänen, V., Nevalainen, T.J., 2000. Level of circulating phospholipase A2 in prediction of the prognosis of patients with suspected myocardial infarction. Basic Res. Cardiol. 5, 413–417.

Prasad, M., Popescu, L., Moraru, I., Liu, X., Maity, S., Engelman, R., 1991. Role of phospholipases A2 and C in myocardial ischemic reperfusion injury. Am. J. Physiol. 260, H877–H883.

Pruzanski, W., Bogoch, E., Wloch, M., Vadas, P., 1991. The role of phospholipase A2 in the physiopathology of osteoarthritis. J. Rheumatol. 18, 117–119.

Pruzanski, W., Stefanski, E., Beer, F.C.d., Beer, M.C.d., Vadas, P., Ravandi, A., Kuksis, A., 1998. Lipoproteins are substrates for human secretory group IIA phospholipase A2: preferential hydrolysis of acute phase HDL. J. Lipid Res. 39, 2150–2160.

Pruzanski, W., Stefanski, E., Kopilov, J., Kuksis, A., 2001. Mitogenic effect of lipoproteins on human vascular smooth muscle cells: the impact of hydrolysis by gr IIA phospholipase A2. Lab Invest. 81, 757–765.

Rader, D.J., Jaye, M., 2000. Endothelial lipase: a new member of the triglyceride lipase gene family. Curr. Opin. Lipid 11, 141–147.

Rantapää-Dahlqvist, S., Wållberg-Jonsson, S., Dahlén, G., 1991. Lipoprotein(a), lipids, and lipoproteins in patients with rheumatoid arthritis. Ann. Rheum. Dis. 50, 366–368.

Romano, M., Romano, E., Björkerud, S., Hurt-Camejo, E., 1998. Ultrastructural localization of secretory type II phospholipase A2 in atherosclerotic and nonatherosclerotic regions of human arteries. Arterioscler. Thromb. Vasc. Biol. 18, 519–525.

Rothblat, G., Llera-Moya, M.d.l., Atger, V., Kellner-Weibel, G., Williams, D., Phillips, M., 1999. Cell cholesterol efflux: integration of old an new observations provides new insights. J. Lipid Res. 40, 781–796.

Rufini, S., de, V.P., Balestro, N., Pescator, M., Luly, P., Incerpi, S., 1999. PLA2 stimulation of Na(+)/H(+) antiport and proliferation in rat aortic smooth muscle cells. Am. J. Physiol. 277, C814–C822.

Rye, K.-A., Clay, M.A., Barter, P.J., 1999. Remodelling of high density lipoproteins by plasma factors. Atherosclerosis 145, 227–238.

Sartipy, P., Johansen, B., Camejo, G., Rosengren, B., Bondjers, G., Hurt-Camejo, E., 1996. Binding of human phospholipase A2 type II to proteoglycans: differential effect of glycosaminoglycans on enzyme activity. J. Biol. Chem. 271, 26307–26314.

Sartipy, P., Bondjers, G., Hurt-Camejo, E., 1998. Phospholipase A2 type II binds to extracellular matrix biglycan: modulation of its activity on LDL by colocalization in glycosaminoglycan matrixes. Arterioscler. Thromb. Vasc. Biol. 18, 1934–1941.

Sartipy, P., Camejo, G., Svensson, L., Hurt-Camejo, E., 1999. Phospholipase A2-modification of low density lipoproteins forms small, high density particles with increased affinity for proteoglycans and glycosaminoglycans. J. Biol. Chem. 274, 25913–25920.

Sartipy, P., Hurt-Camejo, E., 1999. Modification of plasma lipoproteins by group IIA phospholipase A2: possible implications for atherogenesis. Trends Cardiovas. 9, 232–238.

Sartipy, P., Johansen, B., Gåsnik, K., Hurt-Camejo, E., 2000. Molecular basis for the association of group IIA phospholipase A2 and decorin in human atherosclerotic lesions. Cir. Res. 86, 707–714.

Sawada, H., Murakami, M., Enomoto, A., Shimbra, S., Kudo, I., 1999. Regulation of type V phospholipase A2 expression and function by proinflammatory stimuli. Eur. J. Biochem. 263, 826–833.

Schiering, A., Menschikowski, M., Mueller, E., Jaross, W., 1999. Analysis of secretory group II phospholipase A2 expression in human aortic tissue in dependence on the degree of atherosclerosis. Atherosclerosis 144, 73–78.

Schissel, S.L., Tweedie-Hardman, J., Rapp, J.H., Graham, G., Williams, K.J., Tabas, I., 1996. Rabbit aorta and human atherosclerotic lesions hydrolyze the sphingomyelin of retained low-density lipoprotein. J. Clin. Invest. 98, 1455–1464.

Schissel, S.L., Jiang, X.-C., Tweedie-Hardman, J., Jeong, T.-S., Hurt-Camejo, E., Najib, J., Rapp, J.H., Williams, K.J., Tabas, I., 1998. Secretory sphingomyelinase, a product of the acid sphingomyelinase gene, can hydrolyze atherogenic lipoproteins at neutral pH. J. Biol. Chem. 273, 2738–2746.

Schwenke, D.C., Carew, T.E., 1989. Initiation of atherosclerosis lesions in cholesterol-fed rabbits, I: focal increases in arterial LDL concentration precede development of fatty streak lesions. Arterioscler. Thromb. 9, 895–907.

Segrest, J., Jones, M., Loof, H.D., Dashti, N., 2001. Structure of apolipoprotein B-100 in low density lipoproteins. J. Lipid Res. 42, 1346–1367.

Six, D.A., Dennis, E.A., 2000. The expanding superfamily of phospholipase A2 enzymes: classification and characterization. Biochim. Biophys. Acta. 1488, 1–19.

Sparrow, C.P., Parthasarathy, S., Steinberg, D., 1988. Enzymatic modification of low density lipoprotein by purified lipoxygenase plus phospholipase A2 mimics cell-mediated oxidative modification. J. Lipid Res. 29, 745–753.

Stemler, K., Stafforini, D., Prescott, S., McIntyre, T., 1991. Human plasma platelet-activating factor acetyhydrolase. Oxidatively fragmented phospholipids as substrates. J. Biol. Chem. 266, 11095–11103.

Sugiyama, S., Okada, Y., Sukhova, G.K., Virmani, R., Wheinecke, J., Libby, P., 2001. Macrophage myeloperoxidase regulation by granulocyte macrophage colony-stimulating factor in human atherosclerosis and implications in acute coronary syndromes. Am. J. Pathol. 158, 879–891.

Suzuki, N., Ishizaki, J., Yokota, Y., Higashino, K., Ono, T., Ikeda, M., 2000. Structures, enzymatic properties, and expression of novel human and mouse secretory phospholipase A(2)s. J. Biol. Chem. 275, 5785–5793.

Tailleux, A., Torpier, G., Caron, B., Fruchart, J.-C., Fievet, C., 1993. Immunological properties of ApoB-containing lipoprotein particles in human atherosclerotic arteries. J. Lipid Res. 34, 719–728.

Tamminen, M., Mottino, G., Qiao, J.H., Breslow, J.L., Frank, J.S., 1999. Ultrastructure of early lipid accumulation in apoE-deficient mice. Arterioscler. Thromb. Vasc. Biol. 19, 847–853.

Tangt, J., Kriz, R., Wolfman, N., Shaffer, M., Seehra, J., Jones, S., 1997. A novel cytosolic calcium-independent phospholipase A2 contains eight ankyrin motifs. J. Biol. Chem. 272, 8567–8575.

Tew, D.G., Southan, C., Rice, S.Q.J., Lawrence, M.P., Haodong, L., Boyd, H.F., Moores, K., Gloger, I.S., Macphee, C.H., 1996. Purification, properties, sequencing and cloning of a lipoprotein-associated, serine-dependent phospholipase involved in the oxidative modification of low-density lipoproteins. Arterioscler. Thromb. Vasc. Biol. 16, 591–599.

Theilmeier, G., De, B.G., Van, V.P., Stengel, D., Michiels, C., Lox, M., Landeloos, M., Chapman, M., Ninio, E., Collen, D., Himpens, B., Holvoet, P., 2000. HDL-associated PAF-AH reduces endothelial adhesiveness in apoE−/− mice. FASEB J. 14, 2032–2039.

Tischfield, J., Xia, Y., Shih, D., Klisak, I., Chen, J., Engle, S., Siakotos, A., Winstead, M., Seilhamer, J., Allamand, V., Gyapay, G., Lusis, A., 1996. Low-molecular-weight, calcium-dependent phospholipase A2 genes are linked and map to homologous chromosome regions in mouse and human. Genetics 32, 328–333.

Tjoelker, L.W., Stafforini, D.M., 2000. Platelet-activating factor acetylhydrolases in health and disease. Biochim. Biophys. Acta. 1488, 102–123.

Triggiani, M., Granata, F., Oriente, A., Gentile, M., Petraroli, A., Balestrieri, B., Marone, G., 2002. Secretory phospholipase A2 induce cytokine release from blood and synovial fluid monocytes. Eur J. Immunol. 32, 67–76.

Uhl, W., Nevalainen, T.J., Büchler, M.W., 1997. Phospholipase A2. Basic and clinical aspects in inflammatory diseases, First. KARGER, Basel.

Vadas, P., Grouix, B., Stefanski, E., Wolch, M., Pruzanski, W., Schroeder, J., Gauldie, J., 1997. Coordinate expression of group II phospholipase A2 and the acute-phase proteins haptoglobin (HP) and alfa1-anti-chymotrypsin (ACH) by HepG2 cells. Clin. Exp. Immunol. 108, 175–180.

Valentin, E., Ghomashchi, F., Gelb, M.H., Lazdunski, M., Lameau, G., 1999. On the diversity of secreted phospholipases A2. J. Biol. Chem. 274, 31195–31202.

Valentin, E., Ghomashci, F., Gelb, M.H., Lazdunski, M., Lambeau, G., 2000. Novel human secreted phospholipase A2 with homology to the group III bee venom enzyme. J. Biol. Chem. 275, 7492–7496.

Valentin, E., Lambeau, G., 2000. Increasing molecular diversity of secreted phospholipases A2 and their receptors and binding proteins. Biochim. Biophys. Acta. 1488, 59–70.

Valentin, E., Singer, A., Ghomashchi, F., Lazdunski, M., Gelb, M., Lambeau, G., 2000. Cloning and recombinant expression of human group IIF-secreted phospholipase A2. Biochem. Biophys. Res. Commun. 279, 223–228.

van, M.B., der, v.G.J.V., Willemsen, P., Coumans, W., Roemen, T., Reneman, R., 1990. Effects of nicotinic acid and mepacrine on fatty acid accumulation and myocardial damage during ischemia and reperfusion. J. Mol. Cell. Cardiol. 22, 155–163.

van, G.J.V.d., Glatz, J.F., Stam, H.C., Reneman, R.S., 1992. Fatty acid homeostatsis in the normoxic and ischemic heart. Physiol Rev 72, 881–940.

Van, L.B., Wagner, A., Nayak, D., Hama, S., Navab, M., Fogelman, A., 2001. High-density lipoprotein loses its anti-inflammatory properties during acute influenza A infection. Circulation 103, 2283–2288.

Wållberg-Jonsson, S., Johansson, H., Öhman, M.-L., Rantapää-Dahlqvist, S., 1999. Extent of inflammation predicts cardiovascular disease and overall mortality in seropositive rheumatoid arthritis. A retrospective cohort study from disease onset. J. Rheumatol. 26, 2562–2571.

Weglicki, W.B., Owens, K., Urschel, C.W., Serur, J.R., Sonnenblick, E.H., 1973. Hydrolysis of myocardial lipids during acidosis and ischemia. Recent Adv. Stud. Cardiac Struct. Metab. 3, 781–793.

Wery, J.P., Schevitz, R.W., Clawson, D.K., Bobbitt, J.L., Dow, E.R., Gamboa, G., Goodson, T., Hermann, R.B., Kramer, R.M., McClure, D.B., Mihelich, E.D., Putman, J.E., Sharp, J.D., Stark, D.H., Teater, C., Warrick, M.W., Jones, N.D., 1991. Structure of recombinant human rheumatoid arthritic synovial fluid phospholipase A2 at 2.2 A resolution. Nature 352, 79–82.

Weyer, C., Yudkin, J.S., Stehouwer, C.D.A., Schalkwik, C.G., Pratley, R.E., Tataranni, P.A., 2002. Humoral markers of inflammation and endothelial dysfunction in relation to adiposity and in vivo insulin action in Pima Indians. Atherosclerosis 161, 233–242.

Wiklund, O., Hulthe, J., Bondjers, G., Hurt-Camejo, E., 2001. Cell adhesion molecules and secretory type II phospholipase A2 in relation to carotid atherosclerosis in patients with hypercholesterolemia. J. Int. Med. 249, 441–449.

Williams, K.J., Tabas, I., 1995. The response-to-retention hypothesis of early atherogenesis. Arterioscler. Thromb. Vasc. Biol. 15, 551–561.

Williams, S., Gottlieb, R., 2002. Inhibition of mitochondrial calcium-independent phospholipase A2(iPLA2) attenuates mitochondrial phospholipid loss and is cardioprotective. Biochem. J. 362, 23–32.

Yedgar, S., Lichtenberg, D., Schnitzer, E., 2000. Inhibition of phospholipase A2 as a therapeutic target. Biochim. Biophys. Acta. 1488, 182–187.

Zambon, A., Hokanson, J.E., Brown, G., Brunzell, J.D., 1999. Evidence for a new pathophysiological mechanism for coronary artery disease regression. Hepatic lipase-mediated changes in LDL density. Circulation 99, 1959–1964.

**Advances in
Cell Aging and
Gerontology**

Retinal docosahexaenoic acid, age-related diseases, and glaucoma

Nicolas G. Bazan and Elena B. Rodriguez de Turco

*Louisiana State University Health Sciences Center, Neuroscience Center of Excellence and
Department of Ophthalmology, New Orleans, LA, USA.
Corresponding Author: Nicolas G. Bazan, M.D., Ph.D., LSU Health Sciences Center,
Neuroscience Center of Excellence, 2020 Gravier Street, Suite D, New Orleans, LA 70112, USA.
Tel.: + 1-504-599-0831; fax : + 1-504-568-5801.
E-mail address: nbazan@lsuhsc.edu*

Contents

Abbreviations

AA: arachidonic acid; AEA: *N*-arachidonoylethanolamide; DHA: docosahexaenoic acid; DHEA: docosahexaenoylethanolamine; FABP: fatty acid-binding proteins; Hx: hepoxilin; IFN: interferon; IRBP: interphotoreceptor retinoid-binding protein; IsoP: isoprostanes; LC-PUFA: long-chain polyunsaturated fatty acid; LOX: lipoxygenase; LPS: lipopolysaccharide; NarPE: *N*-arachidonoylphosphatidylethanolamine; NDHPE: *N*-docosahexaenoylphosphatidylethanolamine; NO: nitric oxide; NP: neuroprostane; PE: phosphatidylethanolamine; PGF: prostaglandin F; PI: phosphatidylinositol; PLA$_2$: phospholipase A$_2$; PPAR: peroxisome proliferator-activated receptor; PS: phosphatidylserine; ROS: rod outer segments; RPE: retinal pigment epithelium; RXR: retinoid X receptor; TAG: triacylglycerol; TGN: trans-Golgi network; TNF: tumor necrosis factor.

Advances in Cell Aging and Gerontology, vol. 12, 205–222

1. Introduction

Docosahexaenoic acid (DHA, 22:6n−3) and arachidonic acid (AA, 20:4n−6) are the most abundant long-chain polyunsaturated fatty acids (LC-PUFA) of the human body. Major differences in their dietary precursors, tissue distribution, and metabolism are summarized in Table 1. These fatty acids are components of cellular membranes, and are esterified preferentially at the C2 position of the phospholipid-glycerol backbone. Due to selectivity in metabolic pathways, DHA is mainly present in amino phospholipids (phosphatidylethanolamine [PE] and phosphatidyl-serine [PS]) while AA is the primary fatty acid in phosphatidylinositol (PI). Both fatty acids contribute to the modulation of membrane fluidity and function by conforming micro-environments for active membrane proteins, such as receptors, enzymes, and ion channels (Spector and Yorek, 1985; Yeagle, 1989; Murphy, 1990; Clandinin et al., 1991). Moreover, AA- and DHA-phospholipids are reservoirs of bioactive lipid molecules released by activation of phospholipases A_2, C, and D (PLA$_2$, PLC, PLD), which, in turn, modulate responses to extracellular stimulation (Bazan et al., 1995) and gene expression (Long and Pekala, 1996; Duplus et al., 2000; Risé and Galli, 1999; Mata de Urquiza et al., 2000).

DHA synthesis, uptake, and metabolism display unique features reflected in the selective enrichment of DHA in excitable membranes of the retina and brain (Bazan et al., 1993). This contrasts with the more even distribution of AA in all organs of the human body. In photoreceptor outer segments, DHA accounts for 50% of the fatty acids (Aveldaño de Caldironi and Bazan, 1980; Fliesler and Anderson, 1983). In mammals, DHA and AA are synthesized by elongation and

Table 1

DHA and AA: two molecules, unrelated in their biochemistry and metabolism

	Arachidonic acid	Docosahexaenoic acid
Structure	AA, 20:4n−6	DHA, 22:6n−3
Family of essential FA	n−6 fatty acid	n−3 fatty acid
Dietary precursor	18:2n−6	18:3n−3
Synthesis	liver, other tissues	mainly in the liver
Tissue distribution	all tissues	excitable membranes (retina, synaptic terminals)
Esterification into PLs	turnover > de novo	de novo > turnover
PUFA-PLs	inositol PLs (PI, PPI)	amino PLs (SPG, EPG)
Oxidized metabolites	eicosanoids	docosanoids
COX-1/COX-2	PGE$_2$, PGD$_2$, PGI$_2$,TXA$_2$	
5-LOX	5-HETE, LTB$_4$, LTC$_4$, LTD$_4$	
12-LOX/15-LOX	Hydroxy FA (HPETEs, HETEs)	Hydroxy FA (HDHE)
12-LOX/Hx synthase	Hx (HxA$_3$, HxB$_3$)	HxA$_5$
15-LOX/5-LOX	Lipoxins (LXA, LXB)	
PUFA non-enzymatic peroxidation/PLA$_2$	IsoPs (F$_2$, D$_2$, E$_2$)	NPs (F$_4$, D$_4$, E$_4$)
Endocannabinoids	AEA 2-AA-monoglyceride (2-AG)	N-DHA ethanolamide 2-DHA-monoglyceride

desaturation of their dietary precursors linolenic acid (18:3n−3) and linoleic acid (18:2n−6), respectively (Birkle and Bazan, 1986). The content of 18:3n−3 and DHA in plasma lipids is very low, compared to those of n−6 fatty acids, and the liver is the primary organ involved in active DHA synthesis and delivery to tissues for incorporation into lipoproteins (Nelson and Ackman, 1988; Scott and Bazan, 1989; Li et al., 1992; Martin et al., 1994). Because photoreceptors (Wetzel et al., 1991) and neurons (Moore et al., 1991) display a low capacity for DHA synthesis, the long-loop, liver-to-nervous system, efficiently supports brain and retinal DHA requirements. Therefore, the centralized delivery of DHA, combined with its selective uptake and tenacious retention (Bazan et al., 1992; Gazzah et al., 1994), maintains the high DHA levels essential for retinal and brain functions (Wheeler et al., 1975; Neuringer 1986, 1991; Bourre et al., 1989; Bazan, 1990).

Oxidative stress, involved in the aging process and neurodegeneration, targets those organs such as brain and retina that are highly enriched in PUFA-phospholipids with dysfunctional consequences (Urano et al., 1998). Oxidative damage of excitable membranes in the brain and retina leads to the loss of PUFA (Rotstein et al., 1987; Horrocks and Yeo, 1999; Reich et al., 2001), which correlates with impairments in learning ability and cognitive deficits (Favreliere et al., 2000; Lim and Suzuki, 2000) in aged mice and rats. Furthermore, dietary supplementation of aged animals with DHA reduces the hippocampal accumulation of lipid peroxides (Gamoh et al., 2001) and improves learning disabilities (Lim and Suzuki, 2000; Gamoh et al., 2001). Alterations in electroretinograms, reflecting a senile decay in retinal functions both in humans (Kergoat et al., 2001; Jackson et al., 2002) and in rodents (Cavalotti et al., 2001; Katano et al., 2001; Li et al., 2001), may be caused, at least in part, by reduced retinal levels of DHA (Rotstein et al., 1987), as observed in n−3 PUFA food-deprived animals (Wheeler et al., 1975; Neuringer et al., 1986, 1991; Bourre et al., 1989; Bazan, 1990).

Because of the high functional relevance of DHA in the retina, the present review summarizes studies on (a) dynamic interactions between the retinal pigment epithelium (RPE) and photoreceptors in terms of DHA acquisition, supply to photoreceptors, and conservation; (b) DHA polarized trafficking in photoreceptors and functional significance; and (c) DHA-derived messenger generation via enzymatic and nonenzymatic generation of oxidized metabolites (hepoxilins [Hx], neuroprostanes [NPs]) and docosahexaenoylethanolamine (DHEA). Finally, the possible involvement of docosanoids in glaucoma is briefly discussed.

2. Retrieval and supply of DHA to photoreceptors through the interphotoreceptor matrix

RPE interacts through its basal membrane with the choriocapillaris and through its apical surface with photoreceptor outer segments. This strategic localization allows RPE cells to (a) be actively involved in the retrieval of nutrients from the bloodstream for subsequent delivery to photoreceptors; (b) support the visual process, since the RPE stores, isomerizes, and recycles vitamin A; and (c) participate in the daily renewal of rod photoreceptor outer segments (ROS). RPE phagocytizes disc membranes at the tip of ROS, which are constantly replaced by the addition of new membranes at the base of outer segments, thus maintaining a constant ROS length (Gordon and Bazan, 1997).

Disc membranes are unique because DHA is the most prevalent fatty acid in their structure. Moreover, DHA is also essential for visual transduction (Bazan, 1990). The daily renewal of disk membranes results in a high demand for DHA. Because photoreceptors have a very limited capacity for DHA synthesis (Wetzel et al., 1991; Wang and Anderson, 1993a), they rely on other sources and the RPE is the central player in maintaining this supply. RPE supports photoreceptors' DHA needs by way of three pathways. First, RPE elongates and desaturates 18:3n–3 to DHA (Wang and Anderson, 1993a). However, the relatively low plasma levels of DHA precursors limit this potential source. Second, DHA, the most abundant n–3 PUFA of lipoproteins secreted by the liver, is actively and selectively taken up by RPE from lipoproteins circulating through choriocapillaris (Scott and Bazan, 1989; Li et al., 1992; Wang et al., 1992; Gordon and Bazan, 1993; Wang and Anderson, 1993b; Bazan et al., 1985a) (Fig. 1). Moreover, DHA uptake by RPE cells is coupled to efficient and selective delivery to photoreceptors. The *in vivo* trafficking of labeled DHA in the frog retina clearly shows diffuse distribution in RPE cells and preferential delivery to photoreceptors, reaching ROS that is gradually filled with dense labeling as new membranes are formed at the base (Gordon and Bazan, 1990; Bazan et al., 1994). Finally, RPE actively retrieves

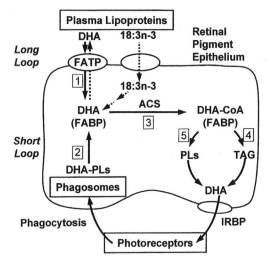

Fig. 1. Retinal pigment epithelial cells in retinal docosahexaenoic acid conservation: uptake and recycling. The RPE avidly take up DHA from lipoproteins circulating through choriocapillaris. 18:3n–3 can also reach RPE by this route and be elongated and desaturated to DHA within RPE. DHA uptake from plasma (arrow 1) or retrieved from phagosomes (arrow 2) is activated to DHA-CoA by the enzyme acyl CoA synthase (ACS, arrow 3). Cytosolic FABPs contribute to the mobilization of both free DHA and DHA-CoA. Activation of DHA prevents its efflux from RPE cells and favors its esterification into TAG (arrow 4) and phospholipids (PLs, arrow 5), transient reservoirs for the subsequent release of free DHA to the interphotoreceptor matrix, where IRBP carries DHA to photoreceptors.

and recycles DHA from phagosomes back to the ROS by way of a short loop through the interphotoreceptor matrix (Gordon et al., 1992; Bazan et al., 1993). This source of DHA, within RPE cells, is critical for conservation of retinal DHA (Bazan et al., 1993; Bazan et al., 1985b), mainly when animals are deprived of dietary n−3 fatty acids. In fact, RPE cells activate the recycling of DHA by way of the short loop when levels of n−3 fatty acids become relatively low in plasma (Stinson et al., 1991), thus conserving DHA in ROS membranes (Wiegand et al., 1991). This suggests an as-yet-unidentified signaling that tightly controls DHA uptake, trafficking, and metabolism in RPE, and a crossroad where RPE balances systemic DHA supply and phagosomal-derived DHA to efficiently support photoreceptor demands. An overall hypothesis to account for this has been suggested (Bazan et al., 1985a; Bazan, 1989; Scott and Bazan, 1989; Bazan, 1990).

Alterations in RPE function during aging may contribute to a deficient DHA supply to the retina. Age-related accumulation of lipofuscin contributes to RPE oxidative damage and dysfunction (Boulton and Dayhaw-Barker, 2001; Davies et al., 2001) and could contribute to the pathogenesis of age-related macular degeneration (AMD), one of the primary causes of blindness in developed countries (Beatty et al., 2000).

The conservation of DHA in retinal cells may be favored by its preferential incorporation into phospholipids by *de novo* synthesis, while other PUFA, such as AA, are actively esterified by turnover (Chen and Anderson, 1993; Bazan and Rodriguez de Turco, 1994). In addition, DHA is actively esterified into triacylglycerols (TAG), a transient reservoir of DHA both in the retina (Rodriguez de Turco et al., 1990, 1991, 1994) and RPE (Bazan et al., 1992; Chen and Anderson, 1993; Gordon and Bazan, 1993). Phagosomal-derived DHA is actively incorporated into TAG from RPE cells both *in vivo* (Gordon et al., 1992) and *in vitro* and released into the media as free DHA (Chen and Anderson, 1993; Rodriguez de Turco et al., 1999). This transient formation of DHA-TAG underlies efficient conservation of DHA in RPE cells for subsequent recycling to photoreceptors (Chen and Anderson, 1993). In studies using rat RPE cells in culture, we have observed that phagosomal-derived DHA is rapidly esterified into TAG and probably into other RPE cell phospholipids (Rodriguez de Turco et al., 1999). In contrast, the release of free DHA into the medium displays a lag period, peaking 8 hours after the phagocytic event, in parallel with a decrease in the DHA-TAG pool. This delay in the release of DHA to the medium suggests that a phagocytosis signaling, involving peroxisome proliferator-activated receptor γ (PPARγ) (Ershov and Bazan, 1998), may activate the expression of genes that have a PPAR-response element in their promoters, including fatty acid-binding proteins (FABP) (Schoonjans et al., 1996a, 1996b; Martin et al., 1997) that are necessary for the efficient mobilization and trafficking of phagosomal-derived DHA. It has been reported recently that DHA is a ligand for the nuclear retinoid X receptor (RXR) (Mata de Urquiza et al., 2000). Thus, the release of phagosomal DHA in RPE cells could be the activator for the RXR signaling pathway and its subsequent heterodimerization with PPARγ, leading to transcriptional activation (Duplus et al., 2000) and thereby fostering the efficient recycling and conservation of retinal DHA.

The inter- and intracellular trafficking of DHA is supported by FABP, which can contribute to its retention in the RPE and its subsequent delivery to photoreceptors. In monkey RPE cells, a FABP that is highly selective for DHA has been found (Lee et al., 1995a). In the interphotoreceptor matrix there are proteins, including the interphotoreceptor retinoid-binding protein (IRBP), that bind DHA and that may play a role in the trafficking of DHA coming from the bloodstream and/or from phagosomal lipids and targeting photoreceptor cells by way of the interphotoreceptor matrix (Bazan et al., 1985a; Bazan et al., 1985b; Chen et al., 1996). Photoreceptors also contain DHA-FABP (Sellner, 1993; Sellner, 1994; Lee et al., 1995b). Thus, it appears that the modulation of FABP levels in retinal cells, especially after the phagocytic event, may be critical to efficient DHA recycling and conservation within the retina. In this context, it is interesting to mention that rat RPE cells grown in Minicell culture inserts and fed with ROS showed a 1.6-fold higher release of DHA toward the apical as compared to the basal compartments (Rodriguez de Turco et al., 2000b). This suggests that RPE cells have the inherent capacity to mobilize DHA toward the apical membrane, favoring its recycling to photoreceptors. Retinal DHA levels decrease during aging, while the mechanism for DHA uptake and incorporation into retinal lipids *in vitro* remains unaffected or stimulated (Rotstein et al., 1987). This may indicate an age-related impairment in the systemic transport of DHA, including its uptake and recycling by RPE cells, which are known to deteriorate with aging as the phototoxic pigment lipofucsin accumulates (Boulton et al., 2001; Davies et al., 2001).

3. Metabolism in photoreceptor cells

There is a selective, high uptake of DHA by photoreceptors (Gordon and Bazan, 1990, 1993; Rodriguez de Turco et al., 1990, 1991, 1994). Ganglion cells in the retina also display very active uptake of DHA (Gordon and Bazan, unpublished observations). DHA is actively esterified into phospholipids of photoreceptors that migrate apically toward the base of outer segments and basally toward synaptic terminals of photoreceptors in frog, rabbit, and primate retinas (Bazan et al., 1993). It has been demonstrated in frog retina that new basal discs actively take up DHA (Gordon and Bazan, 1990, 1993; Gordon et al., 1992) and the labeled discs increase apically, comigrating with [³H]leucine-labeled rhodopsin, until reaching the tips of photoreceptors. At this time, more than 55% of retinal [³H]DHA is located in ROS (Bazan et al., 1994). How photoreceptors regulate this polarized delivery of DHA-lipids to ROS is poorly understood. We have recently reported that vesicles that bud from the trans-Golgi network (TGN) and that are involved in the vectorial transport of opsin to the base of outer segments (Deretic and Papermaster, 1991) also transport newly synthesized DHA-phospholipids, especially choline and ethanolamine phospholipids, which are the major components of photoreceptor membranes (Rodriguez de Turco et al., 1992). The high labeling in this post-Golgi vesicle fraction attained by newly synthesized DHA-phospholipids and rhodopsin suggests that these vesicles are budding from specific microdomains in the TGN, and that the close association of rhodopsin with DHA-phospholipids observed in ROS membranes (Aveldaño, 1988; Rodriguez de Turco et al., 2000a) occurs early on, as they are synthesized in the endoplasmic reticulum. Thus, the assembly of new disc membranes at the base of the ROS is accomplished by the addition of both newly synthesized rhodopsin and DHA-phospholipids. In fact, in cultures of rat retinal cells, DHA protects the cells from degeneration and apoptotic death (Rotstein et al., 1996; Rotstein et al., 1997) and favors cell differentiation and the formation of outer segments (Rotstein et al., 1998). Furthermore, we have found that DHA is necessary for the renewal of rod photoreceptors (Gordon et al., 1995), suggesting that the association of DHA and rhodopsin is essential for rhodopsin-rich membrane biogenesis.

4. Are docosanoids the mediators of the essentiality of DHA for retinal and neural function?

DHA plays essential roles in cellular functions inside and outside the central nervous system (summarized in Table 2). The important role of DHA in retinal function is highlighted by studies in animals deprived of dietary n−3 fatty acids. Under these experimental conditions, both rats (Wheeler et al., 1975; Wiegand et al., 1991) and nonhuman primates (Neuringer et al., 1986, 1991) develop reduced levels of retinal DHA, as well as altered electroretinograms and visual acuity. Moreover, use of milk formulas low in n−3 fatty acids for premature infants results in impairments in retinal functions (Uauy et al., 1990; Birch et al., 1992a, 1992b). In fact, membrane functions depend upon their structure (Clandinin et al., 1991), because alterations in lipid composition affect transmembrane events mediated by receptors, ion channels, and enzymes (Clandinin et al., 1985; Spector and Yorek, 1985; Bourre et al., 1989). DHA is essential for rhodopsin function (Brown, 1994), and reduced levels of DHA-phospholipids in ROS membranes correlate with reduced capacity of rhodopsin for photon absorption (Bush et al., 1994). This occurs despite the compensation for reduced DHA levels by increased 22:5n−6 content, and clearly demonstrates that the function of DHA cannot be fully replaced by other 22-carbon PUFA.

How DHA fulfills its functions in excitable membranes is not defined at present (Bazan, 1990; Mitchel et al., 1998; Fernstrom, 1999). It has been suggested that the six double bonds are necessary for folding the molecule into a slightly spiral shape, thereby reducing the molecular volume of DHA-phospholipids, favoring tight interactions, and influencing acyl-chain packing in the membrane and interactions with rhodopsin (Crawford et al., 1999). The possibility that 18:0- and DHA-PS are the key molecules that, in close association with membrane proteins, underlie the DHA essentiality in excitable membranes has also been proposed (Salem and Niebylski, 1995). The biological significance of DHA was explored in C-6 glioma cells in culture (Kim et al., 2000). DHA promotes the synthesis of DHA-PS, and chronic treatment with ethanol selectively affects the accumulation of DHA-PS, thus implicating its involvement in the pathophysiologic effects of ethanol (Kim and Hamilton, 2000). Similarly, in neuronal cells, DHA promotes

Table 2

Docosahexaenoic acid: implications in specific cellular events

	DHA n−3	References
Phototransduction	+	Bazan, 1990; Brown, 1994; Bush et al., 1994
Rod cell recovery	−	Jeffrey et al., 2002
Photoreceptor apoptosis	−	Rotstein et al., 1996; Rotstein et al., 1997
Neuronal apoptosis	−	Kim et al., 2000
Photoreceptor differentiation	+	Rotstein et al., 1998
Neurite outgrowth	+	Ikemoto et al., 1997
Cellular proliferation/PKC	−	Holian and Nelson, 1992
Neuroprotection (epilepsy, ischemia)	+	Leaf et al., 1999; Lauritzen et al., 2000
Neurologic disorders (bipolar)	−	Calabrese et al., 1999
Neuronal activation ® CBF aging	+	Tsukada et al., 2000
Cholinergic neurotransmission	+	DeGeorge et al., 1991; Jones et al., 1997
NMDA receptor	+	Miller et al., 1992; Nishikawa et al., 1994
GABA receptor	−	Hamano et al., 1996
Long-term potentiation/PKC	−	Mirnikjoo et al., 2001
Long-term depression	−	Young et al., 1998
Voltage-dependent K channels	−	Poling et al., 1995; Poling et al., 1996
background K^+ channels	+	Lauritzen et al, 2000
Na/Ca currents (CA1), excitability	−	Vreugdenhil et al., 1996
Ventricular fibrillation/ sudden death	− − −	Kang and Leaf, 1996; Leaf et al., 1999; Connor, 2000
Atherogenesis, thrombosis	−	Needleman et al., 1979; Kinsella et al., 1990
Vasodilatation	+	Kinsella et al., 1990
Inflammation, arthritis	−	Curtis et al., 2000
Tumor formation	−	Kinoshita et al., 1994; Iigo et al., 1997; Noguchi et al., 1997; Calviello et al., 1998; Narayanan et al., 2001
Gene expression	+	Risé and Galli, 1999; Samples et al., 1999; Narayanan et al., 2001
RXR receptor activation	+	Mata de Urquiza et al., 2000
In vitro NO production	−	Khair-El-Din et al., 1996; Lu et al., 1998; Jeyarajah et al., 1999

the increase in DHA-PS, and this increase correlates with neuronal protection from apoptotic death (Kim et al., 2000). In addition to its unique structural role, the possibility that the importance of DHA involves modulation of AA-mediated regulation of cellular functions has been widely explored (Bazan et al., 1984).

A more dynamic and direct participation of DHA and/or DHA metabolites in the modulation of cell signaling has been reported during recent years, including its involvement in cholinergic signal transduction (DeGeorge et al., 1991; Jones et al., 1997), in long-term potentiation and inhibition of protein kinase C (Young et al., 1998; Mirnikjoo et al., 2001), in interaction with and blockage of neuronal voltage-gated potassium channels (Poling et al., 1995; Poling et al., 1996), in the activation of the RXR signaling pathway (thus underlying transcriptional activation (Mata de Urquiza et al., 2000)), and in the *in vitro* inhibition of the production by macrophages of nitric oxide (NO), a potent injury mediator generated in response to stimulation by lipopolysaccharide (LPS), interferon γ (IFN-γ) and tumor necrosis factor-α (TNF-α) (Khair-El-Din et al., 1996; Lu et al., 1998; Jeyarajah et al., 1999).

Membrane phospholipids are a rich reservoir of AA and DHA that can be released upon agonist-induced PLA_2 activation, thereby generating potent second messengers, including eicosanoids, docosanoids, and platelet-activating factor. Figure 2 summarizes the pathways invoked in DHA metabolism and generation of bioactive molecules. DHA released by PLA_2 can be metabolized by lipoxygenases (LOXs), in pathways analogous to AA lipoxygenation (Birkle and Bazan, 1986). The term "docosanoids" was coined in 1984 (Bazan et al., 1984) to contrast DHA metabolites with eicosanoids, the 20-carbon oxygenated metabolites of AA. In that study, the production of 11-hydroxy-docosahexaenoic acid (and other minor mono-, di-, and trihydroxy isomers) in the retina was reported. Docosanoid synthesis has since been described in other tissues (Birkle and Bazan, 1986), including the pineal body (Sawazaki et al., 1994). An active DHA peroxidation may also lead to racemic mixtures of hydroxylated products, as reported for brain cortex (Kim et al., 1991). The 12-LOX product of 20:4n−6:12S-HPETE can be further metabolized by Hx synthase to hydroxy epoxide metabolites, generating HxA_3 and HxB_3 (Pace-Asciak, 1994). These metabolites display biological functions in many tissues. In the brain, HxA_3 is involved in the modulation of synaptic functions (Pace-Asciak, 1994). It has been reported recently that the products of DHA lipoxygenation, 11- and 14-HPDHE, can generate α-dihomo HxA_5 in the pineal gland and hippocampal slices (Sawazaki et al., 1994; Reynaud

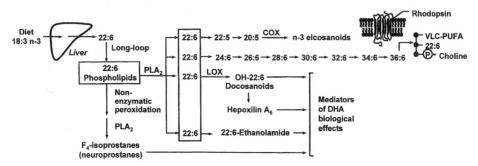

Fig. 2. Docosahexaenoic acid: incorporation into retinal phospholipids and the source of second messengers. DHA, formed by elongation and desaturation of its dietary precursor 18:3n−3 mainly in the liver, reaches the retina by way of a systemic long loop. DHA-phospholipids are actively synthesized in photoreceptors and delivered with high selectivity to ROS and synaptic terminals. DHA can be released from membrane DHA-phospholipids by activation of PLA_2. Free DHA can be further metabolized following different pathways. Its retroconversion to 20:5n−3 favors the synthesis of triene (n-3) eicosanoids by way of a COX pathway. Its successive elongation and desaturation generates very long-chain (VLC) PUFA found in close association with rhodopsin in disc membranes of ROS. DHA can generate by way of the LOX pathway a series of docosanoids or generate DHA-ethanolamine, metabolites that can mediate its biological effects. Nonenzymatic peroxidized DHA-phospholipids, acted upon by PLA_2, can generate NPs, markers of oxidative injury and potential modulators of neuronal function and/ or injury mediators.

and Pace-Asciak, 1997): these metabolites may be modulators of neuronal functions. Exacerbation of these pathways due to age-related oxidative stress may contribute to reduction of retinal DHA levels (Rotstein et al., 1987) and to visual deterioration occurring with age, in humans (Kergoat et al., 2001; Jackson et al., 2002) and in rodents (Cavallotti et al., 2001; Katano et al., 2001; Li et al., 2001).

The amide of 20:4n−6 with ethanolamine (*N*-arachidonoylethanolamide, anandamide, AEA) has been described as an endogenous ligand for cannabinoid receptors with neuromodulatory actions (Axelrod and Felder, 1998; DiMarzo et al., 1998). In bovine retina, the precursor (*N*-arachidonoylphosphatidylethanolamine, NArPE) and the product, AEA, were identified (Bisogno et al., 1999). The authors also reported the presence of *N*-docosahexaenoylphosphatidylethanolamine (NDHPE) and its potentially derived metabolite N-DHEA. Moreover, the other endocannabinoids, 2-arachidonoylglycerol and 2-docosahexaenoylglycerol, are also present in the retina, raising the possibility of their involvement in the modulation of synaptic function in the retina. Therefore, the enzyme-mediated synthesis of DHA metabolites, docosanoids, may be one of the manners through which DHA elicits its functions. A great deal of work remains to be performed in this area. It will be important to clearly define the metabolites produced, refine the experimental conditions, and identify the triggering modulatory events. Once this is done, the identification of putative receptors and signaling will be approachable. Moreover, the functions of specific docosanoids and their pathophysiologic significance in various disease states will be important to assess.

The synthetic docosanoid unoprostone appears as a new, potent therapeutic agent that elicits retinal neuroprotection in experimental models (Hayami et al., 2000; Lafuente et al., 2000; Kimura et al., 2000; Questel et al., 2000). Recent studies using ganglion cell cultures are highly suggestive that unoprostone is effective in protecting against glutamate-induced calcium influx and cell death (unpublished results). The possible connection between the C22 synthetic unoprostone action and DHA docosanoids in the retina is an intriguing and potentially important one in terms of understanding the pathophysiology of glaucoma.

5. F4 isoprostane nonenzymatic peroxidation products of DHA: potential neurotoxicity in retina?

F2 isoprostanes (F2 IsoPs) are a family of prostaglandin $F_{2\alpha}$ (PGF$_{2\alpha}$) isomers generated by free radical peroxidation of cell membrane arachidonoyl-phospholipids (Morrow et al., 1990) and released by PLA$_2$ (Fig. 3). Differing from eicosanoids generated by cyclooxygenase (COX) from free AA, F2 IsoPs are formed *in situ* from esterified AA and subsequently released by PLA$_2$. F2 IsoPs are found in plasma and urine and their content is a direct index of *in vivo* AA peroxidation (Nourooz-Zadeh et al., 1995; Reilly et al., 1996). Peroxidation of DHA generates F4 IsoPs both *in vitro* (Nourooz-Zadeh et al., 1998) and *in vivo* (Roberts et al., 1998). Because of the high concentration of DHA in the brain, these IsoP-like metabolites were named F4 neuroprostanes (NPs) and are

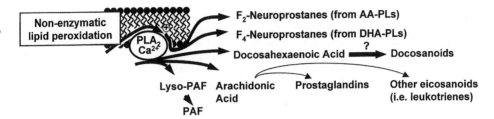

Fig. 3. Phospholipase A_2 and the generation of membrane-derived second messengers. Activation of calcium-dependent cPLA$_2$ leads to the release of PUFA (AA, DHA) esterified at the C2-position of membrane phospholipids. AA and DHA can be metabolized to eicosanoids and docosanoids, respectively, with potent second-messenger functions. PLA$_2$ can also release nonenzymatic peroxidized PUFA from membrane phospholipids, generating F2 IsoPs (from 20:4n−6) and F4 NPs (from DHA).

considered an index of brain oxidative injury (Roberts et al., 1998). There is a high concentration of F4 NP in the cerebrospinal fluid of Alzheimer's patients, which correlates with higher levels of free and esterified NPs found in the brains of patients with this neurodegenerative disease (Nourooz-Zadeh et al., 1999).

Because photoreceptor cells and synaptic terminals are enriched in DHA, the generation of NP in conditions such as retinal light damage, ischemic injury, and glaucoma could contribute to the pathophysiology of retinal damage. These molecules could alter retinal membrane structure and function and, when released by PLA$_2$, generate potent aberrant signaling molecules. This could be a new target for therapeutic intervention.

In glaucoma, an increase in intra-ocular pressure and other factors lead to hypoxic/ischemic oxidative stress and damage of ganglion cells and their axons (Caprioloi, 1997). This includes synaptic signaling dysfunctions, with early alterations in calcium uptake, activation of PLA$_2$, and release of DHA and docosanoids, bioactive molecules that may lead either to ganglion cell degeneration and apoptotic death, or activate pathways of regeneration and repair.

References

Aveldaño, M.I., 1988. Phospholipid species containing long and very long polyenoic fatty acids remain with rhodopsin after hexane extraction of photoreceptor membranes. Biochemistry 27, 1229–1239.

Aveldaño de Caldironi, M.I., Bazan, N.G., 1980. Composition and biosynthesis of molecular species of retina phosphoglycerides. Neurochem. Int. 1, 381–392.

Axelrod, J., Felder, C.C., 1998. Cannabinoid receptors and their endogenous agonist, anandamide. Neurochem. Res. 23, 575–581.

Bazan, N.G., 1989. The identification of a new biochemical alteration early in the differentiation of visual cells in inherited retinal degeneration. In: Inherited and environmentally induced retinal degenerations. New York, Alan R. Liss, pp. 191–215.

Bazan, N.G., 1990. Supply of n−3 polyunsaturated fatty acids and their significance in the central nervous system. In: Wurtman, R.J., Wurtman, J.J., Eds. Nutrition and the Brain. New York; Raven Press, pp. 1–24.

Bazan, N.G., Rodriguez de Turco, E.B., 1994. Pharmacological manipulation of docosahexaenoic-phospholipid biosynthesis in photoreceptor cells: implications in retinal degeneration. J. Ocular. Pharmacol. 10, 591–604.

Bazan, N.G., Birkle D.L., Reddy, T.S., 1984. Docosahexaenoic acid (22:6, n−3) is metabolized to lipoxigenase reaction products in the retina. Biochem. Biophys. Res. Commun. 125, 741–747.

Bazan, N.G., Birkle, D.L., Reddy, T.S., 1985a. Biochemical and nutritional aspects of the metabolism of polyunsaturated fatty acids and phospholipids in experimental models of retinal degeneration. In: LaVail, M.M., Anderson, R.E., Hollyfield, J.G., Eds. Retinal Degeneration: Experimental and Clinical Studies. New York; Alan R. Liss, Inc. 159187.

Bazan, N.G., Gordon, W.C., Rodriguez de Turco, E.B., 1992. The uptake, metabolism and conservation of docosahexaenoic acid (22:6ω-3) in brain and retina: alterations in liver and retina 22:6 metabolism during inherited progressive retinal degeneration. Am. Oil Chem. Soc. 107–115.

Bazan, N.G., Reddy, T.S., Redmond, T.M., Wiggert, B., Chader, G.J., 1985b. Endogenous fatty acids are covalently and noncovalently bound to interphotoreceptor retinoid-binding protein. J. Biol. Chem. 260, 13677–13680.

Bazan, N.G., Rodriguez de Turco, E.B., Gordon, W.C., 1994. Docosahexaenoic acid supply to the retina and its conservation in photoreceptor cells by active retinal pigment epithelium-mediated recycling. In: Galli, C., Simopoulos, A.P., Tremoli, E. (Eds). Fatty Acids and Lipids: Biological Aspects. Basel, Karger, pp. 120–123.

Bazan, N.G., Rodriguez de Turco, E.B., Allan, G., 1995. Mediators of injury in neurotrauma: intracellular signal transduction and gene expression. J. Neurotrauma. 12, 791–814.

Bazan, N.G., Rodriguez de Turco, E.B., Gordon, W.C., 1993. Pathways for the uptake and conservation of docosahexaenoic acid in photoreceptors and synapses: biochemical and autoradiographic studies. Can. J. Physiol. Pharmacol. 71, 690–698.

Beatty, S., Koh, H., Phil, M., Henson, D., Boulton, M., 2000. The role of oxidative stress in the pathogenesis of age-related macular degeneration. Surv. Opthalmol. 45, 115–134.

Birch, D.G., Birch, E.E., Hoffman, D.R., Uauy, R.D., 1992a. Retinal development in very-low-birth-weight infants fed diets differing in omega-3 fatty acids. Invest. Ophthalmol. Vis. Sci. 33, 2365–2376.

Birch, E.E., Birch, D.G., Hoffman, D.R. Uauy, R.D., 1992b. Dietary essential fatty acid supply and visual development. Invest. Ophthalmol. Vis. Sci. 33, 3242–3253.

Birkle, D.L., Bazan, N.G., 1986. The arachidonic acid cascade and phospholipid and docosahexaenoic acid metabolism in the retina. In: Osborne, N., Chader, J., Eds. Progress in Retina Research. New York, Pergamon Press, pp. 309–335.

Bisogno, T., Delton-Vandenbroucke, I., Milone, A., Lagarde, M., Di Marzo, V., 1999. Biosynthesis and inactivation of n-arachidonoylethanolamine (Anandamide) and n-docosahexaenoylethanolamine in bovine retina. Arch. Biochem. Biophys. 370, 300–307.

Boulton, M., Dayhaw-Barker, P., 2001. The role of retinal pigment epithelium: topographical variation and aging changes. Eye 2001. 15, 384–389.

Bourre, J.M., Francois, M., Youyou, A., Dumont, O., Piciotti, M., Pascal, G., Durand, G., 1989. The effect of dietary a-linolenic acid on the composition of nerve membranes, enzymatic activity, amplitude of electrophysiological parameters, resistance to poisons and performance of learning tasks in rats. J. Nutr. 119, 1880–1892.

Brown, M.F., 1994. Modulation of rhodopsin function by properties of the membrane bilayer. Chem. and Physics of Lipids. 73, 159–180.

Bush, R.A., Malnoë, A., Remé, C.E., Williams, T.P., 1994. Dietary deficiency of N-3 fatty acids alters rhodopsin content and function in the rat retina. IOVS 35, 91–100.

Calabrese, J.R., Rapport, D.J., Shelton, M.D., 1999. Fish oils and bipolar disorder: a promising but untested treatment. Arch. Gen. Psychiatry 56, 413–414.

Calviello, G., Palozza, P., Piccioni, E., Maggiano, N., Frattucci, A., Franceschelli, P., Bartoli, G.M., 1998. Dietary supplementation with eicosapentaenoic and docosahexaenoic acid inhibits growth of Morris hepatocarcinoma 3924A in rats: effects on proliferation and apoptosis. Int. J. Cancer 75, 699–705.

Caprioli, J., 1997. Neuroprotection of the optic nerve in glaucoma. Acta Ophthalmol. Scand. 75, 364–367.

Cavallotti, C., Artico, M., Pescosolido, N., Feher, J., 2001. Age-related changes in rat retina. Jpn. J. Ophthalmol. 45, 68–75.

Chen, H., Anderson, R.E., 1993. Differential incorporation of docosahexaenoic acid and arachidonic acid in frog retinal pigment epithelium. J. Lipid Res. 34, 1943–1955.

Chen, Y., Houghton, L.A., Brenna, J.T., Noy, N., 1996. Docosahexaenoic acid modulates the interaction of the interphotoreceptor retinoid-binding protein with 11-*cis* retinal. J. Biol. Chem. 271, 20507–20515.

Clandinin, M.T., Field, C.J., Hargreaves, K., Morson, L., Zsigmond, E., 1985. Role of diet fat in subcellular structure and function. Can. J. Physiol. Pharmacol. 63, 546–556.

Clandinin, M.T., Cheema, S., Field, C.J., Garg, M.L., Venkatraman, J., Clandinin, T.R., 1991. Dietary fat: exogenous determination of membrane structure and cell function. FASEB J. 5, 2761–2769.

Connor, W.E., 2000. Importance of n-3 fatty acids in health and disease. Am. J. Clin. Nutr. 71(Suppl.), 171S–175S.

Crawford, M.A., Bloom, M., Broadhurst, C.L., Schmidt, W.F., Cunnane, S.C., Galli, C., Gehbremeskel, K., Linseisen, F., Lloyd-Smith, J., Parkington, J., 1999. Evidence for the unique function of docosahexaenoic acid during evolution of the modern hominid brain. Lipids 34, S39–S47.

Curtis, C.L., Hughes, C.E., Flannery, C.R., Little, C.B., Harwood, J.L., Caterson, B., 2000. *n*–3 fatty acids specifically modulate catabolic factors involved in articular cartilage degradation. J. Biol. Chem. 275, 721–724.

Davies, S., Ellitt, M.H., Floor, E., Truscott, T.G., Zareba, M., Sarna, T., Shamsi, F.A., Boulton, M.E., 2001. Phototoxicity of lipofuscin in human retinal pigment epithelial cells. Free Rad. Biol. Med. 31, 256–265.

DeGeorge, J.J., Nariai, T., Yamazaki, S., Williams, W.M., Rapoport, S.I., 1991. Arecoline-stimulated incorporation of intravenously administered fatty acids in unanesthetized rats. J. Neurochem. 56, 352–355.

Deretic, D., Papermaster, D.S., 1991. Polarized sorting of rhodopsin on post-Golgi membranes in frog retinal photoreceptor cells. J. Cell. Biol. 113, 1281–1293.

DiMarzo, V., Melck, D., Bisogno, T., De Petrocellis, L., 1998. Endocannabinoids: endogenous cannabinoid receptor ligands with neuromodulatory action. Trends Neurosci. 21, 521–528.

Duplus, E., Glorian, M., Forest, C., 2000. Fatty acid regulation of gene transcription. J. Biol. Chem. 275, 30749–30752.

Ershov, A.V., Bazan, N.G., 1998. Selective induction of peroxisome proliferator-activated receptor gamma (PPARγ) by ROS phagocytosis in RPE cells. Invest. Ophthalmol. Vis. Sci. 39(Suppl.), 189.

Favreliere, S., Stadelmann-Ingrand, S., Huguet, F., De Javel, D., Piriou, A, Tallineau, C., Durand, G., 2000. Age-related changes in ethanolamine glycerophospholipid fatty acid levels in rat frontal cortex and hippocampus. Neurobiol. Aging 21, 653–660.

Fernstrom, J.D., 1999. Effects of dietary polyunsaturated fatty acids on neuronal function. Lipids 34, 161–169.

Fliesler, S.J., Anderson, R.E., 1983. Chemistry and metabolism of lipids in the vertebrate retina. Prog. Lipid Res. 22, 79–131.

Gamoh, S., Hashimoto, M., Hossain, S., Masumura, S., 2001. Chronic administration of docosahexaenoic acid improves the performance of radial arm maze task in aged rats. Clin. Exper. Pharmacol. Physiol. 28, 266–270.

Gazzah, N., Gharib, A., Bobillier, P., Lagarde, M., Sarda, N., 1994. Evidence for brain docosahexaenoate recycling in the free-moving adult rat: implications for measurement of phospholipid synthesis. *Neurosci. Lett.* 177, 103–106.

Gordon, W.C., Bazan, N.G., 1990. Docosahexaenoic acid utilization during rod photoreceptor cell renewal. J. Neurosci. 10, 2190–2202.

Gordon, W.C., Bazan, N.G., 1993. Visualization of [^{3}H]docosahexaenoic acid trafficking through photoreceptors and retinal pigment epithelium by electron microscopic autoradiography. Invest. Ophthalmol. Vis. Sci. 57, 2402–2411.

Gordon, W.C., Bazan, N.G., 1997. Retina. In: Harding, J., Ed. *Biochemistry of the Eye*. London, Chapman and Hall, pp. 144–275.

Gordon, W.C., Rodriguez de Turco, E.B., Bazan, N.G., 1992. Retinal pigment epithelial cells play a central role in the conservation of docosahexaenoic acid by photoreceptor cells after shedding and phagocytosis. Current Eye Res. 11, 73–83.

Gordon, W.C., Rodriguez de Turco, E.B., Bazan, N.G., 1995. Rod photoreceptor renewal depends on docosahexaenoic acid supply. Invest. Ophthalmol. Vis. Sci. 36(Suppl.), 514.

Hamano, H., Nabekura, J., Nishikawa, M., Ogawa, T., 1996. Docosahexaenoic acid reduces GABA response in substantia nigra neuron of rat. J. Neurophysiol. 75, 1264–1270.

Hayami, K., Unoki, K., Ohba, N., 2000. Photoreceptor protection from constant light-induced damage by isopropyl unoprostone, a prostaglandin F2 metabolite-related compound. Invest. Ophthalmol. Vis. Sci. 41(Suppl.), 21.

Holian, O., Nelson, R., 1992. Action of long-chain fatty acids on protein kinase C activity: comparison of omega-6 and omega-3 fatty acids. Anticancer Res. 12, 975–980.

Horrocks, L.A., Yeo, Y.K., 1999. Health benefits of docosahexaenoic acid (DHA). Pharmacol. Res. 40, 211–225.

Iigo, M., Nagakawa, T., Ishikawa, C., Iwahori, Y., Asamoto, M., Yazawa, K., Araki, E., Tsuda, H., 1997. Inhibitory effect of docosahexaenoic acid on colon-carcinoma-26 metastasis to the lung. Brit. J. Cancer. 75, 650–655.

Ikemoto, A., Kobayashi, T., Watanabe, S., Okuyama, H., 1997. Membrane fatty acid modifications of PC12 cells by arachidonate or docosahexaenoate affect neurite outgrowth but not norepinephrine release. Neurochem. Res. 22, 671–678.

Jackson, G.R., Ortega, J., Girkin, C., Rosenstiel, C.E., Owsley, C., 2002. Aging-related changes in the multifocal electroretinogram. J. Optical. Soc. Am. 19, 185–189.

Jeffrey, B.G., Mitchell, D.C., Gibson, R.A., Neuringer, M., 2002. n−3 Fatty acid deficiency alters recovery of the rod photoresponse in Rhesus monkeys. Invest. Ophthalmol. Vis. sci. 43, 2806–2814.

Jeyarajah, D.R., Kielar, M., Penfield, J., Lu, C.Y., 1999. Docosahexaenoic acid, a component of fish oil, inhibits nitric oxide production *in vivo*. J. Surg. Res. 83, 147–150.

Jones, C.R., Arai, T., Rapoport, S.I., 1997. Evidence for the involvement of docosahexaenoic acid in cholinergic stimulation signal transduction at the synapse. Neurochem. Res. 22, 663–670.

Kang, J.X., Leaf, A., 1996. Prevention and termination of arrhythmias induced by lysophosphatidyl choline and acylcarnitine in neonatal rat cardiac myocytes by free omega-3 polyunsaturated fatty acids. Eur. J. Pharmacol. 297, 97–106.

Katano, H., Ishihara, M., Shiraishi, Y., Kawai, Y., 2001. Effects of aging on the electroretinogram during ischemia-reperfusion in rats. Jpn. J. Physiol. 51, 89–97.

Kergoat, H., Kergoat, M.J., Justino, L., 2001. Age-related changes in the flash electroretinogram and oscillatory potentials in individuals age 75 and older. J. Am. Geriatr. Soc. 49, 1212–1217.

Khair-El-Din, T., Sicher, S.C., Vazquez, M.A., Chung, G.W., Stallworth, K.A., Kitamura, K., Miller, R.T., Lu, C.Y., 1996. Transcription of the murine iNOS gene is inhibited by docosahexaenoic acid, a major constituent of fetal and neonatal sera as well as fish oil. J. Exp. Med. 183, 1241–1246.

Kim, H.-Y., Hamilton, J., 2000. Accumulation of docosahexaenoic acid in phosphatidylserine is selectively inhibited by chronic ethanol exposure in C-6 glioma cells. Lipids. 35, 187–195.

Kim, H.-Y., Sawazaki, S., Salem, N., 1991. Lipoxygenation in rat brain? Biochem. Biophys. Res. Commun. 174, 729–734.

Kim, H.-Y., Akbar, M., Lau, A., Edsall, L., 2000. Inhibition of neuronal apoptosis by docosahexaenoic acid (22:6n−3). J. Biol. Chem. 275, 35215–35223.

Kimura, I., Shinoda, K., Kawashima, S., Tanino, T., Mashima, Y., 2000. Effect of topical isopropyl unoprostone on optic nerve head and retinal circulation. Invest. Ophthalmol. Vis. Sci. 41(Suppl.), 553.

Kinoshita, K., Noguchi, M., Earashi, M., Tanaka, M., Sasaki, T., 1994. Inhibitory effects of purified eicosapentaenoic acid and docosahexaenoic acid on growth and metastasis of murine transplantable mammary tumor. *In Vivo* 8, 371–374.

Kinsella, J.E., Lokesh, B., Stone, R.A., 1990. Dietary n−3 polyunsaturated fatty acids and amelioration of cardiovascular disease: possible mechanisms. Am. J. Clin. Nutr. 52, 1–28.

Lafuente, M.P., Mayor-Torroglosa, S., Villegas-Pérez, M.P., 2000. Retinal ganglion cell survival after transient ligation of the ophthalmic vessels and administration of unoprostone isopropyl. Invest. Ophthalmol. Vis. Sci. 41(Suppl.), 15.

Lauritzen, I., Blondeau, N., Heurteaux, C., Widmann, C., Romey, G., Lazdunski, M., 2000. Polyunsaturated fatty acids are potent neuroprotectors. EMBO J. 19, 1784–1793.

Leaf, A., Kang, J.X., Xiao, Y.-F., Billman, G.E., Voskuyl, R.A., 1999. The antiarrhythmic and anticonvulsant effects of dietary N-3 fatty acids. J. Membrane Biol. 172, 1–11.

Lee, J., Jiao, X., Chader, G.J., 1995a. Cultured monkey pigment epithelial (PE) cells contain a fatty acid binding protein (FABP) that binds docosahexaenoic acid (DHA). Invest. Ophthalmol. Vis. Sci. 36(Suppl.),138.

Lee, J., Jiao, X., Gentleman, S., Wetzel, M.G., O'Brian, P., Chader, G.J. 1995b. Soluble-binding proteins for docosahexaenoic acid are present in neural retina. Invest. Ophthalmol. Vis. Sci. 36, 2032–2039.

Li, C., Cheng, M., Yang, H., Peachey, N.S., Naash, M.I., 2001. Age-related changes in the mouse outer retina. Optom. Vis. Sci. 78, 425–430.

Li, J., Wetzel, M.G., O'Brien, P.J., 1992. Transport of $n-3$ fatty acids from the intestine to the retina in rats. J. Lipid Res. 33, 539–548.

Lim, S.Y., Suzuki, H., 2000. Intakes of dietary docosahexaenoic acid ethyl ester and egg phosphatidylcholine improve maze-learning ability in young and old mice. J. Nutr. 130, 1629–1632.

Long, S.D., Pekala, P.H., 1996. Regulation of GLUT4 gene expression by arachidonic acid. J. Biol. Chem. 271, 1138–1144.

Lu, C.Y., Penfield, J.G., Khair-El-Din, T., Sicher, S.C., Kielar, M.L., Vazquez, M.A., Che, L., 1998. Docosahexaenoic acid, a constituent of fetal and neonatal serum, inhibits nitric oxide production by murine macrophages stimulated by IFN gamma plus LPS, or IF gamma plus Listeria monocytogenes. J. Reprod. Immunol. 38, 31–53.

Martin, G., Schoojans, K., Lefebvre, A.M., Staels, B., Auwerx, J., 1997. Coordinate regulation of the expression of the fatty acid transport protein and acyl-CoA synthetase genes by PPARα and PPARγ activators. J. Biol. Chem. 272, 28210–28217.

Martin, R.E., Rodriguez de Turco, E.B., Bazan, N.G., 1994. Developmental maturation of hepatic $n-3$ polyunsaturated fatty acid metabolism: supply of docosahexaenoic acid to retina and brain. J. Nutr. Biochem. 5, 151–160.

Mata de Urquiza, A., Liu, S., Sjöberg, M., Zetterström, R.H, Griffiths, W., Sjövall, J., Perimann, T., 2000. Docosahexaenoic acid, a ligand for the retinoid X receptor in mouse brain. Science 290, 2140–2144.

Miller, B., Sarantis, M., Traynelis, S.F., Attwell, D., 1992. Potentiation of NMDA receptor currents by arachidonic acid. Nature Lond. 355, 722–725.

Mirnikjoo, B., Brown, S.E, Seung, Kim H.F., Marangell, L.B., Sweatt, J.D., Weeber, E.J., 2001. Protein kinase inhibition by omega-3 fatty acids. J. Biol. Chem. 276, 10888–10896.

Mitchel, D.C., Gawrisch, K., Litman, B.J., Salem, N., 1998. Why is docosahexaenoic acid essential for nervous system function? The molecular structure of phospholipids and the regulation of cellular function 26, 365–370.

Moore, S.A., Yoder, E., Murphy, S., Dutton, G.R., Spector, A.A., 1991. Astrocytes, not neurons, produce docosahexaenoic acid (22:6ω3) and arachidonic acid (20:4ω6). J. Neurochem. 56, 518–524.

Morrow, J.D., Hill, K.E., Burk, R.F., Tarek, M., Nammour, T.M., Badr, K.F., Robert, L.J. II., 1990. A series of prostaglandin F_2-like compounds are produced *in vivo* in human by a non-cyclooxigenase, free radical-catalyzed mechanism. Proc. Natl. Acad. Sci. 87, 9383–9387.

Murphy, M.G., 1990. Dietary fatty acids and membrane protein function. J. Nutr. Biochem. 1, 68–79.

Narayanan, B.A., Narayanan, N.K., Reddy, B.S., 2001. Docosahexaenoic acid regulated genes and transcription factors inducing apoptosis in human colon cancer cells. In. J. Oncol. 19, 1255–1262.

Needleman, P., Raz, A., Minkes, M.S., Ferrendelli, J., Sprecher, H., 1979. Triene prostaglandin prostacycline and thromboxane synthesis and unique properties. Proc. Natl. Acad. Sci. USA 76, 944–949.

Nelson, G.J., Ackman, R.G., 1988. Absorption and transport of fat in mammals with emphasis on $n-3$ polyunsaturated fatty acids. Lipids. 23, 1005–1014.

Neuringer, M., Connor, W.E., Lin, D.S., Anderson, G.J., Barstad, L., 1986. Biochemical and functional effects of prenatal and postnatal ω3 fatty acid deficiency on retina and brain in rhesus monkeys. Proc. Natl. Acad. Sci. USA 83, 4021–4025.

Neuringer, M., Connor, W.E., Lin, D.S., Anderson, G.J., Barstad, L., 1991. Dietary omega-3 fatty acids: effects on fatty acids: effects on retinal lipid composition and function in primates. In: Hollyfield, J.G., Anderson, R.E., LaVail, M.M. (Eds). Retinal Degenerations. Boca Raton, CRC Press.

Nishikawa, M., Kimura, S., Akaike, N., 1994. Facilitatory effect of docosahexaenoic acid on N-methyl-D-aspartate response in pyramidal neurones of rat cerebral cortex. J. Physiol. Lond. 475, 83–93.

Noguchi, M., Minami, M., Yagasaki, R., Kinoshita, K., Earashi, M., Kitagawa, H., Taniya, T., Miyazaki, I., 1997. Chemoprevention of DMBA-induced mammary carcinogenesis in rats by low-dose EPA and DHA. Brit. J. Cancer. 75, 348–353.

Nourooz-Zadeh, J., Gopaul, N.K., Barrow, S., Mallet, A.I., Anggård, E.E., 1995. Analysis of F_2-isoprostanes as indicators of nonenzymatic lipid peroxidation in vivo by gas chromatography-mass spectrometry: development of solid-phase extraction procedure. J. Chromatogr. B. Biomed. Appl. 667, 199–205.

Nourooz-Zadeh, J., Liu, E.H.C., Änggård, E.E., Halliwell, B., 1998. F_4-isoprostanes: a novel class of prostanoids formed during peroxidation of docosahexaenoic acid. Biochem. Biophys. Res. Commun. 242, 338–344.

Nourooz-Zadeh, J., Liu, E.H.C., Yhlen, B., Änggård, E.E., Halliwell, B., 1999. F_4-isoprostanes as specific marker of docosahexaenoic acid peroxidation in Alzheimer's disease. J. Neurochem. 72, 734–740.

Pace-Asciak, C.R., 1994. Hepoxilins: a review on their cellular functions. Biochim. Biophys. Acta 1215, 1–8.

Poling, J.S., Karanian, J.W., Salem, N., Vicini, S., 1995. Time- and voltage-dependent block of delayed rectifier potassium channels by docosahexaenoic acid. Molec. Pharmacol. 47, 381–390.

Poling, J.S., Vicini, S., Rogawski, M.A., Salem, N., 1996. Docosahexaenoic acid block of neural voltage-gated K^+ channels: subunit selective antagonism by zinc. Neuropharmacol. 135, 969–982.

Questel, I., Pages, C., Lambrou, G.N., Percicot, C.P., 2000. Effects of unoprostone isopropyl in ocular blood flow decrease induced by systemic vasoconstrictor administration in cynomolgus monkeys. Invest. Ophthalmol. Vis. Sci. 2000, 41(Suppl.), 560.

Reich, E.E., Montine, K.S., Gross, M.D., RobertsII, L.J., Swift, L.L., Morrow, J.D., Montine, T.J., 2001. Interactions between apolipoprotein E gene and dietary alpha-tocopherol influence cerebral oxidative damage in aged mice. J. Neurosci. 21, 5993–5999.

Reilly, M., Delanty, N., Lawson, J.A., FitzGerald, G.A., 1996. Modulation of oxidant stress in vivo in chronic cigarette smokers. Circulation 94, 19–25.

Reynaud, D., Pace-Asciak, C.R., 1997. Docosahexaenoic acid causes accumulation of free arachidonic acid in rat pineal gland and hippocampus to form hepoxilins from both substrates. Biochim. Biophys. Acta 1346, 305–316.

Risé, P., Galli, C., 1999. Arachidonic and docosahexaenoic acids differentially affect the expression of fatty acyl-CoA oxidase, protein kinase C and lipid peroxidation in HepG2 cells. Prostaglandins, Leukotrienes and Essential Fatty Acids 60, 367–370.

Roberts, L.J. II, Montine, T.J., Markesbery, W.R., Tapper, A.R., Hardy, P., Chemtob, S., Dettbarn, W.D., Morrow, J.D., 1998. Formation of isoprostane-like compounds (neuroprostanes) in vivo from docosahexaenoic acid. J. Biol. Chem. 273, 13605–13612.

Rodriguez de Turco, E.B., Gordon, W.C., Bazan, N.G., 1990. Preferential uptake and metabolism of docosahexaenoic acid in membrane phospholipids from rod and cone photoreceptor cells of human and monkey retinas. J. Neurosci. Res. 27, 522–532.

Rodriguez de Turco, E.B., Gordon, W.C., Bazan, N.G., 1991. Rapid and selective uptake, metabolism, and cellular distribution of docosahexaenoic acid among rod and cone photoreceptor cells in the frog retina. J. Neurosci. 11, 3667–3678.

Rodriguez de Turco, E.B., Gordon, W.C., Bazan, N.G., 1994. Docosahexaenoic acid is taken up by the inner segment of frog photoreceptors leading to an active synthesis of docosahexaenoyl-inositol lipids: similarities in metabolism in in vivo and in vitro. Current Eye Res. 13, 21–28.

Rodriguez de Turco, E.B., Deretic, D., Bazan, N.G., Papermaster, D.S., 1997. Post-Golgi vesicles cotransport docosahexaenoyl-phospholipids and rhodopsin during frog photoreceptor membrane biogenesis. J. Biol. Chem. 272, 10491–10497.

Rodriguez de Turco, E.B., Parkins, N., Ershov, A.V., Bazan, N.G., 1999. Selective retinal pigment epithelial cell lipid metabolism and remodeling conserves photoreceptor docosahexaenoic acid following phagocytosis. J. Neurosci. Res. 57, 479–486.

Rodriguez de Turco, E.B., Jackson, F.R., Parkins, N., Gordon, W.C., 2000a. Strong association of unesterified [^{3}H]docosahexaenoic acid and [^{3}H-docosahexaenoyl]phosphatidate to rhodopsin during *in vivo* labeling of frog retinal rod outer segments. Neurochem. Res. 25, 695–703.

Rodriguez de Turco, E.B., Parkins, N., Jackson, F.R., Ershov, A.V., Bazan, N.G. 2000b. Polarized trafficking of docosahexaenoic acid (DHA) in rat retinal pigment epithelial (RPE) cells favors its delivery to photoreceptors. Invest. Ophthalmol. Vis. Sci. 41(Suppl.), 612.

Rotstein, N.P., Ilincheta de Boschero, M.G., Giusto, N.M., Aveldaño, M.I., 1987. Effects of aging on the composition and metabolism of docosahexaenoate-containing lipids of retina. Lipids 22, 253–260.

Rotstein, N.P., Aveldaño, M.I., Barrantes, F.J., Politi, L.E., 1996. Docosahexaenoic acid is required for the survival of rat retinal photoreceptors *in vitro*. J. Neurochem. 66, 1851–1859.

Rotstein, N.P., Aveldaño, M.I., Barrantes, F.J., Roccamo, A.M., Politi, L.E., 1997. Apoptosis of retinal photoreceptors during development *in vitro*: protective effect of docosahexaenoic acid. J. Neurochem. 69, 504–513.

Rotstein, N.P., Politi, L.E., Aveldaño, M.I., 1998. Docosahexaenoic acid promotes differentiation of developing photoreceptors in culture. Invest. Ophthalmol. Vis. Sci. 39, 2750–2758.

Salem, N. Jr., Niebylski, C.D., 1995. The nervous system has an absolute molecular species requirement for the proper function. Molec. Memb. Biol. 12, 131–134.

Samples, B.L., Pool, G.L., Lumb, R.H., 1999. Polyunsaturated fatty acids enhance the heat induced stress response in rainbow trout (*Oncorhynchus mykiss*) leukocytes. Comp. Biochem. and Physiol. Part B. 123, 389–397.

Sawazaki, S., Salem, Jr. N., Kim, H.-Y., 1994. Lipoxygenation of docosahexaenoic acid by the rat pineal body. J. Neurochem. 62, 2437–2447.

Schoonjans, K., Staels, B., Auwerx, J., 1996a. The peroxisome proliferator activated receptors (PPARs) and their effects on lipid metabolism and adipocyte differentiation. Biochim. Biophys. Acta 1302, 93–109.

Schoonjans, K., Staels, B., Auwerx, J., 1996b. Role of the peroxisome proliferator-activated receptor (PPAR) in mediating the effects of fibrates and fatty acids on gene expression. J. Lipid. Res. 37, 907–925.

Scott, B.L., Bazan, N.G., 1989. Membrane docosahexaenoate is supplied to the developing brain and retina by the liver. Proc. Natl. Acad. Sci. USA 86, 2903–2907.

Sellner, P., 1994. Fatty acid-binding protein from embryonic chick retina resembles mammalian heart acid-binding protein. Invest. Ophthalmol. Vis. Sci. 35, 443–452.

Sellner, P.A., 1993. Retinal FABP principally localizes to neurons and not glial cells. Mol. Cell. Biochem. 123, 121–127.

Spector, A.A., Yorek, M.A., 1985. Membrane lipid composition and cellular function. J. Lipid Res. 26, 1015–1035.

Stinson, A.M., Wiegand, R.D., Anderson, R.E., 1991. Recycling of docosahexaenoic acid in rat retinas during n–3 fatty acid deficiency. J. Lipid Res. 32, 2009–2017.

Tsukada, H., Kakiuchi, T., Fukumoto, D., Nishiyama, S., Koga, K., 2000. Docosahexaenoic acid (DHA) improves the age-related impairment of the coupling mechanism between neuronal activation and functional cerebral blood flow response: a PET study in conscious monkeys. Brain Res. 862, 180–186.

Uauy, R.D., Birch, D.G., Birch, E.E., Tyson, J.E., Hoffman, D.R., 1990. Effect of dietary omega-3 fatty acid on retinal function of very-low-birth-weight neonates. Pediat. Res. 28, 485–492.

Urano, S., Sato, Y., Otonari, T., Makabe, S., Suzuki, S., Ogata, M., Endo, T., 1998. Aging and oxidative stress in neurodegeneration. BioFactors 7, 103–112.

Vreugdenhil, M., Bruehl, C., Voskuyl, R.A., Kang, J.X., Leaf, A., Wadman, W.J., 1996. Polyunsatu-
    rated fatty acids modulate sodium and calcium currents in CA1 neurons. Proc. Natl. Acad. Sci. 93,
    12559–12563.
Wang, N., Anderson, R.E., 1993a. Synthesis of docosahexaenoic acid by retina and retinal pigment
    epithelium. Biochem. 32, 13703–13709.
Wang, N., Anderson, R.E., 1993b. Transport of 22:6n−3 in plasma and uptake into retinal pigment
    epithelium and retina. Exp. Eye Res. 57, 225–233.
Wang, N., Wiegand, R.D., Anderson, R.E., 1992. Uptake of 22-carbon fatty acids into rat retina and
    brain. Exp. Eye. Res. 54, 933–939.
Weber, P.C., 1990. The modification of the arachidonic cascade by n-3 fatty acids. Adv. in
    Prostaglandin, Tromboxane, and Leukotriene Res. 20, 232–240.
Wetzel, M.G., Li, J., Alvarez, R.A., Anderson, R.E., O'Brien, P.J., 1991. Metabolism of linolenic acid
    and docosahexaenoic acid in rat retinas and rod outer segments. Exp. Eye Res. 53, 437–446.
Wheeler, T.G., Benoken, R.M., Anderson, R.E., 1975. Specificity of fatty acid precursors for the
    electrical response to illumination. Science 188, 1312.
Wiegand, R.D., Koutz, C.A., Stinson, A.M., Anderson, R.E., 1991. Conservation of docosahexaenoic
    acid in rod outer segments of the rat retina during n-3 and n-6 fatty acid deficiency. J. Neurochem.
    57, 1690–1699.
Yeagle, P.L., 1989. Lipid regulation of cell membrane structure and function. FASEB J. 3, 1833–1842.
Young, C., Gean, P.-W., Wu, S.-P., Lin, C.-H., Shen, Y.-Z., 1998. Cancellation of low-frequency
    stimulation-induced long-term depression by docosahexaenoic acid in the rat hippocampus.
    Neurosci. Lett. 247, 198–200.

List of Contributors

Christopher Fielding

Neider Professor of Cardiovascular Physiology
Cardiovascular Research Institute
University of California San Francisco 94143-0130
Phone: 415-476-4307 Fax: 415-476-2283
Email: cfield@itsa.ucsf.edu

Kathleen Montine,
Joseph Quinn and
Thomas Montine

Address Correspondence to: Kathleen Montine
Department of Pathology
University of Washington
Harborview Medical Center
325 9th Avenue, Box 359791
Seattle, WA 98104
Phone: 206-341-5244 Fax: 206-341-5249
Email: kmontine@u.washington.edu

Cathy Wolkow

National Institute on Aging
Gerontology Research Center
Laboratory of Neurosciences
5600 Nathan Shock Dr.
Baltimore, MD 21224
Phone: 410-558-8566 Fax: 410-558-8465
Email: wolkowca@grc.nia.nih.gov

Peter P. Ruvolo,
Charlene Johnson and
W. David Jarvis

Address Correspondence to: Peter Ruvolo
University of Texas at Houston Health
Science Center
Institute of Molecular Medicine
2121 W. Holcombe Blvd.
Houston, TX 77030
Phone: 713-500-2400 Fax: 713-500-2420
Email: Peter.P.Ruvolo@uth.tmc.edu

Subroto Chatterjee and
Sergio Martin

Address Correspondence to: Subroto Chatterjee
Lipid Research Atherosclerosis Division
Department of Pediatrics
Johns Hopkins University SOM
600 N. Wolfe St., CMSC 6-124
Baltimore, MD 21287-3654
Phone: 410-614-2518 Fax: 410-614-2826
Email: chatter@welchlink.welch.jhu.edu

Mark P. Mattson and
Roy G. Cutler

Address Correspondence to: Mark P. Mattson
National Institute on Aging
Gerontology Research Center
Laboratory of Neurosciences
5600 Nathan Shock Drive
Baltimore, MD 21224
Phone: 410-558-8463 Fax: 410-558-8465
Email: mattsonm@grc.nia.nih.gov

Hari Manev and
Tolga Uz

Address Correspondence to: Hari Manev
The Psychiatric Institute
University of Illinois at Chicago
1601 West Taylor Street
M/C912, Room 238
Chicago, IL 60612
Phone: 312-413-4558 Fax: 312-413-4569
Email: Hmanev@psych.uic.edu

John P. Incardona

Ecotoxicology and Environmental Fish
Health Program
Environmental Conservation Division
NOAA/National Marine Fisheries Service
Northwest Fisheries Science Center
2725 Montlake Blvd E
Seattle, WA 98112
Phone: 206-860-3347/206-860-3411
Fax: 206-860-3335
Email: John.Incardona@noaa.gov

Miguel A. Pappolla,
Suzana Petanceska,
Lawrence Refolo and
Nicolas Bazan

Address Correspondence to: Nicolas G. Bazan
Boyd Professor
Louisiana State University Health Sciences Center
Neuroscience Center of Excellence
2020 Gravier Street, Ste. D
New Orleans, LA 70112-2234
Phone: 504-599-0831 Fax: 504-568-5801
Email: nbazan@lsuhsc.edu

Eva Hurt-Camejo,
Peter Sartipy,
Helena Peilot,
Birgitta Rosengren,
Olov Wiklund and
German Camejo

Address Correspondence to: Eva Hurt-Camejo
Professor
Associate Director Cell Biology and Biochemistry
Research Area CV & GI
AstraZeneca, Molndal S-431 83
Phone: 46-31-776-2967 Fax: 46-31-776-3736
Email: eva.hurt-camejo@astrazeneca.com

Nicolas G. Bazan and
Elena Rodriguez

Address Correspondence to: Nicolas G. Bazan
Boyd Professor
Louisiana State University Health Sciences Center
Neuroscience Center of Excellence
2020 Gravier Street, Ste. D
New Orleans, LA 70112-2234
Phone: 504-599-0831 Fax: 504-568-5801
Email: nbazan@lsuhsc.edu

Advances in
Cell Aging and Gerontology
Series Editor: Mark P. Mattson
URL: http://www.elsevier.nl/locate/series/acag

Aims and Scope:

Advances in Cell Aging and Gerontology (ACAG) is dedicated to providing timely review articles on prominent and emerging research in the area of molecular, cellular and organismal aspects of aging and age-related disease. The average human life expectancy continues to increase and, accordingly, the impact of the dysfunction and diseases associated with aging are becoming a major problem in our society. The field of aging research is rapidly becoming the niche of thousands of laboratories worldwide that encompass expertise ranging from genetics and evolution to molecular and cellular biology, biochemistry and behavior. ACAG consists of edited volumes that each critically review a major subject area within the realms of fundamental mechanisms of the aging process and age-related diseases such as cancer, cardiovascular disease, diabetes and neurodegenerative disorders. Particular emphasis is placed upon: the identification of new genes linked to the aging process and specific age-related diseases; the elucidation of cellular signal transduction pathways that promote or retard cellular aging; understanding the impact of diet and behavior on aging at the molecular and cellular levels; and the application of basic research to the development of lifespan extension and disease prevention strategies. ACAG will provide a valuable resource for scientists at all levels from graduate students to senior scientists and physicians.

Books Published:

1. P.S. Timiras, E.E. Bittar, *Some Aspects of the Aging Process*, 1996, 1-55938-631-2
2. M.P. Mattson, J.W. Geddes, *The Aging Brain*, 1997, 0-7623-0265-8
3. M.P. Mattson, *Genetic Aberrancies and Neurodegenerative Disorders*, 1999, 0-7623-0405-7
4. B.A. Gilchrest, V.A. Bohr, *The Role of DNA Damage and Repair in Cell Aging*, 2001, 0-444-50494-X
5. M.P. Mattson, S. Estus, V. Rangnekar, *Programmed Cell Death, Volume I*, 2001, 0-444-50493-1
6. M.P. Mattson, S. Estus, V. Rangnekar, *Programmed Cell Death, Volume II*, 2001, 0-444-50730-2
7. M.P. Mattson, *Interorganellar Signaling in Age-Related Disease*, 2001, 0-444-50495-8
8. M.P. Mattson, *Telomerase, Aging and Disease*, 2001, 0-444-50690-X
9. M.P. Mattson, *Stem Cells: A Cellular Fountain of Youth*, 2002, 0-444-50731-0
10. M.P. Mattson, *Calcium Homeostasis and Signaling in Aging*, 2002, 0-444-51135-0
11. T. Hagen, *Mechanisms of Cardiovascular Aging*, 2002, 0-444-51159-8
12. M.P. Mattson, *Membrane Lipid Signaling in Aging and Age-Related Disease*, 2003, 0-444-51297-7